Lecture Notes in Computer Science   1262

Springer

Berlin
Heidelberg
New York
Barcelona
Budapest
Hong Kong
London
Milan
Paris
Santa Clara
Singapore
Tokyo

Michel Scholl Agnès Voisard (Eds.)

Advances in Spatial Databases

5th International Symposium, SSD '97
Berlin, Germany, July 15-18, 1997
Proceedings

 Springer

Series Editors

Gerhard Goos, Karlsruhe University, Germany

Juris Hartmanis, Cornell University, NY, USA

Jan van Leeuwen, Utrecht University, The Netherlands

Volume Editors

Michel Scholl
CNAM/INRIA
F-78153 Le Chesnay Cedex, France
E-mail: Michel.Scholl@inria.fr

Agnès Voisard
Freie Universität Berlin, Institut für Informatik
Takustr. 9, D-14195 Berlin, Germany
E-mail: voisard@inf.fu-berlin.de

Cataloging-in-Publication data applied for

Die Deutsche Bibliothek - CIP-Einheitsaufnahme

Advances in spatial databases : 5th international symposium ;
proceedings / SSD '97, Berlin, Germany, July 15 - 18, 1997. Michel
Scholl ; Agnes Voisard (ed.). - Berlin ; Heidelberg ; New York ;
Barcelona ; Budapest ; Hong Kong ; London ; Milan ; Paris ; Santa
Clara ; Singapore ; Tokyo : Springer, 1997
   (Lecture notes in computer science ; Vol. 1262)
   ISBN 3-540-63238-7

CR Subject Classification (1991): H.2-3, H.5, I.4, I.5, J.2,F.2

ISSN 0302-9743
ISBN 3-540-63238-7 Springer-Verlag Berlin Heidelberg New York

Typesetting: Camera-ready by author
SPIN 10548937   06/3142 – 5 4 3 2 1 0   Printed on acid-free paper

Preface

These proceedings contain the research papers selected for presentation at the Fifth International Symposium on Spatial Databases (SSD'97) held in Berlin, Germany, July 15-18, 1997. Since the first conference in Santa Barbara (1989), SSD has been the premier meeting for researchers, developers, and practitioners interested in the integration of database management systems (DBMS) and geographic information systems (GIS). SSD'97 brought together leading spatial database researchers and GIS experts to explore advances in spatial modeling and query processing, and to discuss the requirements from new application domains.

From fifty-five submissions of full papers by authors from nineteen countries, each of them reviewed by four referees, the program committee selected eighteen outstanding papers for inclusion in these proceedings. The stable number of submissions over the SSD series to date is evidence of the continuing interest in spatial databases and is a sign of an active research community. The technical program also includes two keynote presentations (David Abel, Dick Newell), one panel (John Herring) and four tutorials (Hanan Samet, Max Egenhofer, Stéphane Grumbach, and Jiawei Han).

The papers included in this volume show that the need for efficient spatial query processing still motivates numerous papers on geo-algorithms, access methods, spatial joins, and system implementations. Spatial data modeling is still a core topic, but it has shifted toward the study of error correcting and imprecise object representation. Moreover, the emergence of spatial models and systems based on constraint databases sounds a promising framework for spatial query processing and optimization. Spatial data mining has also been established as a topic with great potential. However, some other emerging issues of primary importance are insufficiently addressed in this volume: spatio-temporal applications, modeling of mobile objects, open GIS and integration of heterogeneous spatial sources, World Wide Web applications are among the new topics that deserve more attention in the future.

We wish to acknowledge the assistance and contribution of the following persons and institutions that made this meeting happen. We would like first to thank the program committee and the external referees for their diligence and their high-quality reviews. Max Egenhofer deserves a special thanks for his faithful assistance throughout the preparation of this event. We also thank Oliver Günther, Hans-Jörg Schek, and Heinz Schweppe for their support. We very much appreciated the assistance of Heike Friedrichs in the processing of electronic submissions. We are in debt to Inge Hellmich, Günter Feuer, Heike Friedrichs, Oliver Grzegorski, Danny Moreau, and Stephan Drooff for the organization of the conference. We also appreciated the assistance of Danièle Charlet from Oracle, France. Finally we wish to thank Alfred Hofmann from Springer-Verlag for his help in the preparation of this volume.

We are very grateful for the support given by the following companies and institutions: Bull AG, Cassini, The ESPRIT TMR Network CHOROCHRONOS, DFG, ESRI Inc., ETAK Inc., Fraunhofer ISST, Freie Universität Berlin, ICSI, INRIA, Oracle France, Oracle Spatial Products (USA), SMALLWORLD Systems GmbH. This effort is a testimony of the real-world significance of this event on advances in spatial databases.

Cooperation with ACM SIGMOD, the National Center for Geographic Information and Analysis (NCGIA) and BWO Marketing Service GmbH is also acknowledged.

Berlin, May 1997

Agnès Voisard, General Chair
Michel Scholl, Program Chair

Acknowledgments

In Cooperation with ACM SIGMOD, NCGIA, BWO Marketing Service GmbH

Sponsors
Bull AG
Cassini (French CNRS GdR 1041 and Université de Provence)
CHOROCHRONOS
DFG (Deutsche Forschungsgemeinschaft)
ESRI, Inc.
ETAK, Inc.
Fraunhofer ISST
Freie Universität Berlin
ICSI
INRIA
Oracle France
Oracle Spatial Products, USA
SMALLWORLD Systems GmbH

Conference Organization

General Chair
Agnès Voisard, Freie Universität Berlin, Germany

Program Committee Chair
Michel Scholl, INRIA and CNAM, France

Local Arrangements
Heike Friedrichs
Oliver Grzegorski
Inge Hellmich

Tutorial Chair
Ralf Hartmut Güting, FernUniversität Hagen, Germany

Panel Co-chairs
Andrew U. Frank, TU Vienna, Austria
John R. Herring, Oracle Corporation, USA

Exhibit Chair
Günter Feuer, Freie Universität Berlin, Germany

Program Committee
David J. Abel, CSIRO, Canberra, Australia
Gustavo Alonso, ETH Zürich, Switzerland
Elisa Bertino, Università degli Studi di Milano, Italy
Rolf A. de By, ITC, The Netherlands
Leila De Floriani, Università di Genova, Italy
Max J. Egenhofer, NCGIA and University of Maine, USA
Christos Faloutsos, University of Maryland, USA
Robin Fegeas, USGS, USA
Andrew U. Frank, TU Vienna, Austria
Volker Gaede, Humboldt-Universität, Berlin, Germany
Kenn Gardels, UCB, USA

Contents

Spatial Access Methods

Keynote Paper

Keynote Paper

Spatial Internet Marketplaces: A Grand Challenge?

David J. Abel

CSIRO Mathematical and Information Sciences
GPO Box 664, Canberra, ACT 2601, Australia
dave.abel@cmis.csiro.au

Abstract. While the field of Spatial Information Systems continues to grow impressively as an Information Technology sector, there remain several applications of very high potential value which require greater access to data and data manipulation tools than can currently be provided. This paper considers extension of the electronic market concept to Spatial Internet Marketplaces. A Spatial Internet Marketplace consists of a large number of providers offering data and specialised processing services and customers who buy those services as required. An infrastructure of a limited set of standards for services and of specialist services to aid in planning and executing tasks alleviates the problems of using a very large and heterogeneous collection of services. Key issues in the design and implementation of a Spatial Internet Marketplace are analysed to identify a number of core research challenges.

1 Introduction

The past two decades have seen the development of an impressive body of concepts and techniques assembled by the research community; the emergence of a specialist Spatial Information Systems sector within the Information Technology industry; and massive investments in spatial data acquisition and database maintenance. Predominantly, spatial applications are still implemented as single-site information systems: databases, software, hardware and procedures are assembled locally to satisfy a targeted set of applications. Relatively recent advances have been corporate applications using centralised databases accessed over a network. These systems allow a user to draw on remote resources but continue to rely on assembly by the enterprise of resources to satisfy a selected set of applications. There remain, however, several applications which are potentially significant at corporate, industry and national scales but which demand greater use of resources from external sources.

Three representative examples demonstrate the challenge:

1. in Earth Observation (EO), some tasks such as near-real-time monitoring of environmental disasters require rapid access to recently-acquired imagery, to specialist software and to high-performance computing systems [SPJ+95]. The cost of assembling the resources locally is a barrier to the deployment of such applications. More widely recent European Space Agency study found that the principal disincentives to wider exploitation of EO data were attraction of customers, availability of EO data and lack of financial resources access to the data and processing facilities [And95];

2. in Spatial Decision Support Systems (SDSS), the ill-defined nature of many policy and planning problems means that frequently extra data and visualisation and modelling facilities are needed as the investigation proceeds [CA96]. The times to acquire these can mean that SDSS can be considered as an aid to decision-making only where there is a generous amount of time for investigation of the problem. In practice, this limits severely the applicability of SDSS;

3. in large organisations, data is typically acquired and managed by specific functional units, but used widely through the organisation. Similarly modelling tools for analysis, predictive simulation, optimisation and inferencing are acquired or developed by specific units and are of corporate value. Often these data and modelling resources have been shared by copying from unit to unit and there are high benefits from managing them as corporate resources and accessed in situ. A monolithic, tightly-integrated corporate information system, however, has a high risk of being unable to keep pace with changes in the organisation's business and with technology changes. Some form of 'future-proofing' [BS95] is desirable.

The common element to these problems is access to data resources and to specialised computational resources on an 'as required' basis. Data and software repositories [DG87; GV97], which simplify the acquisition of data and software, are only partly satisfactory, as some important data sets are volatile and as there can be high costs and delays in local installation of data and software. An attractive solution to the accessibility problem is the electronic marketplace, apparently first proposed in the area of decision support technologies [BKM+95]. The core concept is that there are several providers who undertake to perform, at their site with their software, computational processing such as analysis and optimisation. Customers then buy services from the providers and are not tied to specific providers. Providers then compete for customers and are driven by market forces in the precise design and implementation of their services.

This paper considers Spatial Internet Marketplaces which extend the electronic marketplace concept by facilities needed for spatial applications. Essentially these include the presence of both data and computational services providers and of agents which simplify discovery of services relevant to a task, the development of plans to combine services to retrieve or derive a specified data set, and the execution of plans. To contain the problems of heterogeneity, it is proposed that providers conform to a thin set of standards based on a simple architectural model.

In Section 2, a Spatial Internet Marketplace is defined functionally and related to forms of marketplaces proposed in other areas. The concept is also clarified by comparison with multidatabase systems. A design manifesto is offered in Section 3, and a Spatial Internet Marketplace architecture as the types of services and agents and their interactions is sketched in Section 4. We conclude with a brief analysis of the major research challenges in establishing a Spatial Internet Marketplace.

2 What Would a Spatial Internet Marketplace Be?

Bhargava et al [BKM+95, BKR+97, BKQ95] define an electronic (Internet) marketplace as consisting of customers, providers offering computational services, and an infrastructure assisting customers to buy those services. Crucially the customers can exercise choice in buying services so that market forces drive the relationships between customers and providers. Güenther et al [GMS+97] consider a closely- allied topic in publishing econometric analysis services on the World Wide Web, examining such issues as transformation of data and re-expression of commands in linking services implemented independently using different tools.

The concept for a Spatial Internet Marketplace, at a first level, can be demonstrated by a hypothetical example. An Australian telecommunications company wishes to extend its cellphone network in a city by installing an extra repeater. The city maintains a collection of databases, and offers Internet access to these on a fee-for-service basis. The engineer assigned to the task of selecting the site for the repeater firstly accesses, from the company's own databases, the sites of the existing repeaters and their footprints. He buys, from the city's data services, spatial data for residential, commercial, industrial and retail areas. He combines these data with the coverage data to determine the gaps in the coverage of the current cellphone service. He sees that the gaps are in the North-West and South-East sector of the city, both with mixed residential and retail areas. The next step is to buy, from the data service of the Australian Bureau of Statistics, the demographic profiles (including disposable household income) for these sectors, specifying them geographically. The South-East area has a much lower number of households whose disposable income makes them likely cellphone customers, and the engineer focusses in the North-West sector. His next step is to buy, from the city again, a Digital Elevation Model, for the North-West sector. He sees three suitable hill-top sites for repeaters. He uses a computational service offered by a small specialist company to estimate the footprint of the company's standard repeater configuration for the three sites. For each site, he passes to the computational service the location of the repeater, its operating characteristics (frequency, power. and so on) and the Digital Elevation Model, and receives the repeater's footprint. One location is clearly unsuitable because of extensive shadows. To choose the better of the remaining site options, the engineer uses the city's data service to count the number of households and businesses which are enclosed in the footprint.

The application could, of course, have been established as a single-site application. This would have required the telecommunications company to purchase copies of all the data sets used and a licence for the footprint estimation software. This would have required a considerable investment in time to establish and in licence fees, and the demographic and some other data would need to be periodically updated. Additionally, the marketplace, by enabling access to data and software which are needed rarely, allows the engineer to be more expansive in his investigation, rather than being limited by the data and modeling tools installed by the company.

Analysis of the types of services which might be provided in a spatial marketplace and the types of applications which might be built suggests that a marketplace will have a number of distinctive characteristics. Firstly, there are requirements for both data services and for processing services for data manipulation. The costs of acquiring digital spatial data and the role of government agencies in maintaining core data sets of regional or national interest mean that customers rely extensively on data from external sources with locally-captured data restricted to the specific purposes of their organisation. The data services include both delivery of complete data sets and subsets of data sets, by region or theme. The services might be implemented using fully-fledged database systems, but might also include native file systems (for raster data collections, for example). The computational services of interest cover a very extensive range: transformation between coordinate systems, representations (vector to raster, for example), modelling (analysis, predictive simulation, optimisation and inferencing), image rectification and enhancement, and so on. The set of types of processing services is extensive, and there is apparently no widely-accepted approach to their description and classification.

Secondly, spatial analysis often combines independently-acquired data sets in a process using many atomic operations. Essentially geographic space acts a common domain which allows objects to be related across databases. Fusing data from a number of sources typically requires intermediate processing such as coordinate and representation conversion, and similarly many spatial analyses require complex, multi-step processing. This indicates that a Spatial Internet Marketplace must cater for tasks requiring invocation of several services in a certain sequence, with routing of intermediate results from service to service. (We describe statement of the services in the sequence of execution as a plan.)

Thirdly, Spatial Information Systems technology is applied by very many disciplines. Each discipline has its own set of concepts used to describe the physical phenomena and processes of interest: that is, it has its own underlying models to describe objects and processes. Unlike marketplaces aimed at a specific community of interest, a Spatial Internet Marketplace can be expected to have customers from many communities of interest. While some data and some processing services would be specific to a particular community, many are of interest to several communities. Mapping of semantics and representations is then likely to be needed.

These special characteristics, together with the common elements of infrastructures for distributed systems, suggest that the infrastructure for a Spatial Internet Marketplace desirably has three major elements:

1. standards for customer-provider interaction, in terms of connections between them, command languages (equivalently, invocation of services by customers) and data transfer;

2. catalogue services, which advertise and describe the services available. This provides a resource discovery function;

3. agents, which assist a customer to develop a plan to generate a required data set (specified declaratively) by some combination of invocations of data and processing services and to execute that plan.

These elements are similar to those of many other forms of distributed systems (e.g. [OMG96, BM96]). The Spatial Internet Marketplace is distinctive primarily in the presence of planning and execution agents in the infrastructure which (if used) simplify the design and implementation of applications. Further differences become apparent when the specific design goals for a marketplace are identified. Essentially a Spatial Internet Marketplace is a specialised form of distributed system which can draw extensively on concepts and tools from the more generic models and tools for distributed systems.

Another perspective on Spatial Internet Marketplaces is provided by comparing it to a multidatabase system. A multidatabase system principally provides a uniform interface to an integrated collection of heterogeneous databases. Integration is performed by a database administrator who develops a single schema (held in the Global Data Dictionary) which describes uniformly and which relates the data available from the site databases. The focus is then on a static collection of databases which are integrated prior to access by clients, and the sites are only database systems (see, however, Schek and Wolf's proposal of external operation services [SW93]). In contrast, a spatial marketplace has potentially a large number of data and processing services so that prior integration is not possible. Rather a customer (or an agent) performs integration within an application at run-time. There is no equivalent of a Global Data Dictionary, as the marketplace's catalogue service essentially records the equivalent of the export schemas of the multidatabase system's site databases.

3 A Design Manifesto

Clearly, the design and implementation of a Spatial Internet Marketplace is a complex problem on several grounds. Three aspects appear especially important. Firstly, the problem includes the general issues of the organisation, management and use of a potentially very large and diverse collection of resources. Secondly, the Internet environment, in terms of tools available and the standards in wide use, is changing at a rate possibly unprecedented in Information Technology. A marketplace design which is closely tied to current tools and standards would quickly become obsolescent .Thirdly, and perhaps crucially, if a marketplace is to be an effective resource for applications developers, it must provide a critical mass of services. This suggests that the complexity and cost of publishing a service should be sufficiently low to not deter prospective service providers.

These considerations indicate that research on Spatial Internet Marketplaces should be based on some design principles which limit the complexity of the infrastructure and which provide some capacity to exploit changes in Internet technology:

1. the external interfaces for services and agents should be uniform and simple. This restricts the severity of problems of heterogeneity, and provides the greater opportunity for the development of effective agents for resource discovery and for planning;

2. the implementation of agents and services should not prescribed. The standards should deal only with the external interfaces;

3. invention of new protocols and standards should be minimised. This is particularly intended to allow a marketplace implementation to keep pace with the general Internet environment;

4. extensibility should be a deliberate design objective. Extensibility covers the ability to introduce additional data types, descriptions of services, and the range of services offered;

5. scalability is required to cater for potentially a very large number of services.

4 The SMART Project

The SMART (Spatial Marketplace) Project [AGT+97] has begun to explore the design and implementation of Spatial Internet Marketplaces through consideration of an architectural model and implementation of key elements of the infrastructure.

The goal of uniformity and simplicity is addressed by including only three types of services (the catalogue, data services and for processing services). The planning and execution services are modelled as processing services. SMART also adopts a simple, coarse-grained interaction between customers and providers. That is, a customer invokes a service by passing a request message, and the service responds, on completion of the operation, with a data stream which includes the resultant data, the cost of the service and any diagnostics. The request message, uniformly for the three service types, specifies a target set of objects of interest (as constraints expressed in terms of the elements of catalogue description of the service) and a target list of attribute values to be returned (in terms of data types and any functions to be applied before transfer).

Compatibility with the Internet environment is targeted by use of the HTTP protocol [BFF96] for customer-provider communication and of the MIME data transfer format for request messages and data streams.

Key provisions for extensibility in data types and service descriptions are made in the catalogue. Essentially, this includes a services register, a data type register, and a thesaurus and glossary of high- level concepts used to describe services and the objects present in their external schemas.

5 Some Key Research Issues

The development of Spatial Internet Marketplaces will require solution of several challenging research problems in the infrastructure; in new algorithms for key database and processing operations; and in applications drawing on a marketplace.

There appear to be two major issues in establishing an infrastructure. Firstly, the planning agent (to recommend how a specified data collection can be retrieved or derived) is clearly important as an aid to a customer in exploiting a large and diverse collection of resources. The planning problem is quite different to that of query execution planning in relational or object-oriented database systems (because, again, queries are not expressed in terms of a single schema) and it is possible that the starting point for marketplace planning algorithms will come from fields such as Artificial Intelligence or Operations Research. Secondly, if the design manifesto proposed here is adopted, there is the issue of design of a uniform interface, particularly for the command language. As generality of command languages is often bought at the expense of simplicity or expressiveness, it is possible that investigation will find that this objective is unattainable.

The implementation of processing services in the marketplace will require careful consideration of the applicability of approaches formulated for single-site systems and might lead to formulation of new algorithms. For example, an algorithm might assume the presence of a spatial index on a data set. In a conventional system, while building the index incurs a certain cost, the index allows faster execution of many operations of many types on the data set from its creation to its destruction. In a marketplace, where a customer transfers the data set to the processing service and the service destroys the data set on completion of the operation, the cost of building the index is charged fully as part of the cost of the operation.

Finally, the ultimate test of the value of a marketplace must be in terms of the spatial applications which it enables. This paper commenced with a nomination of three types of applications which appear more readily implemented using a marketplace than as single-site systems. A wider analysis of the types of applications enabled by a marketplace or more effectively implemented using a marketplace is clearly necessary. Investigation would desirably extend to how the use of marketplace services influences the design of the application. Clearly, there is great synergy in the investigation of applications, the infrastructure and services and a "whole systems" approach to the research problems is indicated.

Acknowledgment: I am indebted to my SMART colleagues (Volker Gaede, Stuart Hungerford, Kerry Taylor and Xiaofang Zhou) for their contribution to the evolution of the Spatial Internet Marketplace concept. This paper draws on publications in progress by the team. I also acknowledge many stimulating discussions with colleagues in CSIRO, industry and government which led us to identify marketplaces as a research topic and which have tested initial design proposals.

References

[AGT97] Abel, D.J., Gaede, V.J, Taylor, K.L. and Zhou, X., SMART: Towards Spatial Internet Marketplaces, Technical Report SIS-1997-01, CSIRO Mathematical and Information Sciences, Canberra, 1997.

[And95] Andersen, B.H., A Structural Analysis of the European Earth Observation Value Added Industry, Centre for Earth Observation, Ispra, Document CEO/17-/1995, November 1995, 17pp.

[BFF96] Berners-Lee, T., Fielding, F., and Frystyk, H., Hypertext Transfer Protocol - HTTP/1.0, Technical Report HTTP Working Group, February 1996. See http://ds.internic.net/rfc.

[BKM95] Bhargava, H.K., King, A.S. and McQuay, D.S., DecisionNet: an Architecture for Modelling and Decision Support over the World Wide Web. In Tung X. Bui (Ed.), Proceedings of the Third International Society for Decision Support Systems Conference, Hong Kong, 1995, International Society for DSS, Vol II, pages 541-550. See http://dnet.sm.nps.navy.mil.

[BKR+97] Bhargava, H.K., . Krishnan, R., Roehrig, M.S., Kaplan, D., and Mueller, R., Model Management in Electronic Markets for Decision Technologies: a Software Agent Approach. In Proceedings of the Thirtieth Hawaii International Conference on System Sciences, Maui, Hawaii, January 1997. See http://dnet.sm.nps.navy.mil.

[BKK95] Bhargava, H.K, Krisnnan, R., and Kaplan, D., On Generalized Access to a WWW-based Network of Decision Support Systems, In Tung X. Bui (Ed.), Proceedings of the Third International Society for Decision Support Systems Conference, Hong Kong, 1995, International Society for DSS, Vol II, pages 551-560. See http://dnet.sm.nps.navy.mil.

[BM96 Buehler, K. and McKee (Eds), L., The OpenGIS Guide, Open GIS Consortium, Wayland, Mass., 1996. See http://www.ogis.org/guide.

[BS95] Brodie, M.L. and Stonebraker, M.: Migrating Legacy Systems: Gateways, Interfaces, and the Incremental Approach, Morgan Kaufmann Publishers, 1995.

[CA96] Cameron, M.A. and Abel, D.J., A Problem Model for Decision Support Systems, in M.J. Kraak and M. Molenaar (Eds), Proceedings of the 7th International Symposium on Spatial Data Handling, August 12-16, 1996, Delft, The Netherlands, Faculty of Geodetic Engineering, Delft University of Technology, 3A25-3A36.

[DG87] Dongarra, J.C. and Grosse, E., Distribution of Mathematical Software by Electronic Mail, Communications of the ACM, 30: 403-407, 1987. See http://achille.att.com/netlib/index.html.

[GMS+97] Güenther, R. Mueller, P. Schmidt, H. K. Bhargava, and R. Krishnan, MMM: A WWW-Based Method Management System for Using Software Modules Remotely, to appear in IEEE Internet Computing, 1997. See http://mmm.wiwi.hu-berlin.de/mmm/index2.html.

[GV97] Güenther, O., and Voisard, A., Metadata in Geographic and Environmental Data Management, in W. Klas and A. Sheth (Eds), Managing Multimedia Data: Using Metadata to Integrate and Apply Digital Data, McGraw-Hill, New York, 1997.

[OMG96] Object Management Group, Trading Object Service, OMG Document orbis/96-05-06, OMG, Farmingham, MA, May 1996. See http://www.omg.org/docs/orbos/96-05-06.ps.

[SGH+95] D. Sarrat, G. Pierre, J. Harms, G. Richard, R. Oizel and S. Vizzari, Modelling of the Distributed EO Data Handling Facilities- Plans for Year 2000, European Space Agency Report DUS- MGT-006-TR, 4th December 1995, 93 pp.

[SW93] H.J. Schek and A.Wolf, From Extensible Database Systems to Interoperability between Multiple Databases and GIS Applications. In D.J. Abel and B.C. Ooi (Eds), Advances in Spatial Databases, Proc. SSD'93, Singapore, June 1993, LNCS 692, Springer-Verlag, Heidelberg, pp 207-238.

Spatial Similarities

3D Similarity Search by Shape Approximation[1]

Hans-Peter Kriegel, Thomas Schmidt, Thomas Seidl

Institute for Computer Science, University of Munich
Oettingenstr. 67, D-80538 München, Germany
http://www.dbs.informatik.uni-muenchen.de
{kriegel | schmidtt | seidl}@dbs.informatik.uni-muenchen.de

Abstract. This paper presents a new method for similarity retrieval of 3D surface segments in spatial database systems as used in molecular biology, medical imaging, or CAD. We propose a similarity criterion and algorithm for 3D surface segments which is based on the approximation of segments by using multi-parametric functions. The method can be adjusted to individual requirements of specific applications by choosing appropriate surface functions as approximation models. For an efficient evaluation of similarity queries, we developed a filter function which supports fast searching based on spatial index structures and guarantees no false drops. The evaluation of the filter function requires a new query type with multidimensional ellipsoids as query regions. We present an algorithm to efficiently perform ellipsoid queries on the class of spatial index structures that manage their directory by rectilinear hyperrectangles, such as R-trees or X-trees. Our experiments show both, effectiveness as well as efficiency of our method using a sample application from molecular biology.

1 Introduction

The problem we focus on in this paper is similarity retrieval of 3D surface segments in a spatial database system. We propose a new similarity measure for surface segments, based on the approximation of the segments by multi-parametric functions. Furthermore, we show how to efficiently evaluate similarity queries based on this measure using a spatial index structure.

Motivation. The task of finding similar surface segments arises in many recent application areas of spatial database systems, such as:

Molecular Biology. A challenging problem in molecular biology is the prediction of protein interactions (the molecular docking problem): Which proteins from the database form a stable complex with a given query protein? It is well known that docking partners are recognized by complementary surface regions. In many cases, the active sites of the proteins, i.e. the docking regions are known and can be extracted from the protein surface and subsequently stored in a database [SK 95]. Then, the problem of finding docking partners reduces to finding similar (complementary) surface segments from a large database of segments for a given query segment.

Medical Imaging. Modern medical imaging technology such as CTI or MRI produces descriptions of 3D objects like organs or tumors by a set of 2D images. These images represent slices through the object, from which the 3D shape can be reconstructed. A

---

1. This research was funded by the German Ministry for Education, Science, Research and Technology (BMBF) under grant no. 01 IB 307 B. The authors are responsible for the content of this paper.

method for retrieving similar surface segments would help to discover correlations between shape deformations of organs and certain deceases.

Further application fields include CAD and mechanical engineering. In order to meet the typical requirements of these application domains, our method supports invariance against translation and rotation, because the position and orientation of the objects in 3D space does not affect shape similarity. Since the number of objects in a spatial database typically is very large, efficient query processing is important and will be supported by our algorithm. Our method requires the surface segments to be given as sets of points which can be obtained from all common surface representations.

Related Work. In recent years, considerable work has been done on similarity search in database systems. Most of the previous approaches, however, deal with one- or two-dimensional data, such as time series, digital images or polygonal data. However, they do not handle three-dimensional objects.

Agrawal et al. present a method for similarity search in a sequence database of one-dimensional data [AFS 93]. The sequences are mapped onto points of a low-dimensional feature space using a Discrete Fourier Transform, and then a Point Access Method (PAM) is used for efficient retrieval. This technique was later generalized for subsequence matching [FRM 94], and searching in the presence of noise, scaling, and translation [ALSS 95]. However, it remains restricted to one-dimensional sequence data.

Jagadish proposes a technique for the retrieval of similar shapes in two dimensions [Jag 91]. He derives an appropriate object description from a rectilinear cover of an object, i.e. a cover consisting of axis-parallel rectangles. The rectangles belonging to a single object are sorted by size, and the largest ones serve as retrieval key for the shape of the object. Due to a normalization, invariance with respect to scaling and to translation is achieved. Though this method can be generalized to three dimensions by using covers of hyperrectangles, it has not been evaluated for real world 3D data, and furthermore, it does not achieve invariance against rotations.

Mehrotra and Gary suggest the use of boundary features for the retrieval of shapes [MG 93] [GM 93]. Here, a 2D-shape is represented by an ordered set of surface points, and fixed-sized subsets of this representation are extracted as shape features. All of these features are mapped to points in multidimensional space which are stored using a PAM. This method can handle translation, rotation and scaling invariance, as well as partially occluded objects, but is essentially limited to two dimensions.

For retrieving similar 2D polygon shapes from a CAD database system, previous work is presented in [BKK 97] and [BK 97]. This technique applies the Fourier Transform to shape encoding for retrieving similar sections of polygon contours. The polygon sections are stored as extended multidimensional feature objects and a Spatial Access Method (SAM) is used for efficient retrieval [BKK 96] [Ber+ 97]. This approach, however, is also limited to two dimensions.

Korn et. al. propose a method for searching similar tumor shapes in a medical image database [Kor+ 96]. However, they consider only 2D images, and the similarity measure that they investigate is volume based. This might cause problems when applied to surface segments, because surface segments in general do not enclose a volume. Therefore this method cannot be directly applied to the problem we focus on in this paper.

Last but not least, the QBIC (Querying By Image Content) system [Fal+ 94] contains a component for 2D shape retrieval where shapes are given as sets of points. The method is based on algebraic moment invariants and is also applicable to 3D objects [TC 91]. As an important advantage, the invariance of the feature vectors with respect to rigid transformations (translations and rotations) is inherently given. However, the adjustability of the method to specific application domains is restricted. From the available moment invariants, appropriate ones have to be selected, and their weighting factors may be modified. Whereas the moment invariants are abstract quantities, the approximation models that have to be chosen in our approach are concretely visualizable as 3D surfaces, thus providing an early impression of the suitability for certain application domains.

The paper is organized as follows: In section 2, we present the approach of 3D shape approximation and introduce a shape similarity function which will be used for the specification of similarity queries. Section 3 contains the derivation of an efficient method for similarity query processing. In section 4, we present an algorithm for ellipsoid query processing on spatial access methods. Section 5 shows experimental results, and section 6 concludes the paper.

2 3D Similarity based on Shape Approximation

In this section, we present a generic approximation method to represent the shape of 3D surface segments by multi-parametric surface functions. To adapt the shape similarity criterion to the user's application domain, simply instantiate the similarity search method by using an appropriate approximation model. The basic idea of the similarity criterion is that the approximations of the objects of interest are compared mutually. The quantity of (dis-)similarity is defined in terms of the mutual approximation error. Thus, the more an approximation model fits the characteristics of a specific application domain, the more powerful it is to differentiate between a variety of shapes.

The quality of approximation depends on the appropriateness of the approximation model that has been chosen for an actual application. If the real objects actually were instances of the approximation model, approximation errors would only occur due to errors in measurement and roundoff errors. However in general, the real objects of the application are not derived from an artificial approximation model, but the approximation acts as a simplified model for a variety of complex real objects. Thus, choosing an adequate approximation model for specific applications is an important design decision.

2.1 Approximation of 3D-Segments

The objects our method is designed for are 3D surface segments which are represented by sets of points. Multi-parametric two-dimensional (surface) functions $f_{app}(x,y)$ are used to approximate these objects (cf. figure 1). For example, we use paraboloids and trigonometric polynomials of various degrees as approximation models. The length r of the parameter set $app = \langle a_i \rangle$ is called the dimension of the approximation model.

Definition (*approximation error, approximation, relative approximation error*). Let s be a 3D surface segment given by n points $\{p\}$, and $app = \langle a_i \rangle$ any approximation parameter set for a given approximation model. (i) The *approximation error* is defined in a normal-ized least squares sense to be $d_s^2(app) := \frac{1}{n}\sum_{p \in s}(F_{app}(p_x, p_y) - p_z)^2$. (ii) A parameter set app_s for which the approximation error for s is minimum, $\forall app: d_s^2(app) \geq d_s^2(app_s)$, is called the *approximation* of s. (iii) The *relative approximation error* of an arbitrary approximation parameter set app' is defined to be $\Delta d_s^2(app') = d_s^2(app') - d_s^2(app_s)$.

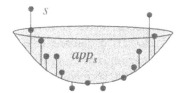

Fig. 1. A 3D surface segment s and its approximation app_s

Notice that in general, the approximation app_s of s is unique. In the worst case, the approximation parameters may vary without affecting the approximation error. Such a case indicates that the approximation model has been chosen inappropriate with respect to the application domain, and should be modified.

In general, even for the approximation app_s of a segment s, the approximation error $d_s^2(app_s)$ will be greater than zero. In order to end up with a similarity function that char-acterizes similarity of an object to itself by the value zero, we introduced the relative approximation error.

Lemma 1. (i) For any 3D surface segment s and any approximation parameter set app', the relative approximation error is non-negative: $\Delta d_s^2(app') \geq 0$. (ii) The relative approx-imation error reaches zero. In particular, $\Delta d_s^2(app_s) = 0$ for all segments s.

Proof. (i) Since the approximation error $d_s^2(app_s)$ is defined to be minimum, the estima-tion $d_s^2(app') \geq d_s^2(app_s)$ holds for all parameter sets app', and thus $\Delta d_s^2(app') \geq 0$. (ii) For $app' = app_s$ we have: $\Delta d_s^2(app_s) = d_s^2(app_s) - d_s^2(app_s) = 0$. ◊

3D-Normalization. In general, the points of a segment s are located anywhere in 3D space and are oriented arbitrarily. Since we are only interested in the shape of s, but not in its location and orientation in 3D space, we transform s by a rigid 3D-transformation into a normalized representation. There are two ways to integrate normalization: (1) *Sep-arate*: First normalize the segment s, and then compute the approximation app_s by least-squares minimization. (2) *Combined*: Minimize the approximation error simultaneously over all the normalization and approximation parameters.

In our experiments, we used the combined normalization approach. For similarity search purposes, only the approximation parameters are used. However, the normaliza-tion parameters may be required later for superpositioning of segments.

Numerical Aspects. Approximation by least squares minimization is a numerical task. When restricting to certain approximation models, efficient solutions are available. An appropriate family of approximation models is given as linear combinations of the approximation parameters a_i with arbitrary two-dimensional functions $f_i(x,y)$. Examples are paraboloids such as $a_1 \cdot x^2 + a_2 \cdot y^2$ for degree two, or trigonometric polynomials such as $a_1 \cdot cos(x) + a_2 \cdot sin(x) + a_3 \cdot cos(y) + a_4 \cdot sin(y)$. The generic formula of linear combination models is as follows:

$$F_{app}(x,y) = \sum_{i=1...r} app_i \cdot f_i(x, y)$$

For such linear combination models, [PTVF 92] recommend to perform least-squares approximation by Singular Value Decomposition (SVD) as the method of choice. Besides the r approximation parameters $app = \langle a_i \rangle$, SVD also returns an r-vector w of condition (confidence) factors as well as an orthogonal $r \times r$-matrix V supporting the computation of relative approximation errors. When denoting the rows of V_s by V_{si}, and $A_s = V_s \cdot diag(w_s)^2 \cdot V_s^T$, the error formula is as follows:

$$\Delta d_s^2(app') = \sum_{i=1...r} w_{si}^2((app' - app_q) \cdot V_{si})^2 = (app' - app_s) \cdot A_s \cdot (app' - app_s)^T$$

2.2 Shape Similarity of 3D-Segments

For 3D surface segments, our shape similarity criterion considers two components, the extension of the segment in 3D space, and the shape of the segments in a narrow sense. We define the extension similarity to be the Euclidean distance of the 3D extension vectors which we obtain as the principle moments of inertia. The shape component of the distance function is based on shape approximation, and we exploit more information than only the approximation parameters. This is recommended, since the confidence of the parameters may substantially vary for single segment approximations. Thus, we introduce the method of *mutual approximation errors*. The basic question for shape similarity quantification is the following: How much will the approximation error increase, if segment q would have been approximated by the approximation of segment s, app_s, instead of its own approximation, app_q, and vice versa (cf. figure 2). This approach leads us to the definition of approximation-based shape similarity.

Definition (*shape similarity*). Let s and q be two 3D surface segments, with ext_s and ext_q being their 3D extension vectors in space. Define the mutual approximation distance to be $d_{app}^2(s, q) := \frac{1}{2}\Delta d_s^2(app_q) + \frac{1}{2}\Delta d_q^2(app_s)$, and the 3D extension distance as the squared Euclidean distance $d_{ext}^2(s, q) := (ext_s - ext_q)^2$. With u_{app} and u_{ext} being non-negative weighting factors, the shape similarity function d_{shape} is defined to be:

$$d_{shape}(s, q) = \sqrt{u_{app}d_{app}^2(s, q) + u_{ext}d_{ext}^2(s, q)}$$

Additionally, for a segment s, we define the $r+3$-dimensional key vector key_s to be the concatenation of the r-dimensional vector app_s and the 3-dimensional vector ext_s: $key_s = (app_s, ext_s)$.

Since we have chosen to combine d_{app}^2 and d_{ext}^2 by the square root over the sum, d_{shape} is related to the Euclidean distance in the following way:

description	symbol	definition
approximation model	$F_{app}(x, y)$	$\sum_{i=1...r} a_i \cdot f_i(x, y) =$ $(a_1, ..., a_r) \cdot (f_1(x, y), ..., f_r(x, y))$
approximation error	$d_s^2(app)$	$\frac{1}{n}\sum_{p \in s} (F_{app}(p_x, p_y) - p_z)^2$
(the) approximation	app_s	$\forall app: d_s^2(app) \geq d_s^2(app_s)$
relative approximation error	$\Delta d_s^2(app')$	$d_s^2(app') - d_s^2(app_s) =$ $(app' - app_s) \cdot A_s \cdot (app' - app_s)^{\mathsf{T}}$
key vector	key_s	$(app_s, ext_s) = (a_1, ..., a_r, e_1, e_2, e_3)$
(simple) shape similarity	$d_{app}^2(s, q)$	$\frac{1}{2}\Delta d_s^2(app_q) + \frac{1}{2}\Delta d_q^2(app_s)$
3D extension similarity	$d_{ext}^2(s, q)$	$(ext_s - ext_q)^2$
(final) shape similarity	$d_{shape}(s, q)$	$\sqrt{u_{app}d_{app}^2(s, q) + u_{ext}d_{ext}^2(s, q)}$ with weighting factors u_{app} and u_{ext}

Table 1. Symbols and definitions used for shape approximation and shape similarity of segments

covering four proteins with a high structural similarity, and the fructose bisphosphatase family (PDB code 1FRP-A) with 18 members.

We queried the database of 6,200 segments with the docking segment of each member of the azurin family. Figure 3 shows how the docking segments of the four azurin molecules rank within the database according to the shape distance d_{shape}. We compare different approximation models that have 2, 4, 8, and 12 parameters, leading to 5-, 7-, 11-, and 15-dimensional *key* vectors, respectively (cf. table 2). In the diagrams, the abscissa axis indicates the four azurin segments, whereas the ordinate axis depicts the minimum, maximum, and average position that has been achieved by the ranking according to d_{shape} within all the 6,200 entries. The experiments support the adequacy of the TRIGO-4 model for the docking-sites application domain. TRIGO-4 is the trigonometric polynomial function $trigo4(x,y) = a_1\cos x + a_2\sin x + a_3\cos y + a_4\sin y$.

model	formula of approximation model
PARAB-2	$(a_1, a_2) \cdot (x^2, y^2) = a_1x^2 + a_2y^2$
TRIGO-4	$(a_1, a_2, a_3, a_4) \cdot (\cos x, \sin x, \cos y, \sin y)$
TRIGO-8	$(a_1, ..., a_8) \cdot (\cos x, \sin x, \cos y, \sin y, \cos 2x, \sin 2x, \cos 2y, \sin 2y)$
TRIGO-12	$(a_1, ..., a_{12}) \cdot (\cos x, \sin x, \cos y, \sin y, ..., \cos 3x, \sin 3x, \cos 3y, \sin 3y)$

Table 2. A sample of approximation models of various dimensionalities

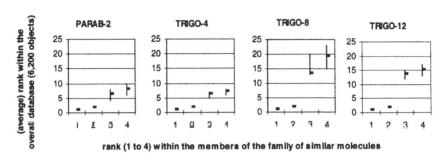

Fig. 3. Ranking by d_{shape} for various approximation models (example: azurin family). The members of the example family rank on top positions, in particular below rank 25 out of 6,200. The best positions are achieved for the TRIGO-4 model.

Additional experiments e.g. for hemoglobin molecules or for trypsin inhibitors show that most of the molecules that were ranked on top positions within the overall database according to d_{shape} also belong to the same family of molecules as the query object.

3 Efficient Similarity Query Processing

There are two types of queries that are relevant with respect to similarity search: range queries and k-nearest neighbor (k-nn) queries. A range query $SIM_{range}(q, \varepsilon)$ is specified by a query object q and a range value ε, defining the answer set to contain all the objects s from the database that have a distance less than ε to the query object q. A k-nearest neighbor query $SIM_{k-nn}(q, k)$ for a query object q and a cardinal number k specifies the retrieval of those k objects from the database that are most similar to q.

Due to the immense and even increasing size of current databases for molecular biology, medical image management, and engineering applications, strong efficiency requirements have to be met. Thus, for the evaluation of complex similarity queries, fast processing is important. As successfully applied for processing of point queries, window queries, and spatial joins, a multi-step query processing architecture is recommended [OM 88] [BHKS 93] [BKSS 94]. Following this paradigm, several filter steps produce and reduce candidate sets from the database, yielding an overall result that contains the correct answer, producing neither false positive nor false negative decisions (no false hits and no false drops). To improve efficiency, filter steps are usually supported by PAMs and SAMs. Refinement steps discard false positive candidates (false hits), but do not reconstruct false negatives (false drops) that have been dismissed by the filter step. Therefore, a basic requirement for any filter step is to prevent false drops.

3.1 Multi-Step Query Processing

For feature-based similarity search, efficient query processing already is available [AFS 93] [Kor+ 96] [BKK 97] . The objects O are transformed to feature vectors $F(O)$ which are efficiently managed by multi-dimensional point access methods (PAM). The algorithms are shown to work correctly, in particular, producing no false drops, if they use a feature distance function $D_{feature}$ that lower-bounds the object distance functions D_{object}, i.e. $D_{feature}(F(O_1), F(O_2)) \leq D_{object}(O_1, O_2)$. The efficiency of the algorithms depends on selectivity and the query processing efficiency provided by the underlying PAM. Figures 4 and 5 show the algorithms $SIM_{range}(q, \varepsilon)$ for range queries (cf. [FRM 94]), and $SIM_{k-nn}(q, k)$ for k-nn queries from [Kor+ 96], where the distance function $D_{feature}$ is a filter function $f(s, q)$, and D_{object} is d_{shape}.

Algorithm SIM$_{range}$ (q, ε)

(1) *Filter Step.* Using the filter function $f(s, q)$, perform a range query on the PAM to obtain the candidate set $\{s \mid f(s, q) \leq \varepsilon\}$.

(2) *Refinement Step.* For each of the candidates s, evaluate the shape similarity $d_{shape}(s, q)$, and report those objects that fulfill $d_{shape}(s, q) \leq \varepsilon$

Fig. 4. Algorithm for range queries wrt d_{shape}, based on a PAM, cf. [FRM 94]

Algorithm SIM$_{k-nn}$ (q, k)

(1) *Primary Candidates.* Perform a k-nn query on the PAM to find the k-nn objects with respect to the filter function $f(x, q)$.

(2) *Range Determination.* For all the candidates s obtained from (1), compute the actual distance $d_{shape}(s, q)$ and return the maximum value d_{max}.

(3) *Final Result.* Perform a range query on the PAM to obtain the candidate set $\{s \mid f(s, q) \leq d_{max}\}$. For all the candidates s, evaluate $d_{shape}(s, q)$, and report the objects having the k smallest values.

Fig. 5. Algorithm for k-nearest neighbors queries wrt d_{shape}, based on a PAM, cf. [Kor+ 96]

To adapt the algorithms to shape similarity query processing, it suffices to provide a filter function $f(x, q)$ that fulfills the correctness and efficiency requirements, i.e.:

(1) Show the correctness of the filter function, i.e. $f(s, q) \leq d_{shape}(s, q)$.

(2) Provide efficient algorithms to perform queries on the PAM that respect the distance function $f(s, q)$.

3.2 A Lower Bound for Shape Similarity

In order to apply the multi-step query processing paradigm to shape similarity search, we derive an appropriate filter function for the shape similarity function d_{shape}. Besides the correctness, which is ensured by a lower-bounding criterion, we look for a solution that provides efficient support by PAMs. Although a PAM has been developed that efficiently manages feature spaces of dimension 10 to 20 or more [BKK 96], the lower dimensionality promises the better performance.

Let us investigate the similarity function d_{shape} and its components with respect to the number of data values that are required for the evaluation. Table 3 illustrates the situation: Observe that for the evaluation of $d_{shape}(s, q)$, r^2+r data values are required concerning the segment q, since the formula contains the r-vector app_q as well as the $r \times r$-matrix A_q. Concerning s, only the r-vector app_s is required. For the extension distance $d_{ext}^{\prime}(s, q) = (ext_s - ext_q)^{\prime}$, it is sufficient to provide the 3D extension vectors ext_q and ext_s.

In order to avoid a dimensionality of the index that is quadratic in the dimensionality of the approximation model, in the filter step, only the components 2 and 3 (cf. table 3) will be evaluated. The overall evaluation of d_{shape} is deferred to the refinement step.

Now, for any given query object q, let us compose the $(r+3) \times (r+3)$-matrix A'_q from the $r \times r$-matrix A_q and the 3×3 unit matrix I_3, resulting in $A'_q = \begin{bmatrix} \frac{1}{2} u_{app} A_q & 0 \\ 0 & u_{ext} I_3 \end{bmatrix}$. Recall the definition of the $r+3$ key vectors $key_s = (app_s,\ ext_s)$, and define the function $f_q: \mathfrak{R}^{r+3} \to \mathfrak{R}$ as follows:

$$f_q(key_s) = \sqrt{(key_s - key_q) \cdot A'_q \cdot (key_s - key_q)^{\mathrm{T}}}$$

	term	formula	data of q	data of s
1	$\Delta d_s^2(app_q)$	$(app_q - app_s) \cdot A_s \cdot (app_q - app_s)^{\mathrm{T}}$	r	$r^2 + r$
2	$\Delta d_q^2(app_s)$	$(app_s - app_q) \cdot A_q \cdot (app_s - app_q)^{\mathrm{T}}$	$r^2 + r$	r
3	$d_{ext}^2(s, q)$	$(ext_s - ext_q)^2$	3	3

Table 3. Number of data values required for the evaluation of the similarity function $d_{shape}(s, q)$

The following lemma states the lower-bounding property of f_q with respect to d_{shape}:

Lemma 3. Given an r-dimensional approximation model with two positive weighting factors u_{app} and u_{ext}, and two arbitrary segments s and q. The filter function $f_q(key_s)$ lower-bounds the shape distance $d_{shape}(s, q)$, i.e. $f_q(key_s) \le d_{shape}(s, q)$.

Proof. Observe that $f_q(key_s)^2 = \frac{1}{2} u_{app} \Delta d_q^2(app_s) + u_{ext} d_{ext}^2(s, q)$, and consider the difference $d_{shape}(s, q)^2 - f_q(key_s)^2 = \frac{1}{2} u_{app} \Delta d_s^2(app_q)$, which is greater or equal to zero, according to Lemma 1. Since $f_q(key_s)^2 \le d_{shape}(s, q)^2$, also $f_q(key_s) \le d_{shape}(s, q)$ due to Lemma 1. ◊

Now, the obvious strategy is as follows: Construct a PAM on the $r+3$-dimensional key vectors key_s of all the segments s in the database. Then provide efficient processing methods for filter queries on the PAM, i.e. range queries $f_q(x) \le \varepsilon$, and k-nn queries using the filter distance function $f_q(x)$.

4 Ellipsoid Queries on Point Access Methods

The efficiency of all the algorithms mentioned above is due to the efficiency provided by the underlying PAM. In the following, we focus on access methods that manage the secondary storage pages by rectilinear hyperrectangles, e.g. minimum bounding rectangles (MBR), for forming higher level directory pages. For instance, this paradigm is realized in the R-tree [Gut 84] and its derivatives, R$^+$-tree [SRF 87], R*-tree [BKSS 90], as well as the X-tree [BKK 96], which has already been used successfully to support query processing in high-dimensional spaces.

4.1 Ellipsoids as Query Objects

Observe that $f_q(x)^2 - c = 0$ with $x \in \Re^{r+3}$, $c \in \Re$, is a quadratic equation that represents an r+3-dimensional ellipsoid for a given level c. Thus, the query region specified by the range criterion $f_q(x) \leq \varepsilon$ which is equivalent to $f_q(x)^2 \leq \varepsilon^2$, is an r+3-dimensional ellipsoid of the level ε^2 (cf. figure 6).

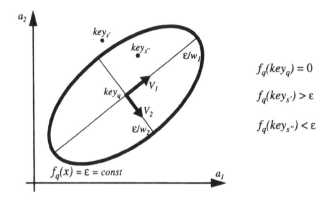

Fig. 6. Ellipsoid for distance range ε in a two-dimensional parameter space

Query processing on MBR-based access methods requires operations for query criterion evaluation of both, data objects (e.g. *query-contains-object*) and MBRs (e.g. *query-intersects-box*). For a query object q, we instantiate the query ellipsoid *ellip$_q$*. A method *eval* is provided for the evaluation of the filter function $f_q(p)$: *ellip$_q$.eval(p)* ≡ $f_q(p)$. In case of a range query, SIM$_{\text{range}}(q, \varepsilon)$, the query object is an ellipsoid around q whose boundary is determined by ε.

In figure 7, we give an illustration of the operations in case of a three-dimensional parameter space. The default way for range query processing would be to compute the MBR of the query ellipsoid, and to perform the standard range query algorithm as filter step. However in general, an MBR is a bad approximation of an ellipsoid, since it introduces a large portion of dead space. The percentage of dead space rapidly grows with increasing dimensionality and affects the selectivity of the index in a negative way.

Therefore, we suggest to perform ellipsoid queries in a direct way: The *ellip-contains-object* operation simply consists of an evaluation of the ellipsoid function, *ellip$_q$.eval(p)* ≤ ε. The *query-intersects-box* operation is more complex: The intersection predicate evaluates to true iff there exists a point that is contained in both the box and the ellipsoid. Our approach is to transform the intersection criterion into an optimization problem: Let p_0 be the point within the box that minimizes the ellipsoid function f_q, i.e. $f_q(p_0) = \min\{f_q(p) \mid p \in \text{box}\}$, and test whether $f_q(p_0) \leq \varepsilon$. Since we are only interested in the fact of intersection but not in the actual minimum value, the algorithm may stop if $f_q(p)$ is below the query range ε, thus improving efficiency.

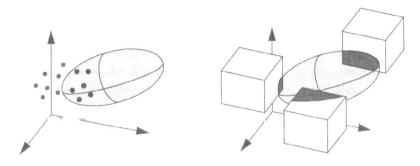

Fig. 7. Problems *ellipsoid contains point* and *ellipsoid intersects box*

For k-nn queries, $\text{SIM}_{\text{k-nn}}(q, k)$, we fall back on the algorithms of [RKV 95] or [HS 95] which provide efficient processing of k-nn queries on PAMs for the Euclidean distance. Both approaches are based on a ranking of distance values. For the processing of ellipsoid queries, we modify the algorithms such that they use the filter distance function f_q instead of the Euclidean distance.

As before, the ellipsoid function, $ellip_q.eval\ (p) = f_q(p)$, returns the distance of the query ellipsoid to a database object p. For the MBRs, the required operation is *mindist*[1], yielding the minimum distance of the box with respect to the query ellipsoid. Thus, once more the ellipsoid function minimization can be applied, since *mindist(ellip, box)* = $\min\{ellip_q.eval(p) \mid p \in box\}$.

Since the two box operations are closely related, we generalize them and provide a single basic operation for ellipsoids concerning boxes, the distance function *ellip.distance(box, limit)* for a (hyper-)rectangle *box* and a float parameter *limit*. The function *distance* returns the minimum distance d_{min} from *ellip* to *box* if $d_{min} \geq limit$, and an arbitrary value below *limit* if $d_{min} < limit$. The relationship of *distance* to the operations *intersect* and *mindist* is as follows:

Lemma 4. The function *ellip.distance(box, limit)* fulfills the following correspondences:

(i) bool *intersect (box, ellip, ε)* $\equiv$ *ellip.distance (box, ε)* $\leq$ ε

(ii) float *mindist (box, ellip)* $\equiv$ *ellip.distance (box, 0)*

Proof. (ii) Since $d_{min} > 0$, *ellip.distance(box, 0)* always returns the actual minimum which is never below the *limit* = 0. (i) By definition, the inequality *ellip.distance(box, ε)* $\leq$ ε holds if and only if the minimum distance d_{min} of *ellip* to *box* is lower or equal to ε. On the other hand, $d_{min} \leq$ ε holds if and only if the (hyper-)rectangle *box* intersects the ellipsoid *ellip* of level ε. ◊

In order to test intersection, Lemma 4 helps to improve runtime efficiency, since the exact value of mindist is not required as long as it is smaller than the limit parameter. In case of intersection, an approximate value suffices and may save iterations.

---

1. In [RKV 95], also a *minmaxdist* operation is investigated. Due to space limitation, we drop the explanation how to adapt minmaxdist to ellipsoid queries.

4.2 Basic Ellipsoid-and-Box Algorithm

The specification of *ellip.distance(box, limit)* is to return the minimum distance $d_{min} = \min\{ellip_q(p) \mid p \in box\}$ from the ellipsoid *ellip* to the (hyper-)rectangle *box* if d_{min} is greater than *limit*, and to return an arbitrary value lower than *limit* otherwise.

For the evaluation of *ellip.distance(box, limit)*, we combined two paradigms, the steepest descent method and iteration over feasible points. The steepest descent as a general optimization technique is applicable to the ellipsoid function since at every location in space, the gradient can be determined. The steepest descent works correctly and stops after a finite number of iterations [PTVF 92]. The feasible points paradigm is adapted from the linear programming algorithm of [BR 85]. The basic idea is that every point that is visited on the way downto the minimum should belong to the feasible domain. In our case, we start within the box, and ensure that our point does not leave the box. Thus, if the algorithm reaches a point that also lies within the ellipsoid of level ε, the ellipsoid and the box intersect (cf. figure 8).

function ellip :: distance (box, limit) —> float;		
1	p_0 := box.closest (*this*.center);	// starting heuristics
2	**loop**	
3	**if** (this.eval (p_i) ≤ limit) **break**;	// ellipsoid is reached
4	g := – this.gradient (p_i);	// descending gradient
5	g := box.truncate (p_i, g);	// truncate by boundary
6	**if** (|g| = 0) **break**;	// no feasible progress
7	s := this.linmin (p_i, g);	// linear minimization
8	p_{i+1} := box.closest (p_i + s*g);	// projection onto box
9	**if** (this.eval (p_i) ≈ this.eval (p_{i+1})) **break**;	// no more progress
10	**endloop**	
11	**return** this.eval (p_i);	
end distance;		

Fig. 8. The basic algorithm *ellip.distance (box, limit)* iterates over feasible points p_i within the *box* until *limit* or the minimum of the ellipsoid *this* is reached

The algorithm *distance* works as follows: The iteration with gradient computation (step 4), linear minimization (step 7), and termination criterion (step 9) performs the steepest descent. The projection in step (8) ensures that *p* does not leave the box, whereas steps (1) and (5) apply heuristics to accelerate the iteration, as we will explain below.

The linear minimization *ellip.linmin (p,g)* in step (7) returns the scaling factor *s* for which $p+s \cdot g$ evaluates to a minimum with respect to the function value of *ellip*, which is fulfilled in case of $\nabla_{ellip}(p + sg) \cdot g = 0$. This equation is immediately solved to $s = -(\nabla_{ellip} p \cdot g)/(\nabla_{ellip} g \cdot g)$.

The feasibility property of the algorithm is ensured by the closest point operation provided by the box type. Every point that is visited by the iteration is projected into the box, or even to the box boundary. Since the box boundary is given by iso-oriented hyperplanes, the projection can be performed in linear time with respect to the number of dimensions (cf. figure 9).

box.closest(p)[d] ≡ **if** (p[d] < box.lower[d]) **then** box.lower[d]
elsif (p[d] > box.upper[d]) **then** box.upper[d]
else p[d];

Fig 9. Projection of a point to a box (closest point) for each dimension d

The algorithm works for any starting point that is feasible, i.e. which is located within the box, e.g. the center of mass or any of the corner points. As a heuristics to save iteration steps, we suggest to start at the point of the box that is closest to the ellipsoid center point. This may be a corner of the box, but in general, it is not. Since the boxes are iso-oriented, the closest point of a box with respect to any given location in space is easily determined by projecting each of the coordinates onto the corresponding box range. If the ellipsoid center lies within the box, it is closest to itself, and iteration will stop immediately at the beginning. Note that the closest point is not yet the global minimum we look for, because in general, the ellipsoid distance function is not compatible with the homogeneous Euclidean distance function.

Another heuristics to save iterations is the gradient truncation (step 5). Obviously, starting at a point from the box boundary, we are not allowed to proceed in any direction that would leave the box. Only directions along the boundary, or towards the box interior, are feasible. Therefore, we truncate the gradient by nullifying components that would leave the box. Formally, we decompose the gradient g into two components $g = g_{remaining} + g_{leaving}$, and proceed only in the direction $g_{remaining}$ that does not leave the box (cf. figure 10).

box.truncate (base, dir)[d] ≡ **if** (base[d] = box.lower[d] **and** dir[d] < 0) **then** 0
elsif (base[d] = box.upper[d] **and** dir[d] > 0) **then** 0
else dir[d];

Fig. 10. truncation of vector dir by box boundary where dir is affixed to base (for each dim. d)

5 Experimental Results

We implemented our algorithms for similarity range query processing in C/C++ and performed experiments on an HP9000/735 under HP-UX 9.0. For our test database, we used the atomic coordinates from the PDB [Ber+ 77] and computed some thousands of protein surfaces [SK 95]. From these surfaces, we extracted 12,000 segments; 6,200 of them are known docking sites. As approximation model, we chose TRIGO-4, leading to a 7D index over the key vectors of the segments. The weighting factors u_{app} and u_{ext} are set to 1. We managed the index using an R*-tree with a pagesize of 4 kbytes. As query objects, we used docking sites from the azurin and the fructose family.

Figure 11 shows performance results for similarity range queries on the database of 12,000 segments where ε ranges from 0 to 2.5 (ε=0 corresponds to a point query). We compare two evaluation strategies: Direct evaluation of ellipsoid queries by the ellip-

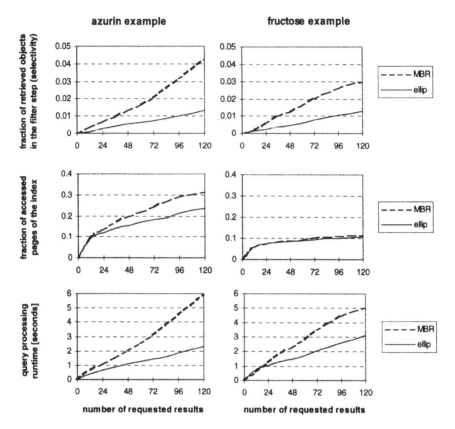

Fig. 11. Performance results for similarity range queries on a database of 12,000 segments using the query examples azurin (left hand side) and fructose (right hand side). The abscissa axis indicates the number of results (up to 120 objects, i.e. 1% of the data). *Top:* Selectivity of the filter step vs. number of results. *Middle:* Fraction of accessed pages vs. number of results. *Bottom:* Overall runtime of shape similarity query processing (in seconds) vs. number of results.

soid-and-box algorithm, and evaluation using a range query where the ellipsoid is approximated by an MBR. In each diagram, the abscissa axis indicates the number of results obtained from the refinement step. In the top diagram, we show the selectivity of the filter step vs. the number of results, i.e. the percentage of the data objects from the database that were selected by the filter step. The selectivity of our ellipsoid method is clearly lower than the selectivity of the standard MBR method. In the middle, the percentage of disk pages that were read by the filter step vs. the number of results is plotted. Again, the ellipsoid method is better than the result of the MBR technique. In the bottom diagram, the overall runtime for shape similarity query processing (in seconds) vs. the number of results is depicted. It shows that the ellipsoid method outperforms the MBR approach if considering both, page accesses and CPU time. The main advantage of the direct ellipsoid method is the reduced number of candidates that are obtained from the filter step.

In figure 12, we present the overall runtime as well as the average runtime per result vs. the size of the database (0 to 12,000 surface segments). These experiments ran for various query ranges (from 0.5 to 2.0). Whereas the number of results increases with a growing query range as expected, the average runtime is nearly constant.

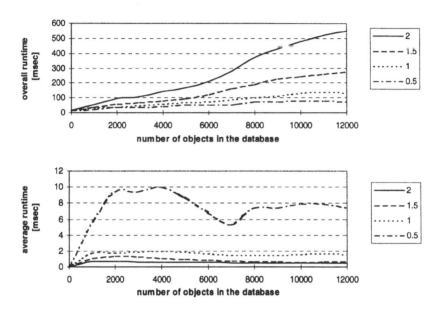

Fig. 12. Overall runtimes (above) and average runtimes per result (below) in seconds vs. database size (0 to 12,000 segments) for various query ranges (from 0.5 to 2.0)

Finally, our experiments have shown that the number of iterations in the *distance* algorithm for ellipsoids and boxes typically ranges from 1 to 8, depending on the data. We never observed an iteration number greater than 10.

6 Conclusion

In this paper, we presented a new approach for similarity search in spatial database systems. First, we introduced the notion of approximation-based similarity for 3D surface segments where similarity is quantified with respect to an application-specific approximation model. Second, to support efficient processing of shape similarity queries, we introduced a new query type, the ellipsoid query. An algorithm for efficient evaluation of ellipsoid queries on PAMs is given.

Experimentally, we could show that our method for ellipsoid query processing outperforms the standard technique using the MBR approximation. Up to now, we mainly investigated databases with known docking segments. We plan to extend our database by arbitrary segments as required for docking prediction systems. This will result in some 100,000 and 1,000,000 data objects which have to be managed by the system.

In our future work, we plan to extend the method of approximation-based similarity from surface segments to overall surfaces of molecules or mechanical parts. Thereby, more complex surfaces will occur which may be managed by two approaches: First, use more complex approximation models. This method leads to an increased dimensionality of the key vectors. In order to efficiently support ellipsoid queries on high-dimensional vectors, techniques to reduce the dimensionality are required. Second, split complex segments into simple pieces. For this purpose, decomposition techniques and new retrieval methods have to be developed. Additionally, we will investigate the promising potential of the concept of ellipsoid queries when applied to other similarity distance functions.

Acknowledgements

We would like to thank the anonymous reviewers who carefully read this paper and gave us helpful hints to improve the presentation of our approach.

References

[AFS 93]   Agrawal R., Faloutsos C., Swami A.: *'Efficient Similarity Search in Sequence Databases'*, Proc. 4th. Int. Conf. on Foundations of Data Organization and Algorithms (FODO'93), Evanston, ILL, in: Lecture Notes in Computer Science, Vol. 730, Springer, 1993, pp. 69-84.

[ALSS 95]   Agrawal R., Lin K.-I., Sawhney H. S., Shim K.: *'Fast Similarity Search in the Presence of Noise, Scaling, and Translation in Time-Series Databases'*, Proc. 21th Int. Conf. on Very Large Databases (VLDB'95), Morgan Kaufmann, 1995, pp. 490-501.

[Ber+ 77]   Bernstein F. C., Koetzle T. F., Williams G. J., Meyer E. F., Brice M. D., Rodgers J. R., Kennard O., Shimanovichi T., Tasumi M.: *'The Protein Data Bank: a Computer-based Archival File for Macromolecular Structures'*, Journal of Molecular Biology, Vol. 112, 1977, pp. 535-542.

[Ber+ 97]   Berchtold S., Böhm C., Braunmüller B., Keim D., Kriegel H.-P.: *'Fast Parallel Similarity Search in Multimedia Databases'*, Proc. ACM SIGMOD Int. Conf. on Management of Data, 1997.

[BHKS 93]   Brinkhoff T., Horn H., Kriegel H.-P., Schneider R.: *'A Storage and Access Architecture for Efficient Query Processing in Spatial Database Systems'*, Proc. 3rd Int. Symp. on Large Spatial Databases (SSD'93), Singapore, 1993, Lecture Notes in Computer Science, Vol. 692, Springer, pp. 357-376.

[BKK 96]   Berchtold S., Keim D., Kriegel H.-P.: *'The X-tree: An Index Structure for High-Dimensional Data'*, Proc. 22nd Int. Conf. on Very Large Data Bases (VLDB'96), Mumbai, India, 1996, pp. 28-39.

[BKK 97]   Berchtold S., Keim D., Kriegel H.-P.: *'Using Extended Feature Objects for Partial Similarity Retrieval'*, accepted for publication in VLDB Journal.

[BK 97]   Berchtold S., Kriegel H.-P.: *'S3: Similarity Search in CAD Database Systems'*, Proc. ACM SIGMOD Int. Conf. on Management of Data, 1997.

[BKSS 90]   Beckmann N., Kriegel H.-P., Schneider R., Seeger B.: *'The R*-tree: An Efficient and Robust Access Method for Points and Rectangles'*, Proc. ACM SIGMOD Int. Conf. on Management of Data, Atlantic City, NJ, 1990, pp. 322-331.

[BKSS 94]   Brinkhoff T., Kriegel H.-P., Schneider R., Seeger B.: *'Efficient Multi-Step Processing of Spatial Joins'*, Proc. ACM SIGMOD Int. Conf. on Management of Data, 1994, pp. 197-208.

[BR 85]   Best, M. J., Ritter K.: *'Linear Programming. Active Set Analysis and Computer Programs'*, Englewood Cliffs, N.J., Prentice Hall, 1985.

[Fal+ 94]   Faloutsos C., Barber R., Flickner M., Hafner J., Niblack W., Petkovic D., Equitz W.: *'Efficient and Effective Querying by Image Content'*, Journal of Intelligent Information Systems, Vol. 3, 1994, pp. 231-262.

[FRM 94]   Faloutsos C., Ranganathan M., Manolopoulos Y.: *'Fast Subsequence Matching in Time-Series Databases'*, Proc. ACM SIGMOD Int. Conf. on Management of Data, 1994, pp. 419-429.

[GM 93]   Gary J. E., Mehrotra R.: *'Similar Shape Retrieval using a Structural Feature Index'*, Information Systems, Vol. 18, No. 7, 1993, pp. 525-537.

[Gut 84]   Guttman A.: *'R-trees: A Dynamic Index Structure for Spatial Searching'*, Proc. ACM SIGMOD Int. Conf. on Management of Data, Boston, MA, 1984, pp. 47-57.

[HS 94]   Holm L., Sander C.: *'The FSSP database of structurally aligned protein fold families'*, Nucl. Acids Res. 22, 1994, pp. 3600-3609.

[HS 95]   Hjaltason G. R., Samet H.: *'Ranking in Spatial Databases'*, Proc. 4th Int. Symposium on Large Spatial Databases (SSD'95), Lecture Notes in Computer Science, Vol. 951, Springer, 1995, pp. 83-95.

[Jag 91]   Jagadish H. V.: *'A Retrieval Technique for Similar Shapes'*, Proc. ACM SIGMOD Int. Conf. on Management of Data, 1991, pp. 208-217.

[Kor+ 96]   Korn F., Sidiropoulos N., Faloutsos C., Siegel E., Protopapas Z.: *'Fast Nearest Neighbor Search in Medical Image Databases'*, Proc. 22nd VLDB Conference, Mumbai, India, 1996, pp. 215-226.

[MG 93]   Mehrotra R., Gary J. E.: *'Feature-Based Retrieval of Similar Shapes'*, Proc. 9th Int. Conf. on Data Engineering, Vienna, Austria, 1993, pp. 108-115.

[OM 88]   Orenstein J. A., Manola F. A..: *'PROBE Spatial Data Modeling and Query Processing in an Image Database Application'*, IEEE Trans. on Software Engineering, Vol. 14, No. 5, 1988, pp. 611-629.

[PTVF 92]   Press W. H., Teukolsky S. A., Vetterling W. T., Flannery B. P.: *'Numerical Recipes in C'*, 2nd ed., Cambridge University Press, 1992.

[RKV 95]   Roussopoulos N., Kelley S., Vincent F.: *'Nearest Neighbor Queries'*, Proc. ACM SIGMOD Int. Conf. on Management of Data, 1995, pp. 71-79.

[SK 95]   Seidl T., Kriegel H.-P.: *'A 3D Molecular Surface Representation Supporting Neighborhood Queries'*, Proc. 4th Int. Symposium on Large Spatial Databases (SSD '95), Portland, Maine, USA, Lecture Notes in Computer Science, Vol. 951, Springer, 1995, pp. 240-258.

[SRF 87]   Sellis T., Roussopoulos N., Faloutsos C.: *'The R+-Tree: A Dynamic Index for Multi-Dimensional Objects'*, Proc. 13th Int. Conf. on Very Large Databases, Brighton, England, 1987, pp 507-518.

[TC 91]   Taubin G., Cooper D. B.: *'Recognition and Positioning of Rigid Objects Using Algebraic Moment Invariants'*, in *Geometric Methods in Computer Vision*, Vol. 1570, SPIE, 1991, pp. 175-186.

Finding Boundary Shape Matching Relationships in Spatial Data

Edwin M. Knorr, Raymond T. Ng*, and David L. Shilvock

Department of Computer Science
University of British Columbia
Vancouver, B.C., V6T 1Z4
Canada

Abstract. This paper considers a new kind of knowledge discovery among spatial objects—namely that of partial boundary shape matching. Our focus is on mining spatial data, whereby many objects called *features* (represented as polygons) are compared with one or more point sets called *clusters*. The research described has practical application in such domains as Geographic Information Systems, in which a cluster of points (possibly created by an SQL query) is compared to many natural or man-made features to detect partial or total matches of the facing boundaries of the cluster and feature. We begin by using an alpha-shape to characterize the shape of an arbitrary cluster of points, thus producing a set of edges denoting the cluster's boundary. We then provide an approach for detecting a boundary shape match between the facing curves of the cluster and feature, and show how to quantify the value of the match. Optimizations and experimental results are also provided. We also describe several orientation strategies yielding significant performance enhancements. Finally, we show how top-k matches can be computed efficiently.

Keywords: spatial knowledge analysis and discovery, pattern matching, GIS

1 Introduction

Knowledge discovery is defined as the nontrivial extraction of implicit, previously unknown, and potentially useful information from data [5]. The purpose of spatial data mining is to efficiently discover relationships or patterns in such spatial data as maps, photographs, and satellite images. Efficiency is a recurring theme in this paper. Although many researchers have investigated relationships that exist among *relational* data [1,2,6,7], a few others have gone beyond the relational model in order to discover relationships in spatial domains [9–12]. To the best of our knowledge, Lu, Han and Ooi were the first ones to develop spatial data mining techniques [10]. They proposed two algorithms that extract high-level relationships between spatial and nonspatial attributes in spatial databases. However, both algorithms require spatial concept hierarchies to be provided.

---

* Person handling correspondence. Email: rng@cs.ubc.ca

This is a serious restriction, because for most applications, it is very difficult to know a priori which hierarchies would be the most appropriate. To overcome this problem, Ng and Han developed the CLARANS algorithm which can find spatial clustering structures implicit in the data [11]. Without relying on any spatial hierarchy, CLARANS may discover, for example, that the most expensive housing units in Vancouver can be grouped spatially into a few clusters. Ester, Kriegel and Xu proposed several focusing techniques that can be used to further optimize the performance of CLARANS [4].

While CLARANS, and for that matter any clustering algorithm, is effective in answering the question of *what* the clusters are, it fails to answer the question, which is more interesting from a knowledge discovery point of view, of *why* the clusters are there spatially. But since we believe that in general it may be too much to expect a computer to be able to pinpoint exactly why a cluster is there spatially, we aim to develop algorithms that can answer a weaker form of the question—namely, what the characteristics and patterns of the clusters are. In [9], we developed efficient algorithms for identifying characteristics based on features that are near to, or far from, the clusters. In [12], we studied how to extract common and discriminating characteristics of clusters from thematic maps. In this paper, we consider another kind of characteristic—namely boundary shape matching relationships among 2-D spatial objects.

1.1 Problem Definition and Contributions

Below, we define boundary shape matching (BSM) and motivate why it is of interest to spatial data analysis and mining. Fig. 1 shows a cluster of 2-D points (x's representing houses), and a man-made feature (a golf course) described by a simple polygon. Note that parts of their boundaries match. We call this a (partial) BSM relationship between the cluster and the golf course. This simple example shows the cluster and the feature adjacent to each other, but in general we do not require proximity for BSM. In this paper, we refer to the point set as a *cluster* and to the geographic entities in a set of maps as *features*. The problem we study in this paper is:

Given a cluster, we find the top-k features, as well as the parts of the matched boundaries, that have the strongest BSM relationships with the cluster.

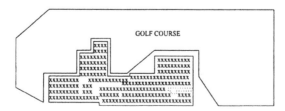

Fig. 1. Boundary shape matching between a housing cluster and a golf course

BSM is useful for spatial data analysis and mining because it can identify some prominent relationships. For example, if part of a housing cluster follows the contour of a golf course, lake, or park, it makes sense to report this fact. In the real estate market, there can be a huge price differential between houses that are across the street from (or simply near) a golf course, versus houses that actually follow the contour (e.g., fairways) of the golf course. (See [9] for a way of analyzing *proximity* relationships among clusters and features.) Note that a relationship can exist even if the cluster and the feature are relatively far apart. For example, if a cluster of expensive houses is a kilometre away from the shoreline, yet follows the shoreline reasonably well, one would suspect a possible relationship. Thus, if the cluster were on a slope, one might be led to believe that some or many of the houses in the cluster yield a splendid view. Relationships such as those described above may be of interest to such groups as the real estate community, Geographic Information Systems (GIS) analysts, demographers, and planners. Although our testbed is GIS, other spatial domains also apply. To show that BSM can be applied concretely, this paper uses real estate examples, and regularly and irregularly shaped geographic features in the city of Vancouver. As a preview, in this paper we provide answers to the following questions key to BSM:

- What is the "boundary" or the "shape" of a collection of points? Our notion of shapes is based on alpha-shapes introduced in [3,8]. Sect. 2 will show how alpha-shapes are adapted for and used in BSM.
- What is a "good fit" between two boundaries? In the ideal case, two boundaries could fit like two adjacent pieces in a jigsaw puzzle. More specifically, if each boundary is represented by a sequence of connected line segments, the characteristics of a good fit are that the peaks and the valleys line up, and that there is a constant separation distance between the two boundaries. However, there is the complication of the orientations of the boundaries. Our solution to this complication is to try many orientations in an intelligent manner. Sect. 3 will define all of these ideas more vigourously.
- How can we quantify the strength of a BSM relationship? Our ranking function takes into account the goodness of fit, partial matching boundaries, the length of the matching sub-boundaries, and a few complications such as hidden segments. Sect. 3 will provide the details.
- How can we efficiently identify the features with the top-k strongest BSM relationships? Our algorithm is based on filtering. In terms of performance, for a pool of 326 features, our algorithm can identify the top-10 features and their matching sub-boundaries in less than 3 seconds of CPU time. Sect. 4 will explain the ideas and show experimental results.

2 Defining Shapes of Clusters by Alpha-Shapes

As we have indicated, BSM will be performed on clusters and features. Each feature is defined by a sequence of points that make up a polygon. Most maps accessible by computer contain features described by an ordered sequence of (x, y) vertices defining the endpoints of a closed curve of line segments (edges). A cluster, on the other hand, is simply a set of unordered points with no specific boundary, perhaps having been produced in response to an SQL query. While clustering algorithms such as CLARANS [4,11] show how to identify one or more clusters from a set of points, no specific boundaries are associated with those clusters. This leads us to our first question: What is the boundary and shape of a cluster? Given a (dense) cluster of points, most humans can easily visualize a polygon whose shape characterizes a cluster. While the shape of such a polygon may be intuitive to humans, it is undefined to machines.

2.1 Basics of Alpha-Shapes

We begin by defining the shape of a cluster as a *convex shape*, which is simply a closed boundary of a set of points such that the line segment connecting any two points in the set always lies on or interior to the boundary. Among convex shapes, the *convex hull* is the most natural choice, because it is the unique, minimum bounding convex shape enclosing a set of points [13]. Convex hulls are a powerful construct for many spatial applications, including data mining [9]; however, they are not very good at providing detailed shape information about nonconvex spatial objects. For example, if we were limited to using a convex hull to describe the cluster in Fig. 1, most of the interesting aspects of the shape would be lost, and BSM would be difficult, if not impossible.

Because the shape of a cluster may be nonconvex, and because convex hulls are too crude for effective BSM, we turn to a generalization of convex hulls, namely *alpha-shapes* [3,8]. While a cluster has a unique convex hull, there is a spectrum of alpha-shapes to consider, each alpha-shape being defined by some α-value. A zero α-value gives the convex hull, positive α-values produce cruder convex shapes for a given set of points, and negative α-values provide finer nonconvex resolutions. The boundary of the cluster in Fig. 1 is the result of a negative α-value. The time complexity for constructing an alpha-shape is the same as that for constructing the convex hull. The complexity is $O(n \log n)$, where n is the number of points in the set.

2.2 Applying Alpha-Shapes to BSM

While certain members of a family of alpha-shapes can describe the intuitive shape of a nonconvex cluster, it is unclear what α-value to use in general, because the choice can vary greatly from cluster to cluster. Furthermore, a fine resolution produced by a large negative α-value may be too fine in the sense that there may be severe, narrow cavities or fissures in the resulting polygon. To illustrate this point, consider Fig. 1 again. In the right side of the cluster, we see a fissure (represented by a dotted area) which may be the result of certain negative α-values. Many such fissures may exist. For BSM, we want alpha-shapes that are neither too crude, nor too fine. In general, it is very difficult to determine the best α-value.

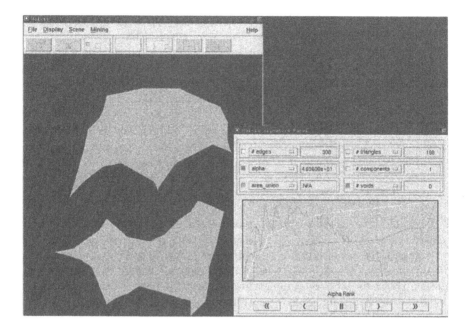

Fig. 2. Matching two alpha-shapes

To play with alpha-shapes empirically, we developed a prototype system, part of which is based on the software called "Shape2D-v1.1" from the National Center for Supercomputing Applications and the Department of Computer Science at the University of Illinois at Urbana-Champaign.[1] Fig. 2 shows a screen session of part of our prototype system. In this session, there are two windows. The left window shows two alpha-shapes brought together for matching. The right window can be used to adjust the alpha-shapes. This can be done by pressing the appropriate buttons at the bottom of the right window. The top part of the window shows how various attributes of the alpha-shape (such as the number of edges, connected components, and voids (holes) in the shape) are changed when the alpha rank is adjusted. The middle part of the window gives a graphical display of the values of these attributes.

We ran many experiments to show how different α-values can affect the shape of a cluster. We observed that an alpha-shape can become disconnected and can change significantly even for small changes in α. For the point sets that we tested, we observed that the best α-values for our application were those for which the number of voids in the shape was zero, and the number of connected

---

[1] This package uses an α-value that differs from the original definition [3]. The package's α-value is the reciprocal of the original α-value. The range of allowable α-values for this package results in a family of shapes ranging from the original, unconnected point set, up to and including the convex hull. Unless indicated otherwise, we use the original definition throughout this paper.

components was one. Even so, a single-component, zero-void alpha-shape can occur for several ranges of α-values. We chose the α-value for a given cluster to be the average of the midpoints of all ranges that yield one component and no void. Using this value resulted in a reasonable alpha-shape for the cluster. Its boundary is thus defined by an ordered sequence of vertices (and hence, line segments), which we use in the BSM algorithm described next.

3 Quantifying BSM Relationships

In this section, we describe how BSM relationships are computed and quantified. We explain the notions of facing and difference curves, give scoring functions for ranking BSM relationships, and describe how the axis of orientation can be selected intelligently.

3.1 Reducing Two Object Boundaries to Their Facing Curves

To develop an effective BSM algorithm, many questions have to be answered. The first question concerns which parts of the boundaries are to be considered. Regardless of the relative positions of the cluster and the feature on a map, it makes sense (particularly for GIS applications) to consider only those parts of the boundaries that face each other. We use outer common tangent lines to delimit the facing curves, as shown in Fig. 3. Intuitively, two chains of line segments face each other in the sense that the endpoints (i.e., vertices touching the tangent lines) and at least one vertex between the endpoints (in the nontrivial case) can "see" some vertex in the other chain. Formally, if x is a vertex in polygon A, and y is a vertex in polygon B, then we say that x and y see each other if the line segment joining x and y does not intersect the interior of either polygon A or polygon B. This definition ignores any objects that may lie between the two polygons.

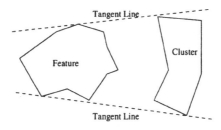

Fig. 3. Identifying facing curves by using tangent lines

A keen reader would observe that for a cluster-feature pair, there may be an infinite number of facing curves and tangent lines, if the feature were to rotate around the cluster. Because we are dealing with features and clusters that have fixed spatial locations, we do not consider the rotation of a feature around a cluster in this paper.[2] Hence, our BSM problem is reduced to determining how well two facing curves match.

3.2 Identifying the Difference Curve

The next question is how to determine whether two shapes or facing curves match. One obvious approach is to translate the curves until they touch, and then calculate the area between them. The smaller the area between them, the greater the likelihood of a match. This naive approach has several problems. First of all, it is biased toward shorter curves. For example, a pair of long curves may match well, but receive a poorer score than a pair of short curves that do not fit quite as well. Secondly, a small local change in one curve (for example, a long, sharp protrusion) can seriously affect the score. So, while we strive for efficiency in spatial data mining, and are often satisfied with approximate answers, those approximate answers still need to be accurate enough to identify good matches rather than weak ones.

The problems described above lead us to consider the notion of "puzzle fit" between two facing curves. Intuitively, if the two facing curves were the appropriate boundaries of two adjacent pieces of a jigsaw puzzle, they would fit snugly, in the sense that the peaks and valleys line up, and that there is a zero separation distance between the two curves. For the kinds of spatial analyses we wish to support, features need not be physically adjacent to a cluster. Thus, we generalize from a zero separation distance to a constant separation distance in the ideal case. We examine the *change* in the distance between the curves as we traverse an appropriate *axis of orientation*. An axis of orientation or "line of scrimmage" separates the curves so that the cluster is on one side of the line, and the feature is on the other. Separating the curves in this manner helps prevent situations in which we have to deal with overlapping curves. As we traverse the relevant part of the axis of orientation, we calculate the separation between the curves. The separation gives rise to a *difference curve*, shown in Fig. 4, which is created by plotting the distance between the curves as the axis is traversed from vertex to vertex. A change in the slope of the difference curve results only when a vertex from either curve is encountered having the same w-coordinate (where w is the axis of orientation). If there are n vertices in the cluster's curve and m vertices in the feature's curve, then $O(n + m)$ distance calculations are required for this scheme.

3.3 Quantifying the Degree of Partial Match

In Fig. 4, if V_A is the set of vertices belonging to Curve A, and V_B is the set of vertices belonging to Curve B, then let $V = V_A \cup V_B$, such that vertices in V are in

---

[2] The work described in this paper can be extended to include rotations. Practical applications include matching pottery fragments, matching fragmented manuscripts, and tracking icefloes.

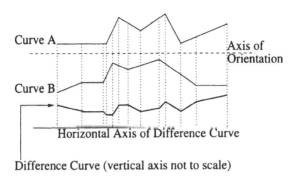

Difference Curve (vertical axis not to scale)

Fig. 4. Computing the difference curve

increasing order along the axis of orientation. Let $dist_i$ be the distance between the two curves (perpendicular to the axis of orientation) at vertex $V_i \in V$, and let L_i be the distance along the axis of orientation between vertices V_i and V_{i+1}. Define θ_i to be the absolute value of the slope of the difference curve (piecewise), that is, $\theta_i = |\ dist_{i+1} - dist_i\ |\ / L_i$. Finally, let Δ_g define the tolerance or maximum acceptable slope of a segment of the difference curve. The following equation shows one simple way to calculate the goodness of fit, where s is the number of segments in the difference curve:

$$\text{score} = \sum_{i=1}^{s} \begin{cases} L_i(\Delta_g - \theta_i) & \text{if } \theta_i < \Delta_g \\ 0 & \text{otherwise} \end{cases} \tag{1}$$

At first glance, (1) seems to satisfy our basic notion of fairness in assigning higher scores to flatter and/or longer difference curve segments. If $\theta_i < \Delta_g$, a larger L_i increases the score, and the closer θ_i is to zero, the larger the score.

Example 1. Let us represent the i-th line segment of a difference curve by the pair $\langle L_i, \theta_i \rangle$. Consider the difference curve: $[\langle 2, 5 \rangle, \langle 20, 0.5 \rangle, \langle 2, 5 \rangle, \langle 2, 2 \rangle, \langle 10, 0.5 \rangle, \langle 2, 5 \rangle]$. If Δ_g is set to 1, (1) only assigns positive scores to the two line segments with $\theta_i = 0.5$. The score is 20 * (1 - 0.5) + 10 * (1 - 0.5) = 15.

To evaluate the effectiveness of the above scoring function, we ran it against a set of geographic features of Vancouver. We observed that the resulting score often did not highlight the strength or quality of an excellent match. In other words, there was no strong distinction between excellent matches and middling matches. This was further complicated by the presence of bad segments within a curve. For example, Fig. 5 shows two pairs of curves that may receive identical scores using the existing linear equation—depending, of course, on the value of Δ_g.

The linear equation presented in (1) was changed in several ways. First of all, we now count both good and bad segments. Good segments are those for which $\theta_i < \Delta_g$; bad segments are those for which $\theta_i > \Delta_b$. Secondly, we show

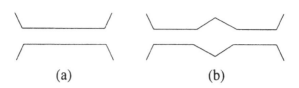

(a)   (b)

Fig. 5. Two pairs of curves that may receive identical scores using the linear equation

favouritism toward flatter difference curves. To achieve this goal, we use an exponential function instead of a linear function. Thirdly, because the difference between a good segment and a bad segment can be highly subjective, we allow for a buffer zone, whereby we assign neither a penalty nor a premium to segments falling into the range $\Delta_g \leq \theta_i \leq \Delta_b$. These three improvements produce the final form of our equation:

$$
\text{score} = \sum_{i=1}^{s} \begin{cases} L_i e^{(\Delta_g - \theta_i)c} & \text{if } \theta_i < \Delta_g \\ -L_i e^{\min[1,\ (\theta_i - \Delta_b)/c]} & \text{if } \theta_i > \Delta_b \\ 0 & \text{otherwise} \end{cases} \tag{2}
$$

The ceiling L_ie is used to bound the penalty for bad segments; otherwise, if θ_i greatly exceeded Δ_b, the penalty could be too severe (even for a relatively small L_i), and weaken the overall score for what might otherwise be a very good match. Furthermore, a constant $c > 1$ is used to help normalize the positive and negative terms because Δ_g and Δ_b are likely to be small compared to θ_i. Without c (and the ceiling), we would typically get small positive terms and large negative ones, making partial matches difficult to detect. We tried numerous values for Δ_g, Δ_b, and c, and found that $\Delta_g = 0.1$, $\Delta_b = 0.3$, and $c = 10$ gave reasonably good results—roughly comparable to the results a human might obtain from a quick visual observation of two facing curves. Although we cannot claim that these are the ideal values for a particular application, our experimental results suggest that these are good heuristics for our testbed.

Equation (2) readily supports the identification of *partial* matches within a pair of facing curves, which is particularly important for this type of knowledge discovery application, because it may be quite unlikely that two facing curves yield a perfect match. We also noted that some curves match quite well except for the final segment at either end of a curve's chain (that is, those segments touching the tangent lines). It seems unfair to penalize an otherwise good curve that just happens to "tail off" at its extremities. For example, although the rightmost segment of Curve B in Fig. 6 falls within the tangent lines, it obviously does not contribute to curve matching. We decided not to count these bad areas in the score. Bad areas lying between good areas are, of course, counted. To be more precise, we can reduce a difference curve to a sequence of the appropriate length from the alphabet {B,G,N}, where B and G stand for bad and good segments respectively, and N stands for neutral segments (i.e., neither a penalty nor a premium is applied). The scoring function presented above quantifies the degree of partial match according to the largest subsequence with G at both ends.[3]

Example 2. Let us consider the difference curve introduced in Example 1. It is: $[\langle 2,5 \rangle, \langle 20, 0.5 \rangle, \langle 2,5 \rangle, \langle 2,2 \rangle, \langle 10, 0.5 \rangle, \langle 2,5 \rangle]$. If Δ_g is set to 1 and Δ_b is set to 4, the difference curve can be reduced to the sequence BGBNGB. Equation (2) then applies only to the subsequence GBNG. More specifically, if the normalization constant c is set to 2, the score is given by: $20 * e^{(1-0.5)*2} - 2 * e^{(5-4)/2} + 10 * e^{(1-0.5)*2} = 30e - 2e^{0.5} \approx 78$.

3.4 Robustness: Trivial Matches and Hidden Segments

Our testbed included a lot of rectangular features such as schools, playgrounds, parks, and shopping centres. We observed that many of the best scores resulted from trivial matches involving one-line segments. To understand why this is a problem, consider a rectangular shopping centre located directly north of a cluster. If one segment in the cluster (especially a long segment) is approximately parallel to a line segment in the feature, then the resulting score can be quite high. We cannot claim, however, that this constitutes a relationship of interest. On the other hand, if many line segments match, such as when an irregularly shaped cluster matches an irregularly shaped feature (e.g., a cluster following a jagged shoreline), it makes sense to report this match as a relationship of interest. We therefore apply the following rule. After determining the tangent lines, if the number of vertices in either the cluster's curve or the feature's curve is ≤ 3, we simply ignore the feature, and move on to the next one. We chose the value 3 because many features in an urban area are rectangular (or nearly rectangular). If an entire curve consists of a corner (that is, three adjacent vertices), we simply skip the calculation, thereby avoiding many trivial matches. Our experiments confirm that this is a useful heuristic.

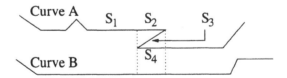

Fig. 6. Curve with hidden segments

Another issue that we had to address concerned overcounting. We encountered this situation when dealing with hidden curves, that is, curves that backtrack. When calculating the score for the pair of curves shown in Fig. 6, we ignore

---

[3] We can use a variant of this scoring function, which returns the highest score of *any* subsequence beginning with, and ending with, a G; however, for the sake of performance, and for returning longer, potential candidates (matches), we use (2), as described above.

the hidden parts of Curve A, which are the line segments in the diagram labelled S_2 and S_3, but we retain S_1 and S_4. If we were to include S_2, for example, an undeservingly high score would result from counting part of Curve A twice.

3.5   Varying the Axis of Orientation Intelligently

As described in Sect. 3.2, a key aspect of constructing the difference curve is the choice of axis of orientation. Using Fig. 7, we describe how to choose an appropriate axis of orientation. First, we calculate a bounding box for each of the facing curves, and then we construct tangent lines to these bounding boxes. The tangent lines are the dotted lines shown. We define our initial axis of orientation to lie in the middle of these two extremes, and we can rotate the axis through any angle ϕ, if ϕ does not go beyond a tangent line. For any angle ϕ, we calculate the separation according to where each curve intersects a line perpendicular to the corresponding axis.

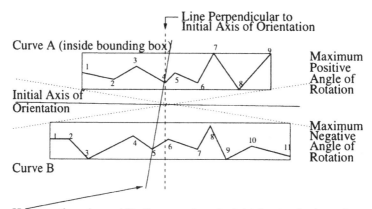

Here, vertices A_4 and B_5 line up when the initial axis of orientation (and hence a line perpendicular to it) is rotated through a negative angle of rotation.

Fig. 7. Axis of orientation

Different axes of orientation result in different matches between peaks and valleys. For example, if $\phi=0$, vertices A_6 and B_7 appear to line up or match, but this is not the situation with vertices A_4 and B_5, as shown by the vertical dashed line. On the other hand, if we rotate the angle of orientation through a negative value of ϕ not exceeding the limit posed by the corresponding tangent line, we see that vertices A_4 and B_5 line up. Different values of ϕ can have considerable effect in determining how well the two facing curves match. Note that different angles of rotation may be more suitable for certain pairs of vertices. There is no way of knowing in advance which angle of ϕ is best. One approach is to take a brute-force approach by exhaustively trying the entire range of allowable ϕ values using a small linear increment, and then choosing the axis that produces the highest score. Although this approach works, it does not scale very well. Efficiency is important because we want to try as many features as possible.

During our analysis of the brute-force results, we plotted each score against its corresponding value of ϕ. We noticed that many graphs contained prominent patterns. For example, a graph often contained several spikes of high scores that stood out from the rest of the graph. This suggested that we might be able to exploit these patterns, and avoid having to test a wide range of ϕ values. Several alternatives were tested:

1. Search for the first local peak.[4] Begin by calculating the score when $\phi=0$, and then either increase or decrease ϕ, depending on which direction produces a score greater than or equal to the previous one. After the direction is determined, continue increasing (or decreasing) ϕ until the new score is less than the previous score, or the limit of the range is reached. At that point, report the peak value as the score for that pair of curves.
2. Divide the range of allowable ϕ values into three equal parts, search for a local peak in each part, and report the highest of the three scores.
3. Use an adaptive technique. Begin by traversing the range of allowable ϕ values using a coarse increment (e.g., 1/30 of the total range). While progressing through these coarse increments, retain the best score and its corresponding ϕ. After calculating the score for a subsequent ϕ_k, if the slope of the graph changes from increasing to decreasing, this indicates that a peak occurs somewhere between ϕ_{k-2} and ϕ_k. In this case, traverse the range defined by ϕ_{k-2} and ϕ_k again but using a finer increment (e.g., 1/80 of the total range) to see if a better score can be obtained. Following this, continue traversing the original range with the coarse increment.

Table 1. Relative effectiveness and efficiency of various orientation strategies

	Relative Effectiveness	Relative Efficiency
Brute-Force	100	100
1 Local Peak	92	58
3 Local Peaks	96	83
Adaptive	100	52

Table 1 shows a representative comparison of the four approaches, with the brute-force method being the baseline. The 1-local-peak method is faster but produces results with lower quality. The 3-local-peaks method is in between

---

[4] Here, we are not referring to a local peak in one of the facing curves, but rather a local peak in the graph of score versus ϕ.

the 1-local-peak method and the brute force approach in both efficiency and effectiveness. The adaptive method is the clear winner. It has an effectiveness of 100% which means that the score returned by this method is equal to that returned by the brute-force method. But the CPU time used is only 52% of that used by the brute-force method.

Our testbed had 326 real-life features in it, with a broad mix of convex, nonconvex, large, small, regularly shaped, and irregularly shaped features of the city of Vancouver. Running the adaptive technique on a Sun SPARCstation LX computer with 24 MB of main memory, we could examine a cluster-feature pair in approximately 0.02 seconds, on average.

3.6 Summary

Having described the various pieces necessary for extracting and quantifying the BSM relationship between a cluster and a feature, we can now stitch all these pieces together to form the algorithm shown in Fig. 8. Throughout the rest of this paper, we call the score computed by Procedure BSM the *BSM score* of the feature.

```
Procedure BSM(Cluster, Feature)
{
  1.1 Set the current maximum score to 0;
  1.2 Extract the facing curves as in Sect. 3.1;
  1.3 Use the adaptive approach and iterate for each axis of orientation
      (switching between coarse and fine increments as shown in Sect. 3.5)
  {
    1.3.1 Identify the difference curve as in Sect. 3.2;
    1.3.2 Check for trivial matches and hidden segments as in Sect. 3.4;
    1.3.3 Compute the score based on (2) in Sect. 3.3;
    1.3.4 If the computed score is higher than the current maximum score,
          replace the current maximum score with the computed score;
  }
  1.4 Output the current maximum score;
}
```

Fig. 8. Algorithm for computing the BSM score of a feature

In summary, here in Sect. 3, we focused on how to identify and quantify the BSM relationship between a given cluster and feature. In the next section, we turn our attention to efficiently finding the top-k features that have the strongest BSM relationships with a given cluster.

We conclude this section by noting that in computer vision and robotics, there are techniques that deal with 2-D and 3-D curve matching (e.g., [14]), but to the best of our knowledge, the techniques are based on physical constraints and do not focus on fast, approximate computation of similarity scores to possible matching objects. As for related work in computational geometry, we are not aware of any work on similarity curve matching based on alpha-shapes.

4 Finding the Top-k Features Efficiently

4.1 SDD vs SDI Rankings

Although a case can be made for applying Procedure BSM only to features lying within a very short distance of the cluster, we did not restrict ourselves in this way because there are applications in which some very interesting items of knowledge discovery are independent of distance. For example, consider a housing cluster that is located on a hill some distance from the shoreline, and which follows the outline of a sheltered bay. We would like to highlight such a spatial relationship. We refer to this approach as the *separation distance independent (SDI)* approach. Under this approach, the top-k features are those that have the k-highest BSM scores.

While there are applications that find the SDI approach appropriate, there are other applications that may prefer to rank features inversely proportional to the separation distance between the cluster and features. That is to say, if the BSM scores of two features are identical, then the one closer to the cluster is ranked higher than the one further away. We refer to this approach as the *separation distance dependent (SDD)* approach. In particular, in this paper we define the SDD score to be the BSM score divided by the separation distance. Under this ranking system, the top-k features are those that have the k-highest SDD scores.

4.2 Efficiency Optimization: Filtering and Upper Bounds

Regardless of whether or not the SDI or SDD approach is used, one way to compute the top-k features is to apply Procedure BSM to every feature one-by-one, and sort the scores (divided by the separation distance, if SDD scores are used) to find the k highest ones. Using the experimental performance figures given in the previous section, for 326 features, the total time taken would be around $0.02 * 326 \approx 6.5$ seconds.

To optimize the performance of finding the top-k features, we use filtering and upper bounds. More specifically, the algorithm keeps track of the k highest scores encountered so far in the search. For the current feature to be tested, instead of applying Procedure BSM directly, which involves a lot of operations as described in Sect. 3, the algorithm applies some lightweight calculation of the best possible score—that is, providing an upper bound on the actual score. If the best possible score attainable for a particular cluster-feature pair is already lower than all the current k highest scores, Procedure BSM need not be applied. Otherwise, the procedure is used to compute the actual, exact score. In this way, as the search for the top-k features goes on, it is expected that more and more features can be screened out by upper bounding and filtering, and many invocations of Procedure BSM can be saved.

The key question is the choice of the upper bound. If the upper bound is too loose, then very few features can be screened away, in which case the entire effort of filtering is pointless. Tight upper bounds are preferrable—if they can be computed easily. If the upper bounds are as complicated to compute as Procedure BSM itself, then again filtering is pointless. Thus, there is the tradeoff between the effectiveness and efficiency of upper bounding.

In our study, we used the following upper bounds. For the SDI approach, we upper bounded the BSM score by the equation $\sum L_i e^{c\Delta_g}$. It is obvious that the score defined by (2) cannot exceed the above upper bound. Notice that in this upper bound, only L_i's are dependent on the feature, as both c and Δ_g are design parameters. In terms of computational effort, the upper bound only requires the extraction of the facing curves, but not any computation and iteration in Procedure BSM that requires the axis of orientation and the difference curve.

The above upper bound on the BSM score applies to the SDD approach as well. But for the SDD approach, more can be done. In particular, we tried the following two optimizations:

- We lower bounded the real separation distance between two curves by the separation between their bounding boxes. This separation distance is easy to compute ($O(1)$ time) after finding the bounding boxes in linear time. Again, this lower bound can be computed based on the facing curves only, but not the axis of orientation and the difference curve.
- Regardless of whether we used the bounding box distance or the exact separation distance, we sorted that distance and tested the features in ascending order. This is because as the distance becomes larger and larger, the chance that a feature will make the top-k SDD list decreases. This makes filtering more effective.

4.3 Experimental Evaluation

To empirically evaluate our upper bounding techniques, we used the dataset introduced at the end of Sect. 3.5. Table 2 shows the run times (in seconds) when applying various optimizations to find the top-10 features. These BSM times include all the computations described in Sects. 3.1 to 4.2, but exclude the time needed to compute the cluster and its alpha-shape. The number in parentheses shows the number of features filtered away, out of a total of 326 features.

If no optimization was applied, the time taken was 6.5 seconds and no feature was filtered out. Once the BSM scores were upper bounded, the time taken reduced to 64%, or 4.2 seconds, and almost half of the features, 150 to be exact, did not require the invocation of Procedure BSM. This indicates that even the SDI approach, which does not involve separation distance, can benefit by filtering and upper bounding.

For the SDD approach, the optimizations are even more pronounced. The third row in Table 2 indicates that by just lower bounding by the bounding box distance, the time taken reduced to 60%, or 3.9 seconds, and 184 features were

Table 2. Efficiently finding the top-k features

	seconds (features filtered out)
no optimization	6.5 (0)
upper bounding BSM scores	4.2 (150)
lower bounding by bounding box distance	3.9 (184)
upper bounding BSM scores + sorting separation distance	3.0 (248)
upper bounding BSM scores + lower bounding by bounding box distance + sorting bounding box distance	2.8 (236)

filtered out. This demonstrates that the bounding box distance optimization is effective.

Furthermore, by comparing the second and fourth rows in Table 2, it is clear that sorting the distance (in this case, the real separation distance) is also useful. The sorting, in conjunction with upper bounding BSM scores, reduced the run time to 46%, or 3.0 seconds, and filtered out 248 features.

The last row in Table 2 shows the performance figures when all the optimizations described above were applied. In this case, the run time was reduced to 43% or 2.8 seconds, and 236 features were filtered away. An interesting point can be made by comparing the number of filtered features in the fourth and last rows. The algorithm corresponding to the last row actually needed to invoke Procedure BSM more times than the algorithm corresponding to the fourth row. But computing bounding box distances is more efficient than computing real separation distances. The savings achieved by this optimization more than offsets the effort required by the extra features.

Fig. 9 shows a screen dump of finding top-k BSM relationships with our prototype system, which implements all the ideas described so far. The cluster and its top-10 features are displayed on a map. The bottom right corner gives the list of the top-10 features and their associated scores. If a user wants to examine the BSM relationship between a feature and the cluster more closely, the user can bring up the screen shown in Fig. 2.

5 Conclusions

We described a problem in spatial knowledge discovery involving boundary shape matching. We began by introducing alpha-shapes as a tool for finding a boundary to describe the shape of a cluster of unordered points. We then described how to perform BSM, and how to quantify the results. Several optimizations and orientation strategies were presented. Our experiments showed that our adaptive approach produces efficient and effective BSM, including partial and total matches of facing curves. Finally, we presented a few optimizations for identifying the top-k features. The optimizations appear to be effective empirically.

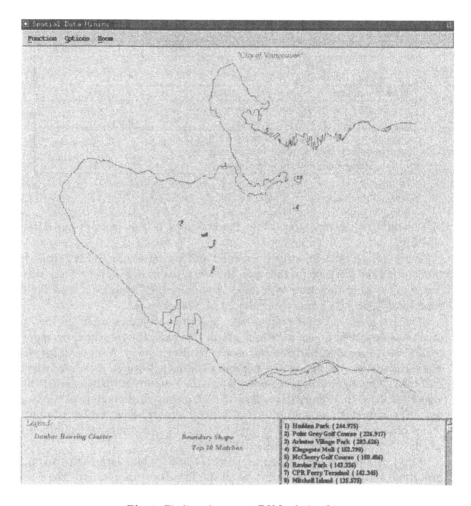

Fig. 9. Finding the top-10 BSM relationships

References

1. R. Agrawal, S. Ghosh, T. Imielinski, B. Iyer, and A. Swami. "An Interval Classifier for Database Mining Applications", *Proceedings of the 18th VLDB Conference*, pp. 560-573, 1992.
2. R. Agrawal, T. Imielinski, and A. Swami. "Mining Association Rules between Sets of Items in Large Databases", *Proceedings of the 1993 SIGMOD Conference*, pp. 207-210, 1993.
3. H. Edelsbrunner, D. Kirkpatrick, and R. Seidel. "On the Shape of a Set of Points in the Plane", *IEEE Transactions on Information Theory*, 29, 4, pp. 551-559, 1983.
4. M. Ester, H. Kriegel and X. Xu. "Knowledge Discovery in Large Spatial Databases: Focusing Techniques for Efficient Class Identification", *Proceedings of the 4th International Symposium on Large Spatial Databases (SSD'95)*, pp. 67-82, 1995.
5. W. J. Frawley, G. Piatetsky-Shapiro, and C. J. Matheus. "Knowledge Discovery in Databases: An Overview", *Knowledge Discovery in Databases*, Piatetsky-Shapiro and Frawley (eds.), AAAI/MIT Press, pp. 1-27, 1991.
6. J. Han, Y. Cai, and N. Cercone. "Knowledge Discovery in Databases: an Attribute-Oriented Approach", *Proceedings of the 18th VLDB Conference*, pp. 547-559, 1992.
7. D. Keim, H. Kriegel, and T. Seidl. "Supporting Data Mining of Large Databases by Visual Feedback Queries", *Proceedings of the 10th International Conference on Data Engineering*, pp. 302-313, 1994.
8. D. Kirkpatrick and J. Radke. "A Framework for Computational Morphology", *Computational Geometry*, G. Toussaint (ed.), The Netherlands: Elsevier Science Publishers B.V., 1985, pp. 217-248, 1985.
9. E. M. Knorr and R. T. Ng. "Finding Aggregate Proximity Relationships and Commonalities in Spatial Data Mining", *IEEE Transactions on Knowledge and Data Engineering*, 8, 6, pp. 884-897, December, 1996.
10. W. Lu, J. Han, and B. C. Ooi. "Discovery of General Knowledge in Large Spatial Databases", *Proceedings of the Far East Workshop on Geographic Information Systems*, Singapore, pp. 275-289, 1993.
11. R. Ng and J. Han. "Efficient and Effective Clustering Methods for Spatial Data Mining", *Proceedings of the 20th VLDB Conference*, pp. 144-155, 1994.
12. R. Ng and Y. Yu. "Discovering Strong, Common and Discriminating Characteristics of Clusters from Thematic Maps", *Proceedings of the Eleventh Annual Symposium on Geographic Information Systems*, pp. 392-394, 1997.
13. J. O'Rourke. *Computational Geometry in C*, Cambridge University Press, New York, 1994.
14. J. T. Schwartz and M. Sharir. "Identification of Partially Obscured Objects in Two and Three Dimensions by Matching Noisy Characteristic Curves", *International Journal of Robotics Research*, Vol. 6, No. 2, pp. 29-44, Summer 1987.

Spatial Data Mining: A Database Approach

Martin Ester, Hans-Peter Kriegel, Jörg Sander

Institute for Computer Science, University of Munich
Oettingenstr. 67, D-80538 Muenchen, Germany
{ester I kriegel I sander} @informatik.uni-muenchen.de
http://www.dbs.informatik.uni-muenchen.de

Abstract. Knowledge discovery in databases (KDD) is an important task in spatial databases since both, the number and the size of such databases are rapidly growing. This paper introduces a set of basic operations which should be supported by a spatial database system (SDBS) to express algorithms for KDD in SDBS. For this purpose, we introduce the concepts of neighborhood graphs and paths and a small set of operations for their manipulation. We argue that these operations are sufficient for KDD algorithms considering spatial neighborhood relations by presenting the implementation of four typical spatial KDD algorithms based on the proposed operations. Furthermore, the efficient support of operations on large neighborhood graphs and on large sets of neighborhood paths by the SDBS is discussed. Neighborhood indices are introduced to materialize selected neighborhood graphs in order to speed up the processing of the proposed operations.

Keywords: Spatial Data Mining, Neighborhood Graphs, Efficient Query Processing.

1 Introduction

Spatial Database Systems (SDBS) are database systems for the management of spatial data. Both, the number and the size of spatial databases are rapidly growing in applications such as geomarketing, traffic control and environmental studies. This growth by far exceeds human capacities to analyze the databases in order to find implicit regularities, rules or clusters hidden in the data. Therefore, automated knowledge discovery becomes more and more important in spatial databases. *Knowledge discovery in databases (KDD)* is the non-trivial extraction of implicit, previously unknown, and potentially useful information from databases [FPM 91].

A wide variety of algorithms have been proposed for KDD. [MCP 93] tries to classify these algorithms and identifies the following generic tasks:

- *class identification*, i.e. grouping the objects of the database into meaningful subclasses.

- *classification*, i.e. finding rules that describe the partition of the database into a given set of classes.

- *dependency analysis*, i.e. finding rules to predict the value of some attribute based on the value of another attribute.

- *deviation detection*, i.e. discovering deviations from the expectations, e.g. outliers in a class of objects.

While a lot of algorithms have been developed for KDD in relational databases, the area of KDD in spatial databases has only recently emerged (see [KHA 96] for an overview). The goal of this paper is to define a set of basic operations for KDD in SDBS which can be used to express many relevant algorithms in the sense that most of the relevant queries in a relational database can be expressed using the five basic operations of relational algebra. [AIS 93] follow a similar approach for KDD in relational databases. The definition of such a set of basic operations and their efficient support by an SDBS will speed up both, the development of new spatial KDD algorithms and their performance.

The rest of the paper is organized as follows. Section 2 discusses a sample geographic information system to illustrate the tasks of KDD in SDBS and to motivate the definition of the basic operations. We present the concepts of neighborhood graphs and paths together with their basic operations in section 3. Section 4 demonstrates the applicability of the proposed basic operations by presenting four spatial KDD algorithms based on these operations, two of them from literature and two new ones. Section 5 discusses efficient database support for neighborhood graphs and paths and their operations. Section 6 summarizes the contributions of this paper and discusses several issues for further research.

2   KDD Tasks in a Geographic Information System

A *geographic information system* is an information system for data representing aspects of the surface of the earth together with relevant facilities such as roads or houses. In this section, we introduce a sample geographic database providing spatial and non-spatial information on Bavaria with its administrative units such as communities, its natural facilities such as the mountains and its infrastructure such as roads. We use an extended relational model and the SAND (Spatial And Non-spatial Database) architecture [AS 91]. The spatial extension of the objects (i.e. polygons or lines) is stored and manipulated using an R*-tree [BKSS 90].

A small part of the relation communities is depicted in figure 1. The geographic database BAVARIA may be used, e.g., by economic geographers to discover spatial rules on the economic power of communities. Some non-spatial attribute such as the unemployment rate is chosen as an indicator of the economic power. In a first step, areas with a locally minimal unemployment rate are determined which are called *centers*, e.g. the city of Munich. The theory of central places [Chr 68] claims that the attributes of central cities influence the attributes of their neighborhood in a degree which decreases with increasing distance. E.g., in general it is easy to commute daily from some community to a close by center implying a low unemployment rate in this community. Thus, in a second step the theoretical distribution of the unemployment rate in the neighborhood of the centers is calculated, e.g.

```
when moving away from Munich,
the unemployment rate increases
```

Due to the general assumption of spatial continuity [IS 89], this distribution is typically continuous. In a third step, deviations from the theoretical distribution are discovererd, e.g.

```
when moving away from Munich towards the north east,
the unemployment rate decreases
```

Communities	name	population	unemploy-ment rate	rate of foreigners	spatial extension
	Munich	1.300.000	0.06	0.15	
	. . .	. . .	. . .	. . .	. . .

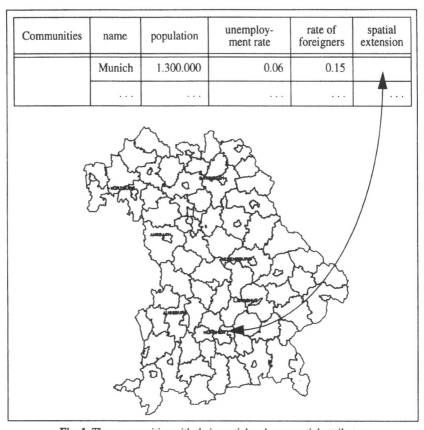

Fig. 1. The communities with their spatial and non-spatial attributes

The goal of the fourth step is to explain these deviations. E.g., if some community is relatively far away from a center but is well connected to it by train, the unemployment rate in this community is not as high as theoretically expected. In our example, the deviation is explained by the location of the Munich airport:

```
when moving away from Munich towards the airport,
the unemployment rate decreases
```

Summarizing, knowledge discovery according to the theory of central places is performed in the following steps:

(1)   discover centers, i.e. local extrema of some non-spatial attribute

(2)   determine the (theoretical) trend of some non-spatial attribute when moving away from the centers

(3)   discover deviations from the theoretical trend

(4)   explain the deviations by other spatial objects e.g. by some existing infrastructure in that area.

Another typical approach for knowledge discovery in geographic databases is to find interesting correlations between different characteristics of certain areas. E.g., we might find that areas with a high value for the attribute rate of retired people are highly correlated with neighboring mountains and lakes. This KDD task is performed in two steps:

(1)   find areas of spatial objects, e.g. clusters or neighboring objects, which are homogeneous with respect to some attribute values

(2)   find associations with other characteristics of these areas, e.g. by correlating them with reference maps or with other attribute values.

We conjecture that these tasks of KDD are representative not only for economic geography but also for a broader class of applications of geographic information systems, e.g. for environmental studies.

Some other approaches to extract knowledge from spatial databases have been proposed in the literature. In [LHO 93] attribute-oriented induction is applied to spatial and non-spatial attributes using (spatial) concept hierarchies to discover relationships between spatial and non-spatial attributes. A spatial concept hierarchy represents a successive merging of neighboring regions into larger regions. In [NH 94] the clustering algorithm CLARANS which groups neigboring objects automatically without a spatial concept hierarchy is combined with attribute-oriented induction on non-spatial attributes. [KH 95] introduces spatial association rules which are discussed in more detail in section 4.1. [Ng 96] and [KN 96] present algorithms to detect properties of clusters using reference maps and thematic maps. E.g. a cluster may be explained by the existence of certain neigboring objects which may "cause" the existence of the cluster.

3 Basic Operations for KDD in SDBS

Roughly speaking, SDBS are relational databases plus a concept of spatial location and spatial extension. The explicit location and extension of objects define implicit relations of spatial neighborhood. We claim that most KDD algorithms for *spatial* databases will make use of those neighborhood relationships, because it is the main difference between KDD in relational DBS and in SDBS, that attributes of the neighbors of some object of interest may have an influence on the object and therefore have to be considered as well. Furthermore, the discussion of the sample applications in the previous section indicates that the efficiency of many KDD algorithms for SDBS depends heavily on an efficient processing of these neighborhood relationships since the neighbors of many objects have to be investigated in a single run of a KDD algorithm.

Therefore, we present in this section a novel approach to KDD in spatial databases aiming at an extension of SDBSs with data structures and operations for efficient processing of implicit relations of spatial neigbhborhoods. This approach allows a tight integration of spatial KDD algorithms with the database management system of a SDBS, speeding up both the development and the execution of spatial KDD algorithms.

3.1 Neighborhood Graphs and Neighborhood Paths

We introduce the concept of neighborhood graphs explicitly representing those implicit neighborhood relations relevant for KDD tasks. [EG 94] follows a similar approach for modeling networks such as roads or telephone lines for the purpose of spatial query processing. While [EG 94] deals with graphs of explicit networks, we use graphs of implicit relations.

A *neighborhood graph* $G_{neighbor}$ for some spatial relation "neighbor" is a graph (N,E) with the set of nodes N and the set of edges E. Each *node* corresponds to an object of the database and two nodes n_1 and n_2 are connected via some *edge* iff neighbor(object(n_1),object(n_2)) holds. The predicate neighbor may be one of the following *neighborhood relations:*

Topological-Relations (c.f. [Ege 91])
                       e.g. {*meet, overlap, covers, covered-by, contains, inside, equal*}
Metric-Relations   e.g. {*distance < d*}
Direction-Relations e.g. {*north, south, west, east*}

Based on the neighborhood graphs, we define a *neighborhood path* in some graph G as a list of nodes of G with an edge of G connecting each pair of successors in the list, e.g. [$n_1, n_2, \ldots, n_k$] where neighbor(n_i, n_{i+1}) holds for each i, $1 \leq i \leq k - 1$. We define the *length* of a path as the number of its nodes.

3.2 The Basic Operations

Now, we present a set of basic operations on neighborhood graphs and paths designed to support KDD tasks such as those discussed in the previous sections. We use the expressions nRelations, nGraphs and nPaths to denote sets of neighborhood relations, neighborhood graphs and neighborhood paths, respectively.

Note that we do not define an explicit domain of Databases. Instead, we use the domain $2^{Objects}$ of all subsets of the set of all objects to define our operations. We assume that standard operations such as select(db:Set-Of-Objects;pred:Predicate) and get-value(o:Object;attr:Attribute) are supported by the SDBS. In the following, we introduce the new operations on neighborhood graphs and paths to be provided by an SDBS. We present both, the signature of the operations and a short description of their meaning.

get_nGraph: $2^{Objects}$ x nRelations -> nGraphs

The operation get_nGraph(db,rel) returns the neighborhood graph representing the neighborhood relation rel on the objects of db. Note that rel may either be one of the primitive neighborhood relations such as intersects or a conjunction or disjunction of two neighborhood relations such as intersects and north.

```
get_neighborhood: nGraphs x Objects x Predicates -> 2^Objects
```

The operation `get_neighborhood(graph,o,pred)` returns the set of all objects o_i directly connected to `o` via some edge of `graph` satisfying the conditions expressed by the predicate `pred`. An additional selection condition `pred` is used if we want to investigate only a specific class of neighbors of object `o` or if we want to exclude explicitly certain types of neighbors of `o`. The definition of `pred` may use spatial as well as non-spatial attributes.

```
create_nPaths: 2^Objects x nGraphs x Predicates x Int -> 2^nPaths
```

The operation `create_nPaths(objects,graph,pred,i)` creates the set of all paths starting from one of the `objects` and following the edges of the neighborhood graph `graph` with `length` $\leq$ `i`. The predicate `pred` expresses further constraints on the paths to be created. This argument of create_nPaths is the most important for the computational complexity of KDD algorithms operating on sets of paths because the number of all paths in a neighborhood graph tends to be very large. Furthermore, most of the neighborhood graphs will contain many cycles because most of the neighborhood predicates are symmetric. However, for the purpose of KDD we are mostly interested in in a certain class of paths, that is to say paths which are "leading *away*" from the starting object in a straightforward sense. We think that a spatial KDD algorithm using a set of paths which are crossing the space in an arbitrary way, leading forward and backwards and contain cycles will not produce understandable patterns (if any will be produced at all). Therefore we assume that the predicate `pred` will in general be defined with respect to a path $p=[n_1,n_2,...,n_k]$ like "direction of the edge $(n_i, n_{i+1}) \cong$ direction of the edge (n_{i-1}, n_i)" or "distance(n_1, n_{i+1}) > distance(n_1, n_i)".

```
extend: 2^nPaths x nGraphs x Pred x Int -> 2^nPaths
```

The operation `extend(set_of_paths,graph,pred,i)` returns the set of all paths extending one of the paths of `set_of_paths` by up to `i` edges of `graph`. The predicate `pred` is assumed to be the same as in the `create_nPaths` operation that was used to create the `set_of_paths`. Note that the members of `set_of_paths` are not contained in the result such that an empty result indicates that none of the elements of `set_of_paths` could be extended.

Finally, we assume some host programming language providing the standard operations on sets of paths and on single paths such as `length(path:Neighborhood-Path)` and iterators such as **for each** path **in** paths.

4 Spatial KDD Algorithms Using the Basic Operations

In this section, we illustrate the applicability of the proposed basic operations by presenting four spatial KDD algorithms based on these operations, two of them from literature (section 4.1 and section 4.2) and two new ones (section 4.4 and section 4.3).

4.1 Spatial Association Rules

[KH 95] proposes a method for mining spatial association rules consisting of five steps. Step 2 (coarse spatial computation) and step 4 (refined spatial computation) involve spatial aspects of the objects and thus are examined in the following. Step 2 computes spatial joins of the target object type (e.g. town) with each of the other specified object types (e.g. water, road, boundary and mine) using the neighborhood relation g_close_to. For each of the candidates obtained from step 2 which passed step 3, in step 4 the exact spatial relation is determined. Finally, a relation such as the one depicted in figure 2 (c.f. [KH 95]) results which is the input of the non-spatial step 5.

Town	Water	Road	Boundary
Victoria	<meet, J.FucaStrait>	<overlap,highway1>, <overlap, highway17>	<g_close_to,US>
Saanich	<meet, J.FucaStrait>	<overlap,highway1>, <g_close_to, highway17>	<g_close_to,US>
PrinceGeorge		<overlap, highway97>	
Petincton	<meet,OkanaganLake>	<overlap, highway97>	<g_close_to,US>
. . .	. . .	. . .	. . .

Fig. 2. sample candidates with exact spatial relations

Both of these spatial steps can be implemented using the operations on neighborhood graphs as follows. Step 2 requires several operations to select the objects of the specified object types. Then, several calls of `get_neighborhood` on the selected sets of objects with the predicate g_close_to yield neighborhood graphs and `create_nPaths` on these graphs with i=2 yields the required pairs of objects. Step 4 is based on the neighborhood graphs for the special neighborhood relations, e.g. spat.int-graph. Then, the refined spatial computation for a pair of objects (o_1,o_2) is equivalent to the test whether o_2 is element of the result of `get_neighborhood(special_graph,o_1,true)`.

4.2 Spatial Clustering

Clustering algorithms group a given set of objects into classes, i.e. clusters, such that objects in one class show a high degree of similarity, while objects in different classes are as dissimilar as possible. In the BAVARIA database (see section 2), e.g., a clustering algorithm can be applied to discover centers of high economic power. Several clustering algorithms for large spatial databases have been designed (e.g. [EKX 95] and [EKSX 96]).

The goal of the algorithm DBSCAN [EKSX 96] is to partition a database into sets of objects, i.e. *clusters*, such that the density of objects inside of each cluster is considerably higher than outside of the cluster. Furthermore, the density within the areas of noise is lower than the density in any of the clusters. DBSCAN discovers all clusters in db with a density of at least MinPts objects in the Eps neighborhood of each object. To find a cluster, DBSCAN uses region queries. Since region queries retrieve a special kind of neighborhood of the center point of the region, DBSCAN can be expressed using some of our proposed basic operations. The neighborhood graph defined by the metric predicate "distance(Object1,Object2) ≤ Eps" is created and then the Eps-neighborhood of each point is retrieved using the operation get_neighborhood(NeighborhoodGraph,Point,true).

4.3 Spatial Trend Detection

A *trend* may be defined as a temporal pattern in some time series data such as network alarms or occurrences of recurrent illnesses (c.f. [BC 96]). In an SDBS, we define a *spatial trend* as a pattern of change of some non-spatial attribute (attributes) in the neighborhood of some database object, e.g. "when moving away from Munich, the economic power decreases".

In the following we introduce an algorithm which discovers trends in SDBS starting from some object o. In each step, the algorithm computes both the local changes of the specified attribute when moving to the neighbors as well as the distance to these neighbors. A linear regression is applied to these pairs of values (change of attribute value, distance). If the resulting correlation coefficient is larger than a specified threshold, the slope of the resulting linear function is returned as the trend for o. If the correlation coefficient is not large enough, no trend is discovered for o. Figure 3 depicts a map of the attribute *average rent* from the BAVARIA database. A significant trend can be observed for the city of Munich: the average rent decreases quite regularly when moving away from Munich.

In the following, we present the algorithm for discovering spatial trends in pseudo code notation. It returns all trends of the non spatial attribute attr in db of length between min_length and max_length with the neighborhood defined by pred. The algorithm incrementally tries to find significant trends of maximal length, i.e. it extends all current neighborhood paths by one step, performs a linear regression on the attr values of the last objects on these paths and continues if the correlation coefficient of the current trend is at least equal to minconf. Note that the algorithm uses a predicate similar_direction, i.e. "direction of the edge $(n_i, n_{i+1}) \cong$ direction of the edge (n_{i-1}, n_i)", to restrict the creation of neighborhood paths.

```
discover_spatial_trends(db:Set_of_Objects;sel:Predicate;
        min_length,max_length:Int;minconf:Real;attr:Attribute;
        pred:NeighborhoodRelation;)
    focus:=select(db,sel);
    graph:=get_nGraph(db,pred);
    for each object in focus do
    all_paths:=create_nPaths({object},graph,
                                similar_direction,min_length);
        local_trends:=EMPTY_LIST;
        trend_list:=EMPTY_LIST;
        correlation:=MAXREAL;
        slope:=MAXREAL;
        new_paths:=extend(all_paths,graph,similar_direction,1);
        current_length:=2;
        current_trend:=NO_TREND;
        while current_length < max_length and new_paths ≠ EMPTY
            and correlation > minconf do
          all_paths:=union(new_paths,all_paths);
          for each path in all_paths do
            last_object:=get_object(path,length(path));
            attr_change:= get_value(object,attr)
            - get_value(last_object,attr);
            distance:=dist(object,last_object);
            insert [attr_change,distance] into list local_trends;
          end for each path in all_paths;
          perform_linear_regression(local_trends,slope,
                                        correlation);
          new_paths:=extend(all_paths,graph,similar_direction,1);
          current_length:= current_length + 1;
          if correlation > minconf then
            current_trend:=[object,slope,correlation];
          end if // correlation > minconf;
        end while // current_length < max_length and . . and . . ;
        if current_trend ≠ NO_TREND then
          insert current_trend into trend_list;
        end if // current_trend ≠ NO_TREND;
    end for // each object in focus;
    return trend_list;
end // discover_trends;
```

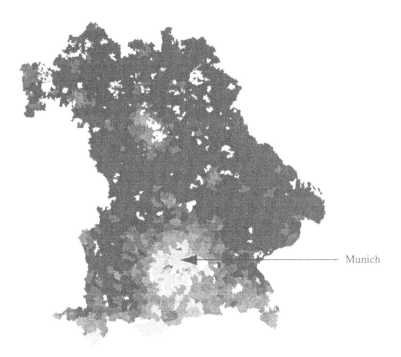

Fig. 3. Average rent for the communities of Bavaria

4.4 Spatial Classification

We assume a database of objects described by a collection of attributes each having a small domain of discrete values. The task of *classification* is to discover a set of classification rules that determine the class of any object from the values of its attributes. A spatial classification algorithm may, e.g., be used to explain the deviations from some discovered or some theoretical spatial trend. The following algorithm is based on the well-known ID3 algorithm [Qui 86] designed for relational DBS. The extension for SDBS is to consider not only attributes of the object o to be classified but to consider also attributes of neighboring objects, i.e. objects of a neighborhood path starting from o. Thus, we define a *generalized attribute* for some neighborhood path $p = [o_1, \ldots, o_k]$ as a tuple (attribute-name, index) where index is a valid position in p representing the attribute with attribute-name of object o_{index}. The generalized attribute (economic-power,2), e.g., represents the attribute economic-power of some (direct) neighbor of object o_1.

Since the influence of neighboring objects and their attributes decreases with increasing distance, we limit the length of the relevant neighborhood paths by an input parameter max-length. Furthermore, the classification algorithm allows the input of a predicate focusing the search for classification rules on the objects of the database fulfilling this predicate. Figure 4 depicts a sample decision tree and two rules derived from it. `Economic power` has been chosen as the class attribute and the focus is on all objects of type city.

In the following, the algorithm for discovering spatial classification rules is presented in pseudo code notation. It discovers all spatial classification rules, i.e. paths from the root to one of the leaves of the decision tree, with all attributes yielding an information gain of at least ε. Note that the algorithm uses a predicate `larger_distance`, i.e. "distance(n_1, n_{i+1}) > distance(n_1, n_i)" to restrict the creation of neighborhood paths.

```
discover_spatial_classification_rules(db:Set_of_Objects;
      sel:NonSpatialPredicate; class_attr:Attribute;
      pred:NeighborhoodRelation; max_length:Int)
   focus:=select(db,sel);
   neighborhood:=get_nGraph(focus,pred);
   paths:=create_nPaths(focus,neighborhood,larger_distance,
                        max_length);
   classify(class_attr,neighborhood,EMPTY_RULE,paths,
            max_length);
end // discover_classification_rules;


classify(class_attr:Attribute; neighborhood:NeighborhoodGraph;
      rule:ClassificationRule; paths:set_of_paths;
      max_length:Int)
   max_info_gain:=0.0;
   max_attr:=NULL;
   for i from 1 to max_length do
      for each generalized attribute (Aj,i) not used in rule do
         info-gain:=calculate_information_gain(Aj,
                                    class_attr,i,paths);
         if info_gain > max_info_gain then
            max_attr:=Aj;
            max_neighbors:=i;
            max_info_gain:=info_gain;
         end if // info_gain > max_info_gain;
      end for // each attribute Aj not yet used;
   end for // i from 0 to max_length - 1;
   if max_attr ≠ NULL and max_info_gain > ε then
      for each value of max_attr do
         extended_rule:=rule + "max_attr,max_neighbors,value";
         classify(class_attr,neighborhood,
                  extended_rule,paths,max_length);
      end for // each value of max_attr;
   else print rule;
   end if // max_attr ≠ NULL and max_info_gain > ε;
end // classify;
```

```
calculate_information_gain(attr,class_attr: Attribute;
     index:Int;paths:set_of_paths);
  for each path in paths do
     consider attr of the index-th object of path and class_attr
     of the first object of path for the calculation of the
     information gain
  end for // each path in paths;
end // calculate_information_gain;
```

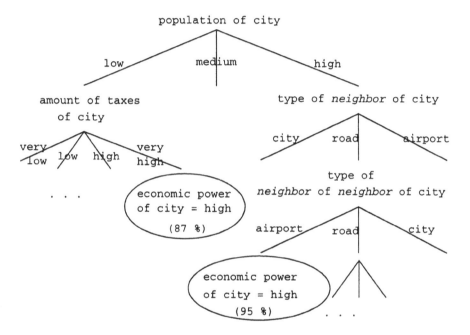

IF population of city = low AND amount of taxes of city = very high
          THEN economic power of city = high (87 %)

IF population of city = high AND type of *neighbor* of city = road
AND type of *neighbor* of *neighbor* of city = airport
          THEN economic power of city = high (95 %)

Fig. 4. sample decision tree and rules discovered by the classification algorithm

5 Efficient SDBS Support for Neighborhood Graphs and Paths

In this section, we discuss the efficient support of the operations on neighborhood graphs and paths by an SDBS. We introduce the concept of neighborhood indices materializing selected neighborhood graphs and show how they can be used to speed up the processing of our basic operations. Furthermore, a cost model is presented which allows to compare the expected execution time of a get-neighborhood operation with vs. without a neighborhood index.

5.1 Neighborhood Indices

First, we discuss some related work. [Rot 91] introduced the concept of *spatial join indices* as a materialization of a spatial join result with the goal of speeding up spatial query processing. This paper, however, does not deal with the questions of efficient implementation of such indices.

[LH 92] extends the concept of spatial join indices by associating their distance with each pair of objects (*distance associated join indices*). Thus, the join index can be used to support not only queries concerning a single spatial predicate but the index is applicable to a large number of queries. In its naive form, however, this index requires $O(n^2)$ space because it needs one entry not only for pairs of neighboring objects but for each pair of objects. Therefore, in [LH 92] a hierarchical version of distance associated join indices is proposed. These indices assume a spatial hierarchy of objects, e.g. countries > cities > houses. Entries in the index are only generated for pairs of objects contained in the same object of the next higher level of the hierarchy, e.g. only for pairs of houses of the same city and only for pairs of cities of the same country. The hierarchical approach significantly reduces the space requirements of the join index but also prevents its application to databases if a spatial hierarchy is either not available or the spatial hierarchy is not relevant for the purpose of KDD. E.g. in the geographic information system on Bavaria, there is a spatial hierarchy of districts > cities etc., but the influence of cities, e.g., to their neighborhood is not restricted to cities inside of the same district. Consequently, we cannot rely on such political hierarchies for the purpose of supporting spatial data mining by neighborhood graphs.

In the following, we present our concept of neighborhood indices. We assume the existence of some spatial index such as an R*-tree [BKSS 90] to support spatial query processing. If many operations are performed on the same neighborhood graph and if this graph is relatively stable, an index should be constructed for this neighborhood graph. This seems to be especially important if neighborhood paths are to be constructed from some neighborhood graph. Note that many SDBS are rather static since there are not many updates on objects such as geographic maps or proteins. We define a *neighborhood index* as an index explicitly representing a neighborhood graph, i.e. a neighborhood index supports the processing of all operations on its corresponding neighborhood graph without accessing the database itself. A simple implementation of a neighborhood index using a B+-tree is illustrated in figure 5. In general, a neighborhood graph is undirected implying a double representation of each edge in the neighborhood index.

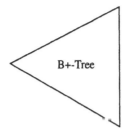

Object	Neighbors
o_1	o_2, o_3, o_7, o_9
o_2	o_1, o_3, o_9
. . .	. . .

Fig. 5. Sample Neighborhood Index

Clearly, it is prohibitive to construct neighborhood indices for each neighborhood graph. It is the task of the database administrator to select some important neighborhood graphs and to create the neighborhood indices for these graphs. The cost model which is presented in the next section may provide support for this task.

5.2 A Cost Model

A cost model is developed to predict the cost of performing a `get_neighborhood(graph,object,filter)` operation with vs. without a neighborhood index. In the database community, usually the number of page accesses is chosen as the cost measue. However, the amount of CPU time required for evaluating a neighborhood relation on spatially extended objects such as polygons may very large so that we model both, the I/O time and the CPU time for an operation. We use t_{page}, i.e. the execution time of a page access, and t_{float}, i.e. the execution time of a floating point comparison, as the units for I/O time and CPU time, resp.

In figure 6, we define the parameters of the cost model and list typical values for each of them (see [BKSS 94] for the values of t_{page} and t_{float}):

name	meaning	values
n	number of nodes in the neighborhood graph	$[10^3 .. 10^5]$
e	number of (directed) edges in the neighb. graph	$[10^3 .. 10^6]$
v	average number of vertices of a polygon	$[10 .. 10^4]$
c_{oid}	capacity of a page in terms of object identificators	1000
c_{pol}	average capacity of a page in terms of polygons	4096 / (4 * v)
t_{page}	execution time for a page access	$1 * 10^{-2}$ sec
t_{float}	execution time for a floating point comparison	$3 * 10^{-6}$ sec

Fig. 6. Parameters of the cost model

In a neighborhood index, there is one entry for each of the e edges of the associated neighborhood graph. The fan-out of the B+-tree is $c_{oid}/2$ since object identificators are used as keys. On the average, object has e/n neighbors which have to be read from data pages of the B+-tree. No refinement step has to be performed. Thus, the expected cost for performing a get_neighborhood(graph,object,filter) operation with a neighborhood index is as follows:

$$\text{cost with neighborhood index} = \left(\log_{c_{oid}/2} e + \left\lceil \frac{e}{n \times c_{oid}} \right\rceil \right) \times t_{page}$$

If no neighborhood index is available for graph, the operation get_neighborhood(graph,object,filter) is evaluated directly on the database, i.e. on the R*-tree. In an R*-tree, there is one entry for each of the n nodes (i.e. for their associated spatial objects) of the neighborhood graph. The fan-out of the R*-tree is $c_{oid}/5$ (in the case of 2D objects) since bounding boxes are used as keys. We assume that the neighbors of object are well clustered on the data pages of the R*-tree and can be read with the minimal number of page accesses. Finally, a refinement step has to be performed. We assume the neighborhood predicate intersects which requires $v * log\ v$ floating point comparisons. Hence, the expected cost for performing a get_neighborhood(graph,object,filter) operation without using a neighborhood index is as follows:

$$\text{cost without neig. index} = \left(\log_{c_{oid}/5} n + \left\lceil \frac{e}{n \times c_{pol}} \right\rceil \right) \times t_{page} + (v \times \log v) \times t_{float}$$

Using the cost model, we performed a simulation of several scenarios. We considered $c_{oid}, c_{pol}, t_{page}$ and t_{float} to be fixed and n, e and v to be variable. In each scenario, a fixed value is chosen for two of the variable parameters and the third one varies over its domain. The results of the simulation of the three scenarios are depicted in figure 7. The implementation using a neighborhood index significantly outperforms the one without using a neighborhood index. The speed-up increases with increasing values of e and v and with decreasing values of n. To conclude, neighborhood indices are recommended for graphs with a high average number of neighbors (e/n) or with large objects (v).

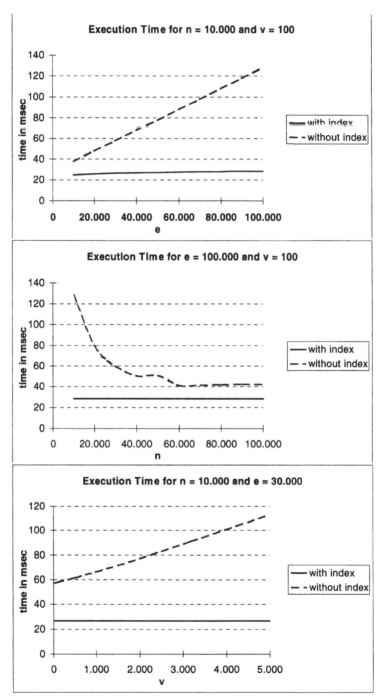

Fig. 7. Simulation of three different scenarios

5.3 Using the Neighborhood Indices

In this section, we discuss the implementation of conjunctions of neighborhood relations, of the operation create_nPaths and of updates on a neighborhood index.

So far, we have only considered neighborhood graphs defined by atomic neighborhood predicates. When the graph is defined by a conjunction of neighborhood relations, the neighborhood index of one of the atomic predicates is used to generate candidates which have to be checked further more in a second step. For this purpose, we define the *selectivity* of a neighborhood graph or index as

$$\text{selectivity} \quad = \quad \frac{n}{e}$$

Thus, a high selectivity implies a low average number of neighbors. The implementation of the operation get_neighborhood(graph,object,filter) for conjunctions of neighborhood relations follows a two-step approach:

- filter step:
  The most selective of the existing neighborhood indices which apply to the graph is determined. The set of all neighbors of object is obtained using this index and is called the set of candidates.

- refinement step:
  For all elements of the set of candidates it is checked whether the other relations of graph hold. This test may either use other neighborhood indices or may be directly performed using the spatial extension of the two objects obtained from the database.

The implementation of the operation create_nPaths(graph,objects,filter,i) is based on the operation get_neighborhood. A call of get_neighborhood(graph,last-object,filter) for the last-object of the current path obtains all one-step extensions of the current path. A depth-first traversal of graph is more efficient than a breadth-first traversal since the main memory requirements are much smaller. This is due to the fact that the depth-first traversal completes the current path before starting to create another one while the breadth-first traversal has to manage a potentially very large number of incomplete paths before the first one is completed. Redundant calls of get_neighborhood(graph,object,filter) for the same arguments can be avoided when providing a buffer of i pages so that all neighbors on the current path are always main memory resident. A lot of work on DBS support for path operations has been reported. However, typically the graphs are assumed to be directed and acyclic (cf. [AK 93], [LD 89]) which is not true for neighborhood graphs.

Clearly, updates of a SDB require updates of all its derived neighborhood indices. Fortunately, updates of the neighborhood indices on the insertion of a new object are restricted to the neighbors of the new object in the database. Therefore, the update of the neighborhood index can be efficiently performed using the operation get_neighborhood.

6 Conclusions

The main contribution of this paper is the definition of a a set of basic operations for KDD in SDBS which should be supported by an SDBS. The definition of such a set of basic operations and their efficient support by an SDBS will speed up both, the development of new spatial KDD algorithms and their performance. We introduce the concepts of neighborhood graphs and paths and a small set of operations for their manipulation. We argue that these operations are sufficient for KDD algorithms considering spatial neighborhood relations by presenting the implementation of four typical spatial KDD algorithms based on the proposed operations. Two of these algorithms are well-known from literature, the other two algorithms are new and are important contributions to clarify the differences between KDD in relational and in spatial databases. Furthermore, the efficient support of operations on large neighborhood graphs and on large sets of neighborhood paths by the SDBS is discussed. Neighborhood indices are introduced to materialize selected neighborhood graphs in order to speed up the processing of the proposed operations.

There are several issues for further research. First, the algorithms for optimizing the performance of the KDD operations using the available indices have to be evaluated. Second, the materialization of neighborhood paths will be investigated. This seems to be feasible if appropriate filters are used to create reasonably small sets of paths. A materialization of relevant paths may further speed-up the overall performance of KDD tasks because different KDD algorithms may use the same set of paths and each algorithm may scan this set of paths many times. Finally, applications in such domains as geography, biology or CAD will be investigated to insure the practical impact of our approach.

Acknowledgements

We thank Xiaowei Xu, Stefan Gundlach and Alexander Frommelt (Institute of Computer Science, University of Munich) for fruitful discussions on draft versions of this paper. Henning Brockfeld (Institute of Economic Geography, University of Munich) introduced us into the KDD problems of economic geographers and kindly provided the BAVARIA data.

References

[AIS 93]   Agrawal R., Imielinski T., Swami A.: *"Database Mining: A Performance Perspective"*, IEEE Transactions on Knowledge and Data Engineering, Vol.5, No.6, 1993, pp. 914-925.

[AK 93]   Agrawal R., Kiernan J.: *"An Access Structure for Generalized Transitive Closure Queries"*, Proc. 9th Int. Conf. on Data Engineering, 1993, pp. 429-438.

[AS 91]   Aref W.G., Samet H.: *"Optimization Strategies for Spatial Query Processing"*, Proc. 17th Int. Conf. VLDB, Barcelona, Spain, 1991, pp. 81-90.

[BC 96]   Berndt D. J., Clifford J.: *"Finding Patterns in Time Series: A Dynamic Programming Approach"*, in Fayyad U., Piatetsky-Shapiro G., Smyth P., Uthurusamy R. (eds.): Advances in Knowledge Discovery and Data Mining, AAAI Press / The MIT Press, 1996, pp. 229-248.

[BKSS 90]   Beckmann N., Kriegel H.-P., Schneider R., Seeger B.: *'The R*-tree: An Efficient and Robust Access Method for Points and Rectangles'*, Proc. ACM SIGMOD Int. Conf. on Management of Data, Atlantic City, NJ, 1990, pp. 322-331.

[BKSS 94]   Brinkhoff T., Kriegel H.-P., Schneider R., Seeger B.: *'Efficient Multi-Step Processing of Spatial Joins'*, Proc. ACM SIGMOD Int. Conf. on Management of Data, Minneapolis, MN, 1994, pp. 197-208.

[Chr 68]   Christaller W.: *"Central Places in Southern Germany"*, (in German), Wissenschaftliche Buchgesellschaft, 1968.

[Ege 91]   Egenhofer M. J.: *"Reasoning about Binary Topological Relations"*, Proc. 2nd Int. Symp. on Large Spatial Databases, Zurich, Switzerland, 1991, pp.143-160.

[EG 94]   Erwig M., Gueting R.H.: *"Explicit Graphs in a Functional Model for Spatial Databases"*, IEEE Transactions on Knowledge and Data Engineering, Vol.6, No.5, 1994, pp. 787-803.

[EKX 95]   Ester M., Kriegel H.-P., Xu X.: *"Knowledge Discovery in Large Spatial Databases: Focusing Techniques for Efficient Class Identification"*, Proc. 4th Int. Symp. on Large Spatial Databases, Portland, ME, 1995, pp.67-82.

[EKSX 96]   Ester M., Kriegel H.-P., Sander J., Xu X.: *"A Density-Based Algorithm for Discovering Clusters in Large Spatial Databases with Noise"*, Proc. 2nd Int. Conf. on Knowledge Discovery and Data Mining, Portland, Oregon, 1996, AAAI Press, 1996.

[FPM 91]   Frawley W.J., Piatetsky-Shapiro G., Matheus J.: *"Knowledge Discovery in Databases: An Overview"*, in: Knowledge Discovery in Databases, AAAI Press, Menlo Park, 1991, pp. 1-27.

[IS 89]   Isaaks E.H., Srivastava R.M.: *"Applied Geostatistics"*, Oxford University Press, New York, 1989.

[KH 95]   Koperski K., Han J.: *"Discovery of Spatial Association Rules in Geographic Information Databases"*, Proc. 4th Int. Symp. on Large Spatial Databases, Portland, ME, 1995, pp.47-66.

[KHA 96]   Koperski K., Adhikary J., Han J.: *"Knowledge Discovery in Spatial Databases: Progress and Challenges"*, Proc. SIGMOD Workshop on Research Issues in Data Mining and Knowledge Discovery, Technical Report 96-08, University of British Columbia, Vancouver, Canada, 1996.

[KN 96]   Knorr E.M., Ng R.T.: *"Finding Aggregate Proximity Relationships and Commonalities in Spatial Data Mining"*, IEEE Transactions on Knowledge and Data Engineering, Vol.8, No.6, 1996, pp. 884-897.

[LD 89]   Larson P.-A., Deshpande V.: *"A File Structure Supporting Traversal Recursion"*, Proc. ACM SIGMOD Int. Conf. on Management of Data, 1989, pp. 243-252.

[LH 92]   Lu W., Han J.: *"Distance-Associated Join Indices for Spatial Range Search"*, Proc. 8th Int. Conf. on Data Engineering, Phoenix, Arizona, 1992, pp. 284-292.

[LHO 93]   Lu W., Han J., Ooi B.C.: *"Discovery of General Knowledge in Large Spatial Databases"*, Proc. Far East Workshop on Geographic Information Systems, Singapore, 1993, pp. 275-289.

[MCP 93]   Matheus C.J., Chan P.K., Piatetsky-Shapiro G.: *"Systems for Knowledge Discovery in Databases"*, IEEE Transactions on Knowledge and Data Engineering, Vol.5, No.6, 1993, pp. 903-913.

[Ng 96]   Ng R.T.: *"Spatial Data Mining: Discovering Knowledge of Clusters from Maps"*, Proc. SIGMOD Workshop on Research Issues in Data Mining and Knowledge Discovery, Technical Report 96-08, University of British Columbia, Vancouver, Canada, 1996.

[NH 94]   Ng R.T., Han J.: *"Efficient and Effective Clustering Methods for Spatial Data Mining"*, Proc. 20th Int. Conf. on Very Large Data Bases, Santiago, Chile, 1994, pp. 144-155.

[Rot 91]   Rotem D.: *"Spatial Join Indices"*, Proc. 7th Int. Conf. on Data Engineering, Kobe, Japan, 1991, pp. 500-509.

[Qui 86]   Quinlan J.R.: *Induction of Decision Trees*, Machine learning 1, 1986, pp. 81 - 106.

Geo-Algorithms

A General and Efficient Implementation of Geometric Operators and Predicates

Edward P.F. Chan & Jimmy N.H. Ng
Department of Computer Science
University of Waterloo
Waterloo, Ontario, Canada N2L 3G1
epfchan@iris.uwaterloo.ca
nhjng@kbl.uwaterloo.ca
http://kbl.uwaterloo.ca

Abstract

Shape and location of objects in a spatial database are commonly represented by geometric data such as points, lines and regions. Numerous geometric operators and predicates have been proposed for spatial database systems. Existing work on their implementation concentrate on individual operators and predicates. This approach makes the realization of geometric operators and predicates in a spatial database system difficult since they are diverse and their implementation are complex. In this paper, we present a plane-sweep algorithm that can be easily modified to realize efficiently a set of frequently used line-region and region-region geometric operators and predicates. The design of this unified algorithm is based on the observation that the covering of elementary regions along the sweep line are updated locally and the implementation of these operators and predicates differ only with the output actions at an intersection point. Any geometric operator or predicate, the output of which can be determined by examining incident edges and covering information at intersection points, can be implemented easily with the algorithm. To demonstrate its generality, extendibility and efficiency, we concentrate on several popular geometric operators and predicates. All these operators and predicates can be realized in $O((N + I) \log N)$ time in the worst case, where N is the number of edges in the operands and I is the number of intersecting pairs. The proposed algorithm has been fully implemented and tested in C++ on a Sun workstation. Although the paper focuses on operators and predicates involving at most two regions, this algorithm can be generalized nicely to r regions, where $r>2$. We describe what changes are needed to make to the basic algorithm to accommodate this generalization.

1 Introduction

In a spatial database, shape and location of objects are denoted by geometric data such as points, lines and regions. Due to the nature of application domain, numerous geometric operators and predicates have been proposed in the literature. See for instance [3, 9, 21]. Existing work on their implementation concentrate on individual operators and predicates [4, 11, 12, 15, 18, 24]. In this paper, we shall present a general and efficient algorithm that implements a set of geometric predicates and operators involving lines and regions. The proposed method is simpler than finding algorithms for individual operators and predicates that are implementable with our approach. To demonstrate that our approach is elegant, we show that popular operators and predicates such as *fusion, set difference, intersection, inside, equal, overlap* and *adjacent* can all be implemented efficiently with this algorithm. Our algorithm can be extended easily to incorporate other operators and predicates such as *superimpose, disjoint, edge_inside, vertex_inside, XOR, overlay, touch* and *meet* [20, 21, 9]. Let us first compare the proposed method with previous work.

1.1 Related Work and Our Approach

1.1.1 Line Intersecting Problem

The *line intersecting* problem has been studied extensively. Shamos and Hoey [22] presented a plane-sweep algorithm for determining if N line segments are free of intersections. Bentley and Ottmann [2] extended Shamos-Hoey algorithm to report all I intersection points of N line segments in time $O((N + I) \log N)$. Both Shamos-Hoey and Bentley-Ottmann algorithms fail to take the degenerate cases such as collinear segments and multi-segment into consideration. Myers [12] gave an $O(N \log N + I)$ expected time algorithm that takes into consideration of all degenerate cases. It has recently been shown that $O(N \log N + I)$ in fact is the lower bound; and an optimal algorithm is found for solving the problem [4]. It is interesting to note that the optimal algorithm in [4] is not based on the plane-sweep paradigm.

1.1.2 Region Intersection Problem

Nievergelt and Preparata [18] extended Bentley-Ottmann algorithm to solve the *region intersection* problem. Nievergelt-Preparata algorithm is based on the plane-sweep paradigm. A vertical sweep line scans from left to right and information on regions intersected is updated according to the types of intersection point encountered. Ottmann, Widmayer and Wood [20] applied the plane-sweep technique to solve the so-called *Boolean Masking Problem*. The problem is to compute resulting disjoint regions given r layers of regions and a Boolean expression. As an illustration to their solution, they highlighted the main ideas (not an algorithm) of how to determine the intersection between two sets of regions. They claimed that the method can be extended to solving the Boolean Masking Problem. As both approaches are based on Bentley-Ottmann algorithm, both are of time complexity of

$O((N + I) \log N)$ and all degenerate cases were not considered. Despite their similarity, their approaches are rather different. Nievergelt-Preparata assume regions are denoted by sequences of vertices while Ottmann et al. accept sets of colored edges as input. As each sweep point, Nievergelt-Preparata uses cyclic lists of vertices to maintain information on regions intersected. On the contrary, Ottmann et al. keep track of region information by color counters in a simple structure called D-list. Kriegel et al. [11] gave a brief description of how to implement a general map overlay operator involving two input maps. Their algorithm is based on and has the same time complexity of Nievergelt-Preparata algorithm. However, the regions assumed are general and holes are allowed. Nevertheless, degenerate cases are not considered explicitly.

1.1.3 Numerical Robustness Problem

Due to computation errors, robustness is an important issue in implementing geometric algorithms. Milenkovic proposed a technique which he calls *double precision geometry* for finding the I intersection points [13]. He showed that if the arithmetic precision of the input is P, then the exact arithmetic for the problem uses $(2P+1)$ precision arithmetic. Mehlhorn and Naher [16] found a robust implementation of Bentley-Ottmann algorithm based on the LEDA library of efficient data types [17]. Their algorithm runs in time $O((N + S) \log N)$, where S is the number of vertices in the input. Since only regions without holes are allowed, each intersection point can have at most six intersecting pairs. In other words, S is of $O(I)$. Hence Mehlhorn-Naher algorithm has the same time complexity as Bentley-Ottmann algorithm. The implementation uses exact arithmetic for the reliable realization of the geometric primitives and it uses floating point filter to reduce the overhead of exact arithmetic. A region intersection and union algorithm is also implemented in LEDA [15]. The algorithm accepts two simple regions without holes as input and returns the intersection or union. The input regions are first processed and a planar graph is produced. Then for each edge in the graph, determine if it is an edge of the resulting region by looking at its orientation and its neighbors. As with the line-intersecting algorithm, the region intersection and union algorithm is robust since the implementation is based on the LEDA library of data types. Nonetheless, the algorithm does not seem to be extended easily to other operators.

A spatial database system has been designed to solve the robustness problem of the implementation of spatial operators and predicates [9, 10]. The approach taken is to convert all geometric objects in a database to be defined on the same *realm*. Informally, a *realm* is a planar graph defined on a discrete grid. Green-Yao algorithm [7] is used in the conversion. Because of the nature of realm, simple, efficient and robust algorithms can be found for implementing most geometric operators and predicates proposed in their query language.

1.2 Our Approach

Unlike previous approaches [18, 20, 11], our problem domain is *general* and *all* degenerate cases that were investigated in the literature such as collinear and vertical edges are explicitly considered. In most existing work based on plane-sweep paradigm [18, 20, 11], information at an intersection point is updated by analyzing its *types*: start, end, bend and intersection. However, we take a different approach by analyzing incident edges and regions in which they are in. As a result, our algorithm is simpler and more comprehensive, and the proof of its correctness is more elegant. Interested readers please compare our approach with those in [18, 20, 11, 15]. Instead of implementing some specific operators, we are interested in finding a *general* and *efficient* algorithm that can implement a largest set of operators and predicates. Since operators such as set difference, union and intersection are implementable with the proposed algorithm, operators should be closed under the regions allowed. Thus, unlike [18, 20], the regions we consider are *general* and holes are allowed in a region [10]. The algorithm that we shall present generalizes and unifies the previous results on implementation of operators involving regions. Instead of considering the so-called *regularized* intersection [10, 11, 20], the proposed algorithm can implement both the *conventional* and *regularized* intersections [23]. Conventional intersection could be a more efficient way of implementing intersection or overlapping operations in a spatial database system. With the exception of [10], existing work concentrate on operators involving regions, the proposed algorithm implements both *operators* and *predicates* involving regions as well as *lines*.[1] Moreover, this method is believed to be simpler than finding algorithms for individual operators and predicates; and can be *extended* easily to accommodate any operator and predicate the output condition of which can be determined locally at intersection points. To implement new operators or predicates, only output actions pertaining to these operators or predicates are added to the program.

To our best knowledge, the only work that is closely related to ours is [10]. Both share the goal of finding efficient algorithms that can implement spatial operators and predicates easily. Nevertheless, there are some important differences. Our work concentrates on finding essential elements for implementing a largest set of spatial operators and predicates while they are interested in solving the robustness problem in implementing spatial operators and predicates. Consequently, the approaches taken are quite different. Firstly, we do not assume realm, a more restricted structure that is not assumed in most existing systems. Secondly, we are interested in finding extendible solution that common to a largest set of operators and predicates; as oppose to finding solutions individually. As will be seen in subsequent discussion, to demonstrate we identify the essential elements, a algorithmic scheme is derived to solve the extendibility problem. Lastly, our approach, as will be demonstrated in Section 9, can be generalized nicely to r layers of geometric objects which does not seem to be the case in their implementation [10]. On the other hand, we do not attempt to address the robustness issue in this work.

In sum, as geometric operators and predicates are diverse and their implementations are complex, spatial query language designer faces the uneasy choice of usability, efficiency and ease of implementation [3]. With the proposed algorithm, many operators and predicates in a spatial database system can be easily and efficiently implemented and thus provides a solution to the above-mentioned problem. The implementation of all operators and predicates as described in this paper have the same time complexity. Each requires $O((N + I) \log N)$ time in the worst case, where N and I are the total numbers of edges in the operands and of intersection pairs, respectively.

In Section 2, we define the notation used throughout the discussion. A plane-sweep algorithm is used in implementing the operators and predicates. In Section 3, we shall show how to update information associated with the sweep line at each sweep point. Basically the intersection points are processed in a left-to-right and bottom-to-top manner; and information required by the algorithm is updated locally at each intersection point. Different operators and predicates can be implemented with different output actions at an intersection point after it is processed. To validate our claim, we focus on the implementation of several important operators and predicates and show what actions are performed at each intersection

---

[1]Our method is applicable to *points* as well. Nevertheless, operators and predicates involving points can be implemented more efficiently with specialized algorithms.

point in Section 4. In Section 5, the pseudocode of the algorithm and the required data structures are described. A brief description of the implementation of various data structures and their primitives are given in Section 6. In Section 7, the time complexity of the algorithm is analyzed. Although the discussion concentrates on operators and predicates involving at most two regions, an intriguing property of the algorithm is that it can be extended to deal with arbitrary number of regions with a minimal modification. We discuss the necessary changes in Section 8. Finally, a summary and future directions are given in Section 9.

2 Definition and Notation

In this Section, we first define some important notation that will be used throughout the discussion. As this paper deals with lines and regions, we first define what they are.

Informally, a region R is an area possibly with a set of holes inside, as is defined in [9]. Holes of R are disjoint (but could be connected at a vertex) with each other and are enclosed in R. An edge is denoted by two endpoints, a color and an orientation. As in [20], the color (either *red* or *green*) of an edge indicates to which set of regions it belongs and the interior of a region is always on the *right side* of an oriented edge. An *edge* e is specified as $e = < ((x_l, y_l), (x_r, y_r)), \text{COLOR}, \text{ORIENTATION} >$, where (x_l, y_l) is the *left-endpoint* and (x_r, y_r) is the *right-endpoint* of e while COLOR and ORIENTATION are the color and direction tags, respectively. All edges of a region will be assigned with the same color. We shall assume that:

- $x_l < x_r$; or
- $x_l = x_r$ AND $y_l < y_r$.[2]

An edge is said to be *left-oriented* (*right-oriented*), if its starting point is its right-endpoint (left-endpoint, respectively). A *segment* is an edge without an orientation. A pair of edges are said to be *collinear* if they share a common sub-segment. A *chain* of edges or simply a chain, is a finite sequence of connected edges such that any two adjacent edges share an endpoint and no endpoint belongs to more than two edges. A chain is said to be *simple* if there is no point other than an endpoint is shared by more than two edges. Informally, a chain is simple if there is no pair of edges crossing over each other and no branching in the chain. A chain is said to be *close* if the two endpoints of the chain are the same.

A *line* is just a set of chains of segments. A *region* is denoted by $<B, S>$, where B and S are a simple closed chain and a set of disjoint simple closed chains, respectively. The closed chain B is assumed to be *clockwise oriented* while chains in S are *counter-clockwise oriented*. Elements in S are enclosed by B. B is said to be the *boundary* of the region and S are *holes* inside. We assume that boundary and holes are at least of length 3 and are not collinear, but some are possibly connected at a vertex. Since holes are allowed, each vertex of a region is incident with an *even number* of edges. Moreover, the number of incoming and outgoing edges at a vertex are the *same*. An *intersection point* is either a point common to two non-collinear edges or an endpoint of the common sub-segment of two collinear edges. The edges involved could come from the same or different lines or regions.

To simplify our discussion, we assume input involves a line, and two regions with different colors. A plane-sweep technique will be used to find out all output points and edges. We sweep a vertical line L (*sweep line*) in the horizontal direction through

---

[2]Edges with the same left-endpoint and right-endpoint is actually a point. and is treated as an invalid input in our algorithm.

the plane, keeping track of all regions under consideration. An edge or a region is said to be *active* if it intersects with the sweep line L. At each sweep-line position (*sweep point*), active edges are totally y-ordered, and so are the intervals between them. The sweep points are exactly the x-coordinates of intersection points. An *elementary region* is an area between two consecutive sweep-line positions and two adjacent active edges [20]. The unbounded regions just above the topmost and just below the bottommost active edges are also regarded as elementary regions. Points in an elementary region are covered by exactly the same set of colored regions. For example, in Figure 1, there are four active edges (i.e., e_1, e_8, e_5 and e_4) and five active elementary regions (i.e., two of them are covered by one region while the others are covered by no region) to the right of L. The colors pertaining to an elementary region is said to be the *covering* of the region.[3] To determine if a point or an edge need to be output, we shall keep track of the covering of all elementary regions to the right of the sweep line L.

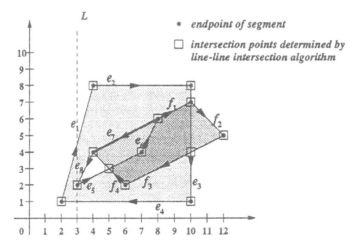

Figure 1: *An Input Example*

At each sweep point (say, x) of L, all active edges, and thus all active elementary regions, will be totally y-ordered according to the following rule:

$$
\begin{aligned}
e_1 < e_2 \quad \text{iff} \quad & y_{e_1}(x) < y_{e_2}(x) \\
\text{or} \quad & y_{e_1}(x) = y_{e_2}(x) \ \text{AND} \ m_{e_1} < m_{e_2} \\
\text{or} \quad & y_{e_1}(x) = y_{e_2}(x), \ m_{e_1} = m_{e_2}, \ \text{AND} \ p_l(e_1) < p_l(e_2) \\
\text{or} \quad & y_{e_1}(x) = y_{e_2}(x), \ m_{e_1} = m_{e_2}, \ p_l(e_1) = p_l(e_2), \ \text{AND} \ p_r(e_1) < p_r(e_2) \\
\text{or} \quad & y_{e_1}(x) = y_{e_2}(x), \ m_{e_1} = m_{e_2}, \ p_l(e_1) = p_l(e_2), \ p_r(e_1) = p_r(e_2), \ \text{AND} \\
& \text{colors of } e_1 \text{ and } e_2 \text{ are red and green, respectively}
\end{aligned}
$$

where $y_e(x)$ is the y-value of e at x, $p_l(e)$ is the left-endpoint of e, $p_r(e)$ is the right-endpoint of e, m_e is the slope of e, and for two points p_1 and p_2,

$$
\begin{aligned}
p_1 < p_2 \quad \text{iff} \quad & x_{p_1} < x_{p_2} \\
\text{or} \quad & x_{p_1} = x_{p_2} \ \text{AND} \ y_{p_1} < y_{p_2}
\end{aligned}
$$

and

---

[3]The number of colors in a covering is called *overlap number* in [10].

$$p_1 = p_2 \quad \text{iff} \quad x_{p_1} = x_{p_2} \quad \text{AND} \quad y_{p_1} = y_{p_2}$$

where x_p and y_p are the x-coordinate and y-coordinate of p, respectively.

To find out all changes in the covering of elementary regions, we have to determine all changes in the total order of active edges since a change in covering can only take place when there is an ordering change of active edges [20]. Changes in the total order of active edges occur exactly at the intersection points. Given all intersection points formed between edges, L is supposed to scan through each of them in an increasing order with respect to the x-coordinates and to keep track of the covering of all regions appropriately.

Since the method in [20] is simpler and can be extended easily to arbitrary number of regions, we maintain the covering of an elementary region in terms of *color counters*, one counter for each colored (input) region [20]. These color counters keep track of the number of red and the number of green regions covering an elementary region. Since each region is assigned with a distinct color, each color counter is either 0 or 1. More precisely, whenever an intersection point (say, p) is reached, the total order of active edges along L as well as the covering of active elementary regions along L are updated appropriately. An important observation is that covering information can be determined and updated *locally* at an intersection point. Based on this observation, all output edges and points can then be decided and reported at p.

Suppose a line-intersecting algorithm has been invoked and all intersection points are found. Assume further that we have a data structure l which stores current ordering of all active edges, as well as all active elementary regions, along L^4. We shall show how l is updated at each predetermined intersection point.

3 Active Elementary Regions Update

As the sweep line L scans from left to right, elementary regions to the right of the sweep line change. In this Section, we shall show how l is updated correctly. We begin with the simple case in which there is no vertical or collinear edges. We then show how the method could be extended easily to handle all degenerate cases.

3.1 Cases with no vertical and collinear edges

In this Subsection, we assume that all edges are non-vertical and no collinear edges exist. We shall show how l is updated correctly and efficiently at all intersection points. Contrary to existing work where intersection points are updated depending on the type of intersection points encountered [18, 20, 11], we take a different approach by analyzing the incident edges. As a result, our algorithm is much simpler and more comprehensive, and the proof of its correctness in updating the sweep line is more elegant. Interested readers may wish to compare our approach with those in [18, 20, 11, 15, 10]. We shall start by looking at several important facts about incident edges and covering of elementary regions.

Observation 1: At an intersection point, edges of the same color are of even number and are alternately incoming and outgoing edges.

As the sweep line L is a vertical edge, incident edges at an intersection point are either to the left or to the right of L.

---

[4] In other words, besides keeping track of the ordering information of the edges, l is also responsible for keeping track of covering of the elementary regions.

Observation 2: At an intersection point, the covering of the elementary regions above the left and right topmost incident edges, if they exist, are the same. The observation is also applicable for the elementary regions just below the bottommost left and right incident edges.

Lemma 3.1 *The color c counter value flip-flops across an edge e of a region of color c.*

[**Proof**]: Follows from **Observation 1** and the fact that the color c appear only to the right of e. □

We are now ready to outline a simple procedure for updating covering of elementary regions at an intersection point.

Suppose that we identify all incident edges at an intersection point and assume further that the covering of elementary regions to the left of the sweep line have been updated correctly. We first process and delete edges, if any, to the left and determine the covering I of elementary region just below the bottommost left edge. Then the edges to the right are inserted. Finally we use the covering I to update covering of elementary regions created by the inserted right edges.

If there is no edge to the left, the covering I is the same as the elementary region in which the intersection point is in. If there is some incident edge to the left, then delete each edge from l and its *just above* elementary region. This has the same effect as retaining the elementary region B just below the bottommost left incident edge and its covering. We then insert edges to the right, if any, into l and keep track of which edge is the *highest* and which is the *lowest*. The action performed for an edge insertion into l is to locate the active elementary region in which it is inserted and divide it into two new elementary regions with the same covering as in the original one. In effect, a new elementary region is created *just above* the inserted edge. By the definition of y-order, when the first right edge is inserted, it subdivides B into two active elementary regions. Each subsequent insertion again subdivides one of these newly created elementary regions. By the way the covering is assigned, all newly created elementary regions have the same color counter values. Having inserted all the right edges, the covering of newly created elementary regions are updated as follows. Process each edge e to the right of the intersection point from the lowest y-order to the highest as follows: if e is of color c, the covering of elementary region just above e is set to be the same as its elementary region just below with color c counter value flip-flopped. The pseudocode of this algorithm is given as the routine **ResolveIP** in Section 5.2.

Theorem 3.2 *Suppose L is swept from left to right. The above algorithm correctly updates the covering of elementary regions to the right of L at an intersection point after the right edges are processed.*

[**Proof**]: Assume the covering of elementary regions to the left of L has been updated correctly. We want to show that the covering of elementary regions to the right of L at an intersection point are correct after they are processed. If there is no edge to the left, the way of finding I is clearly correct. If there is some edge to the left, by **Observation 2**, the initial covering I is again determined correctly. If there is no edge to the right, then I is the covering for the region to the right of the intersection point. If there is some right edge, then I is the correct covering for the region just below the bottommost edge to the right of the intersection point. Initially the covering of all newly created elementary regions due to insertion of right edges are the same. As covering is updated bottom-up, the correctness of updating the covering of elementary regions just above inserted right edges follows from **Observation 1** and from Lemma 3.1. □

3.2 Cases with Collinear Edges

In this Subsection, we shall examine cases with collinear edges. Since edges from the same region are not collinear, collinear edges are of different colors.

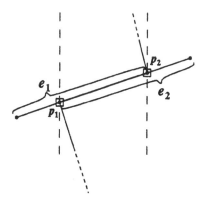

Figure 2: *Example of Collinear Edges.*

Consider two non-vertical collinear edges (say, e_1 and e_2) and the *common sub-segment* formed between them is $[p_1, p_2]$ (see Figure 2). Under our assumption, e_1 and e_2 intersect only at p_1 and p_2. In other words, there is no intersection in the middle of $[p_1, p_2]$. Conceptually, they are viewed as follows:

$[p_1, p_2]$ is split into two edges (say, k_1 and k_2 - one is of green color and the other is of red color) such that k_1 and k_2 are separated by a negligible angle α (where $\alpha \rightarrow 0$) at p_1 and are combined at p_2, as is shown in Figure 3. Moreover, we shall assume that there is an elementary region between k_1 and k_2.

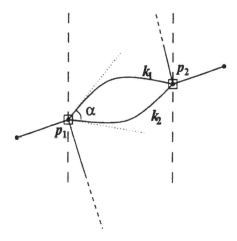

Figure 3: *Split Collinear Edges.*

As a result, collinear edges cease to exist. Therefore, we can use the same argument as described before to update l at all these intersection points.

When splitting $[p_1, p_2]$ into two edges, the edges are ordered according to the total order as defined in Section 2. For instance (see Figure 3), we assume that k_1 is on top of k_2. It is worth noting that the covering of elementary regions just above k_1 and just below k_2 are correct, but not necessarily for the in-between elementary region. Nevertheless, the assignment is arbitrary and care must be given when output is generated.

As long as an intersection point with collinear edges can be converted into one specific intersection point with degree four, collinear edges can be handled as non-collinear edges.

3.3 Cases with Vertical Edges

Consider an imaginary vertical line treated as a slanted segment as follows:

> The imaginary line, with all edges lying on or touching it, is assumed to be rotated clockwisely around its lowest point (where y-value of the lowest point is assumed to be the minimal y-values of the input edges) by a *negligible* amount so that it has a positive slope which is greater than that of *any* non-vertical edge.

Conceptually, the further a point on the imaginary line from the lowest point, the greater its x-value. As a result, we have

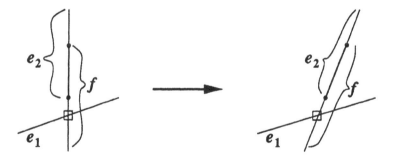

Figure 4: *Instance of Slant Intersection Point and Line.*

- each *slanted intersection point* is assumed to be an intersection formed between two non-vertical edges, while

- each *slanted collinear edge* is assumed to be a collinear edge formed between two non-vertical edges.

With this assumption, we are able to update l at all slanted intersection points and lines as discussed in previous Subsections. As long as l is updated in a left-to-right and bottom-to-top manner, l will be updated correctly at each sweep point.

4 Output Condition Analysis At Intersection Points

Our approach to a general implementation of geometric predicates and operators is based on the observation that once the covering of elementary regions to the right of L are updated correctly at an intersection point, various operators or predicates differ only on the output actions taken, if any, at the intersection point. The output action involves outputting an edge at the intersection point or returning *true* or *false*. The testing that determines if an output action is carried out includes examining the *covering* of elementary regions as well as the *color* and *orientation* of incident edges. As collinear edges are assigned with an arbitrary order when they are split, covering of elementary region between the split collinear edges may not be correct. Consequently, these two types of edges are handled differently in determining if an output action is performed.

As an illustration to this general approach, we analyze what are the output actions for several commonly used predicates and operators found in systems such as those in [3, 5, 9, 21]. The pseudocode and the primitives needed for implementing these predicates and operators are given in Section 5.

Assumption on input lines: As the critical positions are the intersection points and since lines in general may not be closed, the first and last endpoints of a non-closed chain may not be processed. To avoid this problem and to simplify our discussion, we assume from now on all chains in a line are made *closed* and those added segments are *marked* and *ignored* in the output actions.

Throughout this Section, when l is said to be updated successfully at an intersection point, the algorithm **ResolveIP** as described in the Subsection 3.1 on updating the intersection point is executed. That is, all incident edges to the left are removed from l, all right incident edges are added and their newly created elementary regions are updated correctly.

4.1 Predicates Output Conditions

The predicates included in this Subsection are *equal*, *inside*, *overlap* and *adjacent*. These predicates return *true* or *false* as the output.

4.1.1 Equal

Two regions are said to be *equal* if they are identical. Or equivalently, at each intersection point, after l is updated successfully, the two sets of colored edges to the right are collinear and have the same orientation.

4.1.2 Inside

A green region is said to be *inside* a red region, if for every elementary region R, R is covered by green implies R is covered by red. Since edges at an intersection point could be collinear, this condition can be rephrased as follows.

At each intersection point, after l is updated successfully, for each edge e to the right of the sweep line, do the following:

1. If there is an edge collinear with e but with different orientation then return *false*.

2. If there is no edge collinear with e, and if either the region just above or just below e has a green covering but not red, then return *false*.

A green region is inside a red region exactly after all intersection points are processed and no *false* has been returned.

A line c is *inside* a red region R if every segment of c is either collinear with an edge of R or is completely embedded in some red elementary region. This condition can be expressed equivalently as follows.

At each intersection point, after l is updated successfully, for each segment e of c to the right of the sweep line, do the following:

1. If no (red) edge is collinear with e and there is no red elementary region completely embeds e, then return *false*.

A line is inside a red region exactly after all intersection points are processed and no *false* has been returned.

4.1.3 Adjacent

Two regions are said to be *adjacent* to each other, if they share some common boundary segment, but no interior points. Thus, two regions are adjacent if they are disjoint and share some common boundary segment. We express this as follows.

We need a boolean variable called *ShareAnEdge* which is set to *false* initially. At each intersection point, after l is updated successfully, for each edge e to the right of the sweep line, do the following:

1. If e is not collinear with other edges, then if either the region just above or just below is covered by red and green, then return *false*.

2. If e is collinear with another edge then if they have the same orientation then return *false* else set *ShareAnEdge* to *true*.

Two regions are adjacent exactly after all intersection points are processed, no *false* is returned and *ShareAnEdge* is *true*.

A line c is *adjacent* to a red region R if (i) no part of a segment is completely embedded in some red elementary region and (ii) at least a sub-segment is collinear with an edge of R. To test these conditions, we do the following:

A boolean variable called *ShareAnEdge* is set to *false* initially. At each intersection point, after l is updated successfully, for each segment e of c to the right of the sweep line, do the following:

1. If e is not collinear with other edges, and if there is a red elementary region containing e, then return *false*.

2. If e is collinear with another edge then set *ShareAnEdge* to *true*.

A chain is adjacent to a red region exactly when all intersection points are processed and no *false* has been returned and *ShareAnEdge* is set to *true*.

4.1.4 Overlap

Two regions are said to be *overlap* if some elementary region is covered by both red and green. At each intersection point, after l is updated successfully, for each edge e to the right of the sweep line, do the following:

1. If e is not collinear with other edges, then if either the region just above or just below is covered by red and green, then return *true*.

2. If e is collinear with another edge, then if they have the same orientation then return *true*.

Two regions overlap exactly after an intersection point is processed, *true* is returned. Conversely, if no *true* is returned after all intersection points are processed, then the two regions do not overlap.

A line c is said to *overlap* with a red region if a sub-segment of c is completely embedded in a red elementary region. This condition can be tested by inspecting the relative position of a segment to an elementary region at an intersection point.

4.2 Operators Output Conditions

Unlike the predicates, a region-region or line-region operator returns is a list of edges, segments or points. The list of edges denote boundary and holes of regions returned. This list is constructed as intersection points are processed. The operators analyzed are *fusion*, *set difference* and *intersection*. A problem with plane-sweep algorithm is how to generate an output edge. A solution is presented in [20]. In that method, each edge is associated with an output buffer. The output buffer of an edge e, if not empty, stores the left-endpoint of an output edge which is a sub-edge of e. When the right-endpoint of the output edge is reached, the output buffer is examined. If it is non-empty, an output edge is generated and the output buffer is then emptied. From now on, we assume edges are output as described above.

4.2.1 Fusion

The *fusion* of two regions is the union of the operands involved [3]. It returns a set of edges representing a set of regions. An edge is output if it represents part of the boundary of the union. This occurs exactly when one side of the edge is neither red nor green, that is, it has *no color*.

For each edge e to the right of the sweep line, do the following.

1. If e is not collinear, output e if either the region just above or just below has no color covering.

2. If e is collinear and both edges are of the same orientation, output the common sub-edge.

4.2.2 Set Difference

The *set difference* of green region from red $(R - G)$ is the areas that are in red but not in green. It returns a set of edges representing a set of regions. An edge that is part of the boundary or holes of a resulting region if it is part of a red or green edge such that either its just above or just below elementary region is covered by red but not green.

For each edge e to the right of the sweep line, do the following.

1. If e is not collinear, output e if either the region just above or just below is red but not green. However, if e is a green edge, then the orientation is reversed, (that is, from left to right and from right to left) to reflect the edge is part of the boundary of a red region.

2. If e is collinear and both edges are of different orientation, output e if e is a red edge.

The set difference of a line c from a red region R are the sub-segments of c that are neither adjacent to R nor are inside R. This operator could be implemented by looking at each segment of c to the right of the sweep line and see if the condition is satisfied.

82

4.2.3 Intersection

Unlike most intersection operators in the literature, the intersection operator in QL/G returns a set of regions, segments and points. It is referred to the *conventional intersection* in [23]. The result of *intersection* of two regions are the regions, segments and points that are common to both operands [3]. The set of regions returned is precisely the result of the so-called *regularized intersection* of two regions involved [20, 23]. A segment is returned if it is common to both regions but is not part of an edge in the set of regions returned. A point is returned if it is not on any of the segments or regions returned and is a point common to both operands. This occurs exactly when two regions touching each other at a common vertex.

We again look at the actions taken at each intersection point.

To find the edges output for a returned region, perform the following.

For each edge *e* to the right of the sweep line, do the following:

1. If *e* is not collinear, output *e* if either the region just above or just below have both the red and green covering.

2. If *e* is collinear and both edges are of same orientation, output the common sub-edge.

The set of segments returned is determined as follows. For each pair of collinear edges of different orientation, output the common sub-segment.

The points returned are vertices common to both regions and are determined as follows. There are no collinear edges at the intersection point. Moreover, the intersection point must have a positive even number of green as well as red edges. Sort all left and right incident edges. Edges of the same color must occur in consecutive order. Then identify the two pairs of consecutive edges of different colors. Let the two edges in the four edges identified with the same color (as well as orientation) be a *colored edge*. The two colored edges are on the *left-hand side of each other* with respect to their orientation exactly when the intersection point is returned.

The result of *intersection* a line *c* with a red region *R* are the segments and points common to both operands [3]. A segment is returned if it is a sub-segment of *c* and is part of *R*. A point is returned if it is not on any of the segments returned and is a point common to both operands. This occurs precisely when a segment 'touching' a vertex of *R*.

The segments returned can be computed by looking at each segment *e* of *c* to the right of the sweep line as follows:

1. If *e* is not collinear with any red edge, output *e* if it is in a red elementary region.

2. If *e* is collinear with some red edge, output the common sub-segment.

The intersection point is returned if there are some segments as well as some red edges at the intersection point; but none of segments satisfies the two conditions above.

5 The Algorithm

In Section 5.1, we shall describe the data structures and some of their primitives that we are going to use in our algorithm. Then the general algorithm will be presented in Section 5.2.

5.1 Data Structures

5.1.1 Edges

An edge e is specified as $e = < ((x_l, y_l), (x_r, y_r)),$ COLOR, ORIENTATION $>$, where (x_l, y_l) and (x_r, y_r) are the *left-endpoint* and the *right-endpoint* of e, respectively, and COLOR and ORIENTATION are the color and direction tags, respectively.

An edge supports the following set of operations:

orientation(e): Given an edge e, returns LEFT or RIGHT, if e is left- or right-oriented, respectively. If e is a segment then NULL is returned.

color(e): Given an edge e, returns its color.

5.1.2 An Array $\mathcal{L}$

Our algorithm requires first to find out all intersection points from the operands. Any line-intersecting algorithm that returns the set of intersecting pairs that satisfies our input format will do. The output of a line-intersecting algorithm should be an array of 3-tuples. Each 3-tuple is of the form $<i, e_k, e_j >$, where $e_k \neq e_j$ are a pair of intersecting edges and i is their intersection point. Collinear edges are also regarded as intersecting pairs and two 3-tuples $<i_1, e_k, e_j >$ and $<i_2, e_k, e_j >$ are generated for each such pair, where i_1 and i_2 are the endpoints of the common sub-edge. In our implementation, we employ the line-intersecting algorithm by Myers [12]. These intersection points are then sorted ascendingly on the x-axis and then ascendingly on y-axis. These points are stored into an array $\mathcal{L}$ of 3-tuples. The following set of operations is provided for this data structure:

sort($\mathcal{L}$): It sorts the array ascendingly on the x-coordinates and then on y-coordinates of intersection points in the list.

getnext($\mathcal{L}$): Return the next 3-tuple $<i, e_k, e_j >$, if it exists.

5.1.3 D-list

A plane-sweep technique is being used as we process 3-tuples in $\mathcal{L}$ in an ascending order of intersection points. At each sweep point, as was described in Section 2, l is used to keep track of a y-ordered list of all active edges, as well as covering (color counters) of all active elementary regions. Thus, as in [20], a data structure called D-list will be used here to maintain l.

To facilitate the search process, the D-list is kept as an *AVL-tree* so that we are able to add, delete and search edges efficiently. The D-list supports the following set of operations:

D_add(e): Given an active edge e, add it to the D-list. An active elementary region in which e is supposed to be inserted is subdivided into two active elementary regions. The color counters of newly created elementary regions assume the corresponding values of the original elementary region.

D_delete(e): Given an edge e, remove e and its just above active elementary region from the D-list.

D_above(e): Return the edge f just above e in the D-list, if it exists.

We need to update color counter values of elementary regions. The following operation is supported by the D-list:

D_update(e,RED_VALUE,GREEN_VALUE): Assign the red color counter and the green color counter by RED_VALUE and GREEN_VALUE, respectively, to the active elementary region just above e.

We need to query covering of elementary regions in the current D-list. The following are required.

color-above(c,e): Return the color c counter value of the elementary region just above the edge e.

color-below(c,e): Return the color c counter value of the elementary region just below the edge e.

region-in(c,e): Return the color c counter value of the elementary region in which e is in. The edge or segment e is supposed to be active and is in some active elementary region.

5.1.4 Incident Edges at an Intersection Point

Only when all incident edges at an intersection point are identified are we able to update the D-list and output points and edges. Let us create a structure IE which contains an array of records which is used to store all incident edges at a particular intersection point p. Each such record in the array is of format $\langle e, position \rangle$, where e is one of the incident edges at p and *position* indicates on which side (either L stands for left-hand-side or R stands for right-hand-side) of the sweep line that e is lying.

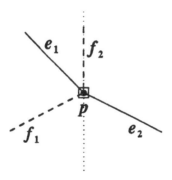

Figure 5: *An Instance for Demonstrating Array IE.*

Consider the example in Figure 5. Suppose that p is an intersection point with four intersecting edges. These four incident edges will be found and the array of records in IE will be

$$\langle f_2, R \rangle, \langle e_2, R \rangle, \langle f_1, L \rangle, \langle e_1, L \rangle$$

Note that vertical edges are treated as slanted with a positive slope. This means that f_2 is lying on the right hand side of p.

In sum, IE is a structure storing the current intersection point and an array of records. An IE supports the following set of operations:

reset(IE): Initialize intersection point and records in IE to NULL.

setIP(IE, p):The current intersection point is set to p.

IP(IE): Return the current intersection point.

sort(IE): Sort edges in IE in clockwise direction starting at 12 o'clock.

removeDuplicate(IE): Remove duplicate records from IE.

appendIE($\langle e, LorR \rangle$): Append a record $\langle e, LorR \rangle$ to the IE. The current **IP**(IE) must be a point on e.

size(IE): Return the number of records currently stored in IE.

collinear(e): e is an edge or segment at **IP**(IE). If there is an incident edge or segment f (that from the other operand) in IE collinear with e, then return f, else return NULL.

right(IE): Return the set of edges or segments that have the position R.

left(IE): Return the set of edges or segments that have the position L.

5.2 The Pseudocode

Our algorithm will be based on the following three major steps:

1. Use a line-intersection algorithm to obtain $\mathcal{L}$.

2. Sort $\mathcal{L}$ with respect to intersection points.

3. Scan the array $\mathcal{L}$ in an ascending order and update l at each intersection point.

When scanning the sorted $\mathcal{L}$, we shall do the following for each set of all 3-tuples with the same intersection point p:

1. Find out the set S_p of edges incident on p.

2. Create IE for S_p with p as the intersection point.

3. Update D-list with edges in IE.

4. Perform output operations at the intersection point.

The following is the pseducode for the main program.

Function GeoOp(R, G, Line, Op, UserDefine): A general algorithm that implements many of the predicates and operators involving two regions R and G and a line *Line*.

Input: R is the given list of edges of the red region; G is the given list of edges of the green region; *Line* is assumed to be a set of *closed* chains for line-region predicates or operators; Op is one of the predicates or operators implemented and *UserDefine* is a union of structures; one for each Op implemented. *UserDefine.Op* contains Op-dependent variables as well as the resulting list of output points and/or output edges for the operation Op involved.

Output: If a geometric predicate is involved, *true* or *false* is returned. If an operator Op is involved, the result is stored in the corresponding structure *UserDefine.Op*.

Method: The algorithm sweeps the plane from left to right. At each sweep point, the intersection points are processed in an increasing y-coordinate values. At each intersection point, the incident edges are collected and analyzed, and appropriate actions are performed to update the *D*-list. This is accomplished with routines **UpdateIE** and **ResolveIP**. Having correctly updated the active elementary regions at the intersection point, *Op*-dependent tests are performed to see if edges, points, *true* or *false* should be returned. All these actions are described in the function **HandleOutput**.

begin
  /* Initialization. */
  $UserDefine.Op \leftarrow$ NULL
  /* Use a line-intersection algorithm to find out all intersecting pairs.
  Store and sort the pairs in $\mathcal{L}$.*/
  $S \leftarrow$ all edges in $R \cup$ all edges in $G \cup$ all segments in *Line*
  /* LineIntersection accepts a set S of segments and returns an array $\mathcal{L}$
  of 3-tuples representing the set of intersecting pairs. */
  LineIntersection$(S, \mathcal{L})$
  sort$(\mathcal{L})$
  /* Scan $\mathcal{L}$ */
  reset(IE)
  $previous = < i_0, f_1, f_2 > \leftarrow$ **getnext**$(\mathcal{L})$
  setIP(i_0)
  /* Accumulate incident edges or segments at an intersection point*/
  UpdateIE$(previous)$
  while $\mathcal{L}$ is not empty **do**
    $current = \langle i, e_1, e_2 \rangle \leftarrow$ **getnext**$(\mathcal{L})$
    if $i \neq i_0$ **then**
      /* All incident edges at an intersection point have been identified*/
      /* Sort records in IE in clockwise direction starting at 12 o'clock */
      /* and remove duplicate records in *IE* */
      sort(IE)
      removeDuplicate(IE)
      ResolveIP(IE)
      $flag \leftarrow$ **HandleOutput**$(IE, i_0, Op, UserDefine)$
      if Op is a predicate **then return**$flag$
      reset(IE)
      setIP(i)
    end if
    UpdateIE$(current)$
    $previous \leftarrow current$
  end while
  /* Resolve the last intersection point */
  sort(IE)
  removeDuplicate(IE)
  ResolveIP(IE)
  $flag \leftarrow$ **HandleOutput**$(IE, i_0, Op, UserDefine)$
  if Op is a predicate **then**
    switch(Op)

/* Since a variable is tested, *adjacent* are handled differently*/
case *RR_adjacent* or *LR_adjacent*
   if *flag* is *false* **or** *ShareAnEdge* in *UserDefine.OP* is *false*
      then return *false* **else return** *true* **end if**
  break
case *default*
  return *flag*
  break
end switch
end if
end /* GeoOp */

UpdateIE(t): Given $t = \langle i, e_1, e_2 \rangle$, where e_1, e_2 are edges or segments and i an intersection point such that i is on both edges. Determine the position of each edge with respect to i and then store the edge and its associated position into IE. $IP(IE)$ should be i.

Input: A 3-tuple $t = \langle i, e_1, e_2 \rangle$ of $\mathcal{L}$.

Output: IE is updated accordingly, as was described in Section 5.1.4.

ResolveIP(IE): Update covering of elementary regions to the right as described in Section 3.

Input: IE contains all incident edges at an intersection point.

Output: D-list got updated by removing incident edges to the left and then inserting edges to the right followed by an update on covering of the newly created elementary regions.

Method: IE may contain segments which are ignored in updating the D-list. Let CIE be IE with those segments removed.

begin
  /* Remove left edges and their just above elementary regions from D-list */
  if size(left(CIE)) > 0 **then do**
    for each edge e **in** left(CIE) **D_delete**(e) **end for**
  end if
  /* Insert edges to the right */
  if size(right(CIE)) > 0 **then do**
    highest − *lowest* ← first edge in CIE
    for each edge e **in** right(CIE) **do**
      if e is y-order higher than *highest* **then** *highest*← e **end if**
      if e is y-order lower than *lowest* **then** *lowest*← e **end if**
      D_add(e)
    end for
  end if
  /* Update covering of newly created elementary regions in a bottom-up manner*/
  e← *lowest*

```
    while(e is not NULL and e is not y-order higher than highest) do
        if color(e) is RED then
            /* flipflop is a function returns 0 (1) when 1 (0) is given */
            RV← flipflop(color_below(RED, e))
            GV← color_below(GREEN, e)
        else
            GV← flipflop(color_below(GREEN, e))
            RV← color_below(RED, e)
        end if
        D_Update(e, RV, GV)
        e← D_above(e)
    end while
end /* ResolveIP */
```

HandleOutput(IE, IP, Op, UserDefine): This handles the output for an *Op*.

Input: *IP* is the current intersection point, *Op* is one of the predicates or operators implemented and *UserDefine* is a union of structures, one for each *Op* implemented.

Output: If a geometric predicate is involved, *true*, *false* or *continue* is returned. *continue* is returned if so far the final result is not known to be *true* or *false*. *UserDefine.Op* is returned which contains *Op*-dependent variables as well as the resulting list of output points and/or output segments for the operation involved.

Method: The output condition for each operator and predicate is given in Section 4. Due to space constraint, not every predicate and operator is included here. Predicates or operators begin with *RR* (*LR*) denote region-region (line-region, respectively).

```
begin
switch (Op)
    /* Test if a green region is inside a red region.*/
    case RR_inside:
        for each edge e in right(IE) do
            if collinear(e) = NULL then
                if (color-above(red,e) = 0 and color-above(green,e) = 1)
                    then return false end if
            else if orientation(e) ≠ orientation(collinear((e)))
                then return false end if
        end for
        return continue
        break

    /* Test if a green region is adjacent to a red region.*/
    case RR_adjacent:
        for each edge e in right(IE) do
            if collinear(e) = NULL then
```

```
            if (color-above(red, e) = 1 and color-above(green, e) = 1)
                then return false end if
        else if orientation(e) = orientation(collinear(e))
                then return false
                else UserDefine.RR_adjacent.ShareAnEdge ← true end if
        end if
    end for
    return continue
    break



/* Perform a set difference of green from red region. */
case RR_set difference:
    for each edge e in right(IE) do
        if collinear(e) = NULL then
            if (color-above(red, e) = 1 and color-above(green, e) = 0)
            or (color-below(red, e) = 1 and color-below(green, e) = 0) then
                if color(e) is GREEN then e orientation is reversed end if
                output e to UserDefine.RR_SetDifference.Result
            end if
        else if orientation(e) ≠ orientation(collinear(e))
            then if color(e) is RED then
                    output e to UserDefine.RR_SetDifference.Result endif
            end if
        end if
    end for
    break

/* Perform an intersection of green and red region. */
/* Only actions on output edges and segments are described here. */
case RR_intersection:
    for each edge e in right(IE) do
        if collinear(e) = NULL then
            if (color-above(red, e) = 1 and color-above(green, e) = 1)
            or (color-below(red e) = 1 and color-below(green e) = 1 )
                then output e to UserDefine.RR_Intersection.RegionResult
            end if
        else if orientation(e) = orientation(collinear((e)))
            then output the common sub-edge to
                    UserDefine.RR_Intersection.RegionResult,
                    if it is not already output
            else output the common sub-segment to
                    UserDefine.RR_Intersection.LineResult,
                    if it is not already output
        end if
    end for
    break
case default:
    Print an error message of undefined operator or predicate
```

end switch
end /* HandleOutput */

6   The Implementation of D-list

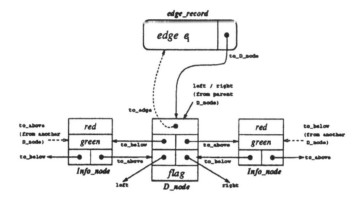

Figure 6: *Relationship Between Edge Records and D-nodes.*

A *D*-list is constructed as an AVL tree. As shown in Figure 6, nodes of a *D*-list, namely *D-nodes*, are not explicitly linked together in a doubly-linked list, but are rather linked alternately with another type of nodes, namely *Info-nodes*. Each *Info-node* is used to store the current covering (in terms of color counters) of an active elementary region. Functions **D_above**, **D_delete**, **D_add D_update**, **flipflop**, **color-above**(c, e), **color-below**(c, e) and **region-in**(c, e) can all be implemented easily. Among all these routines, **D_add**, **D_delete**, **D_update** and **region-in** take $O(\log N)$. All other routines require only constant time.

7   Analysis

In this Section, we analyze the time complexity of the algorithm. The routine **UpdateIE** requires only constant time. For each edge processed, **ResolveIP** requires at most $O(\log(N))$ time. In **HandleOutput**, a **for** loop is used in each individual case which iterates for each record in IE. Since for each segment or edge processed, the body of the **for** loop requires at most $O(\log(N))$ time, the procedure takes only $O(\log(N))$ time for each input segment or edge.

Let N be the number of edges or segments in the operands and I be the number of intersecting pairs from the operands. In the main body of **GeoOp**:

- The line-intersection algorithm we implemented is the Myer's algorithm [12] which requires $O(N \log N + I \log N)$ time in the worst case and $O(N \log N + I)$ time in the average case.

- Sort $\mathcal{L}$ takes $O(I \log I)$ or equivalently $O(I \log N)$, as $I < N^2$.

There is a **while** loop which iterates once for each element in $\mathcal{L}$. Let us consider the total cost of **sort**(IE) and **removeDuplicate**(IE), and the total cost of other statements separately.

- Clearly the total cost of removing duplicate *IE* records over all intersection points is $O(I)$.

- We first observe that the total number of incident edges at all intersection points is $O(I)$. It can be shown that the total cost of sorting *IE* records over all intersection points is $O(I \log I)$ or equivalently $O(I \log N)$.

- Except the **sort**(*IE*) and **removeDuplicate**(*IE*) statements, all routines invoked in the **while** loop require at most $O(\log N)$ time for each incident edge processed. Since the total number of incident edges is of $O(I)$, $O(I \log N)$ time is required for the **while** loop without the **sort** and **removeDuplicate** statements.

Therefore, in the worst case, the time taken by the algorithm is:

$$O(N \log N + I \log N) + O(I \log N) + O(I) + O(I \log N) + O(I \log N)$$

which is equal to

$$O((I + N) \log N).$$

8 Generalization to r Layers of Regions

We assume a red and a green region are involved. The algorithm can trivially be extended to two layers of colored regions, where regions in a layer are of the same color and are neither overlapping nor sharing common edges.

Throughout the discussion, we assume at most two regions are involved in a predicate or an operator. An intriguing property of the proposed method is that it can be generalized easily to cases in which r regions are involved, where $r > 2$. Let us explain what are some of the most important changes that need to be made to the proposed algorithm.

As before, each region is assigned with a distinct color. A total order is imposed on these colors so that active edges can be y-ordered at a sweep point. For each elementary region, there is a color counter for each of the r colors, each counter assumes the value of either 0 or 1. The way the active elementary regions are updated is essentially the same as described here. For p collinear edges, we conceptualize it as p edges splitting in a similar manner as in Figure 3. Thus, as before, it can be reduced to cases in which collinear edges do not exist. Although the ordering of these collinear edges is arbitrary, the covering just above the topmost edge and just below the bottommost edge are the same no matter how we assign ordering to these edges. Thus, the algorithm **ResolveIP** will correctly update the covering information, except possibly for those in-between elementary regions of collinear edges. Some primitives such as **collinear** need to be changed to accommodate more than two colored edges at an intersection point. Other changes may be required in routines such as **HandleOutput** depending on the predicates and operators involved.

9 Conclusion

We presented a general algorithm that implements efficiently a set of region-region and line-region geometric operators and predicates. It has been fully implemented and tested in C++ on a Sun workstation. The algorithm is based on the plane-sweep paradigm. Our problem domain is broad as it includes all degenerate cases identified in the literature. As various operators are implementable with the algorithm, closure of operators is essential. Consequently, regions considered are general and holes are allowed in a region. Existing work focus on either the implementation of individual operators or finding a common set of functions for implementing a set of operators and predicates [18, 20, 11, 10]. We took a different approach by identifying the essential elements that that are common to a largest set of operators and predicates. As a result, an algorithmic scheme can be constructed for implementing all these operators and predicates in a relatively simple and comprehensive manner. The algorithm presented implements both operators and predicates involving regions as well as lines, and it generalizes and unifies existing approches. Although we illustrate our approach with a number of popular predicates and operators, the algorithm can be extended easily to other operators and predicates so long as their output conditions can be determined locally with information available at an intersection point. All operators and predicates investigated are implemented in $O((N + I) \log N)$ time in the worst case, where N be the number of edges or segments in the operands and I be the number of intersecting pairs. We also showed that the algorithm can be generalized to r regions, where $r > 2$.

Currently we are working on extending this work to efficient implementation of spatial joins. Despite its generality, operators such as *maximum distance* between two geometric objects do not seem to be implementable with the proposed algorithm. Another open problem is to characterize the class of operators and predicates that can be realized with this method. As geometric objects are modeled on a finite precision machine, robustness of geometric algorithms is an important issue that has not been addressed in this work. Existing techniques can be classified into:

1. Solution based on inexact floating point arithmetic. See for instance [8, 13, 14].

2. Solution based on an exact arithmetic library. Such as those in [1, 6, 15].

3. Mixed solution. See for example [19].

As no clear consensus has emerged, this important issue deserves further investigation in the near future.

Acknowledgement

We thank other members of Spatial Database Group at the University of Waterloo for their input and assistance in building QL/G. The first author wishes to acknowledge the financial support of the Natural Sciences and Engineering Research Council of Canada.

References

[1] Benouamer, M.O., Jaillon, P., Michelucci, D. and Moreau, J.M., "A Lazy Solution to Imprecision in Computational Geometry," *Proceedings of the Fifth Canadian Conference on Computational Geometry*, 1993, pp. 73-78.

[2] Bentley, J.L. and Ottmann, T.A., " Algorithms for Reporting and Counting Geometric Intersections," *IEEE Transactions on Computer*, Vol. c-28 (1979), pp. 643-647.

[3] Chan, E.P.F. and Zhu, R., "QL/G: A Query Language for Geometric Databases," *Proceedings of the First International Conference on GIS in Urban and Environmental Planning*, Island of Samos, Greece, April 1996, pp.271-286.

[4] Chazelle. B. and Edelsbrunner, H., "An Optimal Algorithm for Intersecting Line Segments in the Plane," *JACM 39:1, January 1992*, pp. 1-54.

[5] DeWitt, D., Kabra, N., Luo, J., Patel, J.M., Yu, J.B., "Client-Server Paradise," *Proceedings of the 20th International Conference on VLDB*, Santiago, Chile, 1994.

[6] Fortune, S. and Van Wyk, C. "Efficient Exact Arithmetic for Computational Geometry," *Proceedings of the Ninth ACM Symposium on Computational Geometry*, San Diego, May 1993, pp. 163-172.

[7] Green, D.H. and Yao, F.F., "Finite-resolution Computational Geometry," *Proceedings of the 27th Annual Symposium on the Foundation of Computer Science*, pp. 143-152, 1986.

[8] Guibas, L., Salesin, D., and Stolfi, J., "Epsilon Geometry: Building Robust Algorithms from Imprecise Computations," *Proceedings of the Fifth ACM Symposium on Computational Geometry*, San Diego, 1989, pp. 208-217.

[9] Guting, R.H. and Schneider, M., "Realms-Based Spatial Data Types: The ROSE Algebra," *VLDB Journal 4(2)*, April 1995, pp. 243-286.

[10] Guting, R.H., Ridder, T. and Schneider, M., "Implementation of ROSE Algebra: Efficient Algorithms for Realm-Based Spatial Data Types," *Proceedings of Advances in Spatial Databases - Fourth International Symposium, SSD '95*, pp. 216-239, Portland, Maine, August 1995. Lecture Notes in Computer Science 951. Springer-Verlag.

[11] Kriegel, H., Brinkhoff, T. and Schneider, R., "An Efficient Map Overlay Algorithm Based on Spatial Access Methods and Computational Geometry" *Proceedings of the International Workshop on DBMS's for Geographic Applications*, Capri, May 12-17, 1991.

[12] Myers, E.W., " An $O(E \log E + I)$ Expected Time Algorithm for the Planar Segment Intersection Problem," *SIAM J. of Computing, Vol. 14 (1985)*, pp. 625-637.

[13] Milenkovic, V.J., "Double Precision Geometry: A General Technique for Calculating Line and Segment Intersections Using Rounded Arithmetic," *Proceedings of the Thirtieth Annual Symposium on the Foundations of Computer Science*, 1989, pp. 500-505.

[14] Milenkovic, V.J. and Li, Z., "Constructing Strongly Convex Hulls Using Exact or Rounded Arithmetic," *Algorithmica (8)*, 1992, pp. 345-364.

[15] Naher, S., *LEDA Manual Version 3.2*, TR MPI-I-95-1-002, Max-Planck-Institut fur Informatik, Saarbrucken, 1995.

[16] Mehlhorn, K. and Naher, S. "Implementation of a Sweep Line Algorithm for the Straight Line Segment Intersection Problem," *TR MPI-I-94-105, Max-Planck-Institut fur Informatik*, Saarbrucken, 1993.

[17] Mehlhorn, K. and Naher, S. "LEDA, A Platform for Combinatorial and Geometric Computing," *CACM (38), Number 1995*, pp. 96-102.

[18] Nievergelt, J. and Preparata, F.P., "Plane-Sweep Algorithms for Intersecting Geometric Figures," *CACM (25), Number 10, 1982*, pp. 739-747.

[19] Ottmann, G., Thiemt, G. and Ullrich, C., "Numerical Stability of Geometric Algorithms," *Proceedings of the Third ACM Symposium on Computational Geometry*, 1987, pp. 119-125.

[20] Ottmann, T.A., Widmayer, P. and Wood, D. "A Fast Algorithm for the Boolean Masking Problem," *Computer Vision, Graphics, and Image Processing (1985)*, pp. 249-268.

[21] Scholl, M. and Voisard, A., "Thematic Map Modelling," *Proceedings of the 1st Symposium, SSD 89*, Santa Barbara, CA., July 1989, pp.167-190. Lecture Notes in Computer Science 409, Springer-Verlag, 1989.

[22] Shamos, M.I. and Hoey, D., " Geometric Intersection Problems," *Proc. 17th ACM Symposium on Foundations of Computer Science* (1976), pp. 208-215.

[23] Tilove, R.B., "Set Membership Classification: A Unified Approach to Geometric Intersection Problems", *IEEE Transaction on Computers*, Vol. C-29, No. 10, 1980, pp. 874-883.

[24] *XYZ GeoBench User Manual*, ETH Zurich. Switzerland, 1994.

Materialization Trade-Offs in Hierarchical Shortest Path Algorithms

Shashi Shekhar, Andrew Fetterer, and Brajesh Goyal

University of Minnesota Department of Computer Science
[shekhar|fetterer|brajesh]@cs.umn.edu
http://www.cs.umn.edu/research/shashi-group/

Abstract. Materialization and hierarchical routing algorithms are becoming important tools in querying databases for the shortest paths in time-critical applications like Intelligent Transportation Systems (ITS), due to the growing size of their spatial graph databases [16]. A hierarchical routing algorithm decomposes the original graph into a set of fragment graphs and a boundary graph which summarizes the fragment graphs. A fully materialized hierarchical routing algorithm pre-computes and stores the shortest-path view and the shortest-path-cost view for the graph fragments as well as for the boundary graph [9]. The storage cost of the fully materialized approach can be reduced by a virtual or a hybrid materialization approach, where few or none of the relevant views are pre-computed. This paper explores the effect of materializing individual views for the storage overhead and computation time of hierarchical routing algorithms. Our experiments with the Twin Cities metropolitan road-map show that materializing the shortest-path-cost view for the boundary graph provides the best savings in computation time, for a given amount of storage and a small number of fragments. Materializing the relevant part of the shortest-path-cost view for the fragment graphs provides the next best savings, followed by materializing the shortest-path view for the boundary graph. Virtual shortest-path-view on fragments can reduce storage costs by an order of magnitude or more for large graphs

1 Introduction

Spatial graph databases [16] form the kernel of many important applications including transportation; water, electric, and gas utilities; telephone networks; urban management; sewer maintenance, and irrigation canal management. Shortest-path queries retrieve a sequence of nodes and edges forming an optimal (e.g. shortest distance or travel time) path between the given node-pair (source,destination). The shortest-path-cost query computes the aggregate cost (e.g. distance) of the shortest-path. Materialization and hierarchical algorithms are becoming important in processing shortest-path queries in time-critical applications such as Intelligent Transportation Systems [2]. Hierarchical algorithms decompose a large spatial graph into a boundary graph and a collection of fragment graphs, each of which is smaller than the original graph. The shortest-path query on the large graph is decomposed into a collection of shortest-path-cost queries and shortest-path queries on the smaller graphs.

1.1 ATIS Architecture

Path-planning is an essential component of ATIS, aiding travelers in choosing the optimal path to their destinations in terms of travel distance, travel time and other criteria. Due to the dynamic nature of the ITS database and the frequency and nature of queries, it is necessary to explore alternatives to traditional disk-resident databases (DRDB). Main-memory databases (MMDB) [4] store most of their content permanently in main memory, resulting in high speed access to that data. Data stored in memory is *hot* if it is frequently accessed, while data kept on disk is *cold* if it is accessed less frequently. Main memory databases can store both types of data. MMDBs differ from DRDBs with large caches, since the MMDB does not have to treat the data as if it were stored on disk (i.e. deal with index structures and the buffer manager). Storing the data permanently in memory eliminates the need for clustering since since random access is no more costly than sequential access. Since disk access time is not a concern for the hot data, performance is also dependent upon processing time.

Some of the issues in the performance tuning of the hierarchical routing system include (i) the choice of views to materialize and (ii) the partitioning of those views between MMDB and DRDB. Other issues relate to the decomposition of graphs into fragments, to the choice of the number of levels in the hierarchy, and to view maintenance strategies [9].

1.2 Related Work and Our Contributions

The single-pair path-computation problem is a special case of single-source path computation and all-pair path computation. Traditional research in database query languages [1] , transitive closure [6] and recursive query processing [13] has approached single-pair path-computation as a special case of more general problems. For example, partial-transitive-closure computation [8] and transitive-closures [7] have been used for single-pair path computations. Previous evaluation of the transitive-closure algorithms examined the iterative, logarithmic, Warren's, Depth-first search (DFS), hybrid, and spanning-tree-based algorithms [14].

Previous evaluation of single-pair path-computation algorithms in a database environment examined the iterative breadth-first and best-first search algorithms [12]. Another study [15] examined A* and other estimator-based algorithms, along with the effect of path lengths and edge costs on the relative performance of search algorithms. Query-language based implementations were found to be inefficient.

Hierarchical path-finding has been explored in the context of computer networks in [11] and in the context of planar graphs in [3]. In ATIS, a hierarchical routing algorithm called HEPV [9] that guarantees optimal solutions has been investigated. That hierarchical routing algorithm divides the base graph into a boundary graph and a set of fragment graphs. It has materialized views for the shortest-path and shortest-path-cost for the fragments and the boundary graph. The storage of these views reaches gigabytes for a graph with 100,000 nodes and may become prohibitively large for ATIS maps with millions of nodes. Selective materialization can be used to manage storage costs. That work also explored performance issues such as the decomposition of graphs into fragments, the number of levels in the hierarchy, clustering of the materialized view to reduce IO, and view maintenance in the face of updates.

Three types of view implementations are possible [18]. The *fully materialized* view stores all relevant information, i.e. all shortest paths, the cost of the shortest paths, or both are pre-computed. A *virtual view* relies on on-line computation and stores neither the shortest paths nor their cost. The *hybrid view* stores part of the relevant information, for example, storing the costs of the shortest paths but not the shortest path itself.

This paper explores hybrid materialization to analyze the storage / computation-time trade-offs in materializing individual views needed for processing shortest-path queries. Our experiments with the Twin Cities metropolitan road-map show that materializing the shortest-path-cost view for the boundary graph provides the best savings in computation time for a given amount of storage when the number of fragments is small. Materializing the relevant part of the shortest-path-cost view for the fragment graphs provides the next-best savings, followed by materializing the shortest-path view for the boundary graph. Virtual shortest-path view for fragments can reduce the storage cost by an order of magnitude or more for large graphs.

1.3 Scope of Paper and Outline

This paper focuses primarily on materialization issues and makes simple choices for resolving other issues. For this study, we work with a hierarchical algorithm with a two level hierarchy. The issue of view maintenance in the face of updates, buffering in DRDB, etc. will be addressed in future work. Min-cut [10] partitioning is used to decompose the base graph into fragments. The shortest-path-cost view and the shortest-path view are clustered by the source-node. The design decisions made to resolve these issues are orthogonal to the issue of materialization and have been studied in the literature.

Section 2 describes the basic concepts in hierarchical routing, giving a description of the algorithm and a proof of correctness. Section 3 introduces the concept of and describes the choices in view materialization. In section 4, we present experimental results for the materializations in terms of CPU operations, IOs required, MMDB size, and storage space used. Section 5 summarizes our work and presents future challenges in hierarchical routing.

2 Basic Concepts

2.1 Flat-Graph Definitions

Graph $G = (N, E, C)$ is a flat graph consisting of a node set N, a cost set C, and an edge set E. The edge set E is a subset of the cross product $N \times N$. Each element (i, j) in E is an edge that joins node i to node j. Each edge (i, j) is associated with a cost $c_{i,j}$. Cost $c_{i,j}$ takes values from the set of positive real numbers. A node i is a neighbor of node j if $E_{i,j} \in E$. The degree of a node is the number of neighboring nodes. A path $P_{s,d}$, in a graph from a source node s to a destination node d, is a sequence of nodes $(N_0, N_1, N_2, ..., N_k)$, where $s = N_0$, $d = N_k$, and the edges $E_{0,1}, E_{1,2}, ..., E_{k-1,k} \in E$. The cost of the path is the sum of the cost of the edges, i.e. $\sum_{i=1}^{k} c_{i-1,i}$. An optimal path from node i to node j is the path with the smallest cost.

2.2 Hierarchical Graph Definitions

The hierarchical graph has a 2-level representation of the original graph G. The lower level is composed of a set of fragments $F_1 \ldots F_k$ of G. The higher-level graph is comprised of the boundary nodes (BN) and is called the boundary graph (BG). Boundary nodes are defined as the set of nodes that have a neighbor in more than one fragment, i.e. $N_i \in BN \iff \exists E_{i,j}, E_{i,k} | FRAG(k) \neq FRAG(j)$. Edges in the boundary graph are called boundary edges, and the boundary nodes of a fragment form a clique, i.e. are completely connected. The cost associated with the boundary edge is the shortest-path cost through the fragment between the boundary nodes. A boundary edge is associated with a fragment identifier. A *boundary path* is the shortest path through the boundary graph.

Theorem 1. $BG.SP(s, d)rewrite[Edge \rightarrow] = G.SP(s, d)rewrite$
[1] $[BoundaryNode \rightarrow BoundaryNode, InteriorNode \rightarrow, Edge \rightarrow]$ if $s, d \in BN$.

Theorem 2. $p_g = (G.SP(s, d)rewrite[BoundaryNode \rightarrow BoundaryNode, InteriorNode \rightarrow, Edge \rightarrow]) \in P_{set}$. where $P_{set} = \{p_{ij}rewrite[Edge \rightarrow] \mid p_{ij} = BG.SP(b_i, b_j), b_i \in BoundaryNodes(Fragment(s)), b_j \in BoundaryNodes(Fragment(d))\}$

The shortest path[2] $G.SP(s, d)$ between nodes s and d in C corresponds directly to the shortest path $BG.SP(s, d)$ between s and d in BG where s and d are boundary nodes, as formalized in Theorem 1. The sequence of boundary nodes in $G.SP(s, d)$ is identical to that in

---

[1] Rewrite [5] in this case drops all the interior nodes and edges in the path and keeps just the boundary nodes

[2] More detailed discussion of the proofs for the theorems and lemmas may be found at http://www.cs.umn.edu/research/shashi-group/paper_list.html

```
1    //BG = boundary graph
2    //CASE 1: Both Source s and Destination d are boundary nodes
3
4    boundaryPath = BG.GetPath(s,d);
5    path = ExpandBoundaryPath(boundaryPath);
6
7    //CASE 2: Source s is a local node and destination d is a
8    //         boundary node
9
10   c = INFINITY;
11   for each bN in BoundaryNodes(Fragment(s)) do
12           if ((cpc = Fragment(s).SPC(s, bN) + BG.(bN,d)) < c) {
13                   c = cpc;
14                   minB = i;
15           }
16   path = Fragment(s).SP(s, minB)
17           + ExpandBoundaryPath(BG.SP(minB, d);
18
19   //CASE 3: Destination is a local node and source is a boundary
20   //         node
21   //Similar to CASE 2, but with source and destination reversed
22
23   //CASE 4: Both source and destination are local nodes
24   c = INFINITY;
25   for each sBN in BoundaryNodes(Fragment(s)) do
26           for each dBN in BoundaryNodes(Fragment(d)) do
27                   if ((cpc=Fragment(s).SPC(s,sBN)
28                           + BG.SPC(sBN,dBN)
29                           + Fragment(d).SPC(dBN,d))< c) {
30                                   c = cpc;
31                                   minSB = sBN;
32                                   minDB = dBN;
33                           }
34   if (Fragment(s) == Fragment(d) && c > Fragment(s).SPC(s,d))
35           path = Fragment(s).SP(s,d);
36   else
37           path = Fragment(s).SP(s, minSB)
38                   + ExpandBoundaryPath(BG.SP(minSB, minDB))
39           + Fragment(d).SP(minDB, d);
```

Fig. 1. Hierarchical Routing Algorithm Template

$BG.SP(s, d)$, due to the construction of boundary edges in BG. This observation can be generalized to the other node pairs in G. The sequence of boundary nodes in the shortest path $G.SP(s, d)$ between nodes s and d in G is a shortest path in BG between a boundary node in $fragment(s)$ and a boundary node in $fragment(d)$. This is stated in Theorem 2. Shortest path $G.SP(s, d)$ can be constructed by choosing a pair $<$ a boundary node in $fragment(s)$, a boundary node in $fragment(d) >$, which minimize the cost of $G.SP(s, sBN) + BG.SP(sBN, dBN) + G.SP(dBN, d)$.

Figure 1 gives a template algorithm for finding a path in a hierarchical graph. The following queries are sent to the database to compute the optimal path:

- **SP(s,d)** : This query returns the shortest path between node s and d.
- **SPC(s,d)** : Query to find the cost of the shortest path from a source s to a destination d.
- **BoundaryNodes(f)** : Query that returns the set of boundary nodes for fragment f.
- **Fragment(n)** : This query returns the fragment identifier for interior node n.
- **ExpandBoundaryPath(boundaryPath)** : Query that expands the path through the boundary graph and returns the corresponding path in G by using the **ExpandBoundaryEdge(boundaryEdge)** query for each edge in the boundary path.
- **ExpandBoundaryEdge(boundaryEdge)** : This query expands a boundary edge and returns a corresponding path in G by computing the shortest path between the endpoints of the edge in the fragment.

The hierarchical algorithm is composed of three steps: finding the relevant boundary-node pair in the boundary graph, computing the boundary path, and expanding the boundary path. The first step in determining the shortest path is to compute the boundary node through which the shortest path leaves the source's fragment and enters the destination's fragment. If both the source and destination are boundary nodes, then it is trivial. If the source is an internal node and the destination is a boundary node, the boundary node through which the shortest path leaves the source's fragment is found by querying the fragment graph for the cost of the path from the source to all boundary nodes of that fragment, and by querying the boundary graph for the cost of the shortest path from all boundary nodes of the source's fragment to the destination. The source-boundary-destination triple with the lowest aggregate cost determines the appropriate boundary node. The case where the source is a boundary node and the destination is an internal node is similar, but the roles of the source and destination are reversed. When both the source and destination are internal nodes, the appropriate boundary node pair is found by querying the fragment graphs to determine the cost of the shortest path from the internal nodes to all boundary nodes of the fragment. Next, the boundary graph is queried to compute the shortest path cost between all pairs of boundary nodes. The path with the lowest aggregate cost determines the boundary-node pair. Once the appropriate boundary-node pair has been determined, the boundary graph is queried to determine the shortest path between those boundary nodes. The final step is to expand the boundary path by querying the fragments for the shortest path through them. Adjacent nodes in the boundary path form source/destination pairs on which the shortest-path query can be run on in a fragment.

Lemma 3. *The hierarchical routing algorithm finds an optimal path from s to d.*

3 Degree of Materialization

3.1 View Materialization Terminology

A cost view (CV) on a graph materializes the cost of the shortest path between all node pairs in the graph. It does not store any path information. A compressed path view CPV stores the set of optimal paths between all points on the graph as a series of "hops". For a fragment graph, a partial materialization of the CV, the $C2B$ or cost-to-boundary-nodes view stores the cost of the shortest path from the interior nodes of the fragment to the boundary nodes of the fragment.

Figure 2 shows CPV and CV for an example graph with 5 nodes and 8 edges. Each node is associated with a table, with one entry for all other nodes. The node table in CV contains the costs of the shortest paths between two nodes. For example, the cost from node C to node A is given as 5. To find the shortest path between those two nodes, the CPV table is used. Starting at the table for node C, the destination is looked up to get the next node on the path. In this case, the next destination is node E, and the CPV table for node E is used. From that table, node A is found to be the next hop and the destination is reached.

3.2 Asymptotic Analysis of Storage Cost Using a Grid Graph

	Adjacency list information		Compressed Path View		Cost View		Cost2-Boundary-Nodes
	per fragment graph	boundary graph	per fragment graph	boundary graph	per fragment graph	boundary graph	
Raw Storage	$F*G_f$	G_b	$N(\frac{N}{F}-1)*2S_{id}$	$B(B-1)*3S_{id}$	$N(\frac{N}{F}-1)*(S_c+S_{id})$	$B(B-1)*(S_c+S_{id})$	$(N-B)K*(S_c+S_{id})$
Asymptotic Storage	$O(N)$	$O(N)$	$O(N^{5/3})$	$O(N^{4/3})$	$O(N^{5/3})$	$O(N^{4/3})$	$O(N^{4/3})$
Real Size F=100	4M	3M	2.3G	32M	2.3G	32M	40M

Table 1. Asymptotic Size and Complexity of the Materialized Views

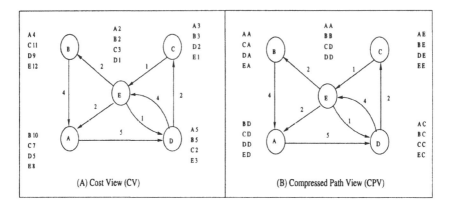

Fig. 2. Path Closure and Cost Closure Examples

Symbol	Explanation		Symbol	Explanation
N	Number of nodes		S_c	size of storing a cost
S_{id}	Size of storing a node identifier		B	Number of boundary nodes
F	Number of fragments		G_b	Size of boundary graph in bytes
K	Average number of border nodes per fragment		L	Length of path through base graph
M	Length of path through boundary graph			

Table 2. Table of Notations

A summary of the parameters used in the remainder of the paper is given in Table 2. A square grid-graph of N nodes is an undirected planar graph. If $\sqrt{N}$ is an integer, the square grid-graph can be embedded in a Cartesian space with nodes at coordinates $(i * \sqrt{N}, j * \sqrt{N})$ with $i, j \in 0, 1, 2, ..., \sqrt{N} - 1$. Edges connect nodes $(i\sqrt{N}, j\sqrt{N})$ to nodes $((i+1)\sqrt{N}, j\sqrt{N})$ and $(i\sqrt{N}, (j+1)\sqrt{N})$. A grid-graph is often used in complexity analysis for transportation algorithms because of its regular structure and similarity to urban road-maps [15].

In a graph decomposed into F fragments, each fragment has approximately N/F nodes. The average number of boundary nodes for each fragment is $K = 4\sqrt{N/F}$. The total number of boundary nodes is approximately $B = \sqrt{NF}$. Table 1 shows the view storage requirements. The storage formulas for the materialized views are shown first and are given in terms of N, B, K, and constants. F can be chosen to minimize the storage of the materialized views, as is studied in [17]. Setting F in terms of N, the asymptotic storage sizes for the views are computed. When $F = N^{1/3}$, the storage that minimizes the size of CV and CPV for the boundary graph, the values $K = N^{1/3}$ and $B = N^{2/3}$ are used to compute the asymptotic storage sizes for the views. The asymptotic analysis shows that the sizes of the CPV and CV views for the fragments is $O(N^{1.66})$, which is larger than the CPV and CV for the boundary graph and $C2B$ for the fragments which grow at $O(N^{1.33})$. This is verified in the real sizes of the views are given for $F = 100$

($F \approx N^{2/5}$). The CPV and CV views for the fragments take a couple of gigabytes of storage, while the other views are two orders of magnitude smaller,
as shown in Table 1. The will not consider the materialization of the CPV on
the fragments further in this paper. The size of the CPV for the base graph
is $O(N^2)$, which would take about 100G to store.

3.3 Levels of Hybrid Materialization

A fully materialized view has all relevant information pre-computed and
stored, while a virtual view has no information pre-computed. Hybrid materialization forms the middle ground between those two approaches – some
of the relevant data is materialized, while some is left for computation. In
terms of the hierarchical routing algorithm, three different levels of hybrid
materialization could be used: table level, graph level, and algorithm level.
Hybrid materialization with respect to a single table means that only part of
the table is kept as a materialized view. An example of this is the $C2B$ view
which materializes only a part of the CV table for the fragment graphs. In
this case, the CV table is fully materialized because it stores cost information
between all pairs, while the $C2B$ table only stores cost information between
internal nodes and boundary nodes. The graph-level materialization chooses
to materialize all views related to either boundary graph or the fragment
graphs, but not both. means that only materialization at the graph level is
considered. The third definition of hybrid materialization is when we only
materialize the tables necessary for the version of the hierarchical algorithm
under consideration. Full materialization of all views for the hierarchical algorithm was researched in [9]. The hybrid materialization of the hierarchical
algorithm (the focus of the remainder of the paper) will materialize a subset
of the CV and CPV for the boundary graph and fragment graphs. Figure 3(a)
shows the possible hybrid materializations for the hierarchical algorithm.

4 Experimental Evaluation of Hybrid Materialization

4.1 Experiment Design

Goals: The goal of the experiments is to study the effects of materializing
individual views on fragments and on the boundary graph on the storage and
computation cost of the hierarchical routing algorithm. Computation cost is
measured by the total number of IO's required and the on-line computation
effort in terms of heap operations and heap size.

The candidate hybrid materializations are evaluated using the experimental setup shown in Figure 4. Benchmark maps are first converted to the
input format of the *graph fragmentizer* and the *work load generator*. The
fragmented graph is used by the *hierarchical routing application*. The *view
manager* is responsible for implementing candidate hybrid materializations.
The materialized views are partitioned between the MMDB and DRDB by

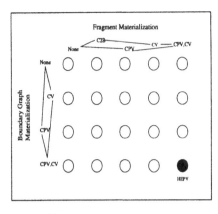

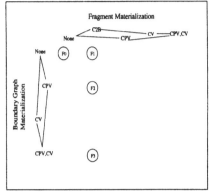

(a) Materialization Matrix

(b) Candidate Hybrid Materializations for Experiments

Fig. 3. Degree of Materialization in Hierarchical Routing Algorithms

the *view partitioner*. The *work-load generator* uses the list of nodes to generate (source,destination) pairs. The *hierarchical routing application* executes the hierarchical routing algorithm using the views stored in the MMDB and DRDB with the (source,destination) pairs. Statistics from the *heap manager* and *IO manager* are collected for *data analysis*.

Hybrid Materialization Candidates: We chose four candidate hybrid materializations for direct comparison, as shown in Figure 3(b). These candidates are chosen to facilitate studying the effects of materializing individual views on the fragment graph and the boundary graph. F0 has no materialization in either the boundary graph or the fragment graphs. F1 only materializes the *C2B* table in the fragment graph. F2 materializes the *C2B* table of the fragment graph and the *CV* table of the boundary graph. F3 materializes the *C2B* table in the fragment graph and both the *CPV* and *CV* tables in the boundary graph. We compare the candidates to determine the effect of materialization on the hierarchical algorithm. We compare F0 and F1 to determine the effect of materializing the *C2B* table in the fragment graph. By comparing F1 and F2, we determine the effect of materializing the *CV* in the boundary graph. Comparing F2 and F3, we determine the effect of materializing the *CPV* view in the boundary graph.

Table 3 shows the patterns of access to the views for the different candidate materializations we have chosen to study. The table is summarized below.

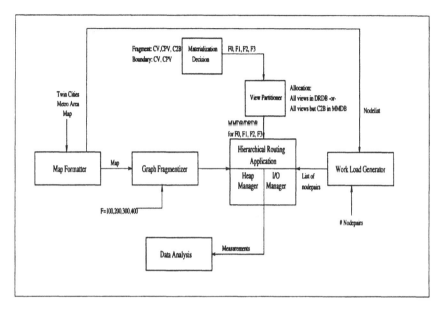

Fig. 4. Experimental Method for Evaluating Shortest Path Algorithms

	Adjacency list information		Compressed Path View		Cost View		Cost2-Boundary-Nodes
	per fragment graph	boundary graph	per fragment graph	boundary graph	per fragment graph	boundary graph	
F0	$2K + M + 2$	$K^2 + 1$					
F1	$M + 2$	$K^2 + 1$					2
F2	$M + 2$	1				K	2
F3	$M + 2$			M		K	2

Table 3. Views and their Access Patterns for Various Algorithms

- **F0:** There will be K^2 A* run in the boundary graph to compute the cross-fragment paths. In the fragment graph, there are $2K$ A* run to determine the cost to the boundary nodes, $M + 2$ A* to expand the boundary path, and 1 A* in the boundary graph to compute the boundary path
- **F1:** This is identical to F0, but the $2K$ A* are replaced with 2 lookups to the *C2B* view.
- **F2:** F2 is identical to F1, but the K^2 A* are replaced with K lookups in the *CV* for the boundary graph.
- **F3:** This case is identical to F2, except that the boundary path is determined using the *CPV* instead of running an A*.

Partitioning of Views Between MMDB and DRDB: In the experiments, we look at two partitioning strategies. In the first partitioning, the *C2B* table was stored in the DRDB and the rest of the views were stored in the MMDB. In the other partitioning, all views are stored in the DRDB with only the adjacency-list of the fragment graphs and border graph are in main memory.

Metrics: The storage cost is measured by the number of bytes needed to store the data-structures i.e. graphs, views. The computation time has two components commonly used by query processors in choosing a query evaluation strategy: i) CPU time and ii) disk IO time. CPU time is determined by the size of the graph and the number of A*s run. One measure of CPU time is the total number of heap operations used by the A* algorithm and the size of the heap when those operations are made. Those parameters give an indication of how many operations need be done and how expensive those operations are. IO cost is dominated by the number of disk IOs incurred to fetch data which cannot be loaded into the MMDB from the views.

4.2 Comparing Hybrid Materialization Candidates

We report the CPU-costs, IO-costs, and storage-cost of the candidate hybrid materialization strategies in our experiments using the Twin Cities metropolitan road map with 123,000 nodes and 313,000 edges. The are analyzed to deduce the effect of materializing individual views on the performance of the hierarchical routing algorithm.

CPU Cost Comparison: We measure the average heap size and average number of heap operations for the hierarchical routing algorithm using alternative materializations. A set of 500 randomly chosen (source,destination) pairs are used for the experiments. The number of heap operations include inserts and deletes only. A count of these operations is measured for each path and an average is reported in Figure 5(b). For each path, the heap-manager measures the heap-size at every operation (insert/delete) and produces and average heap size for a given (source,destination) pair. We report the average of these over 500 paths in Figure 5(b).

We compare the heap results against running A* on the base graph to get a baseline performance measurement. The effect of creating a hierarchy of graphs is apparent, as all hierarchical algorithms have a smaller average heap size than when running A* on the base graph, as shown in Figure 5(a). A* has a large average heap size since it searches the entire breadth of the graph, while the hierarchical algorithms are focused on the fragments or boundary graph. If the $C2B$ table is materialized as in F1, the average heap size increases, since there are $2K$ fewer A* run in the fragment graphs, but the same number of A* run in the boundary graph. The effect of materializing the CV in the boundary graph is apparent in F2. In this case, the average heap size decreases dramatically, since the K^2 A* in the boundary graph are replaced with lookups. The cost of running A* in the boundary graph is further highlighted by materializing the CPV for the boundary graph (F3),

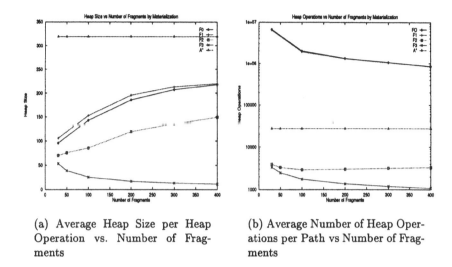

(a) Average Heap Size per Heap Operation vs. Number of Fragments

(b) Average Number of Heap Operations per Path vs Number of Fragments

Fig. 5. CPU Cost vs Number of Fragments

which replaces the A* and finds the path through the boundary graph with a lookup. The average heap size is small, because the only A* are run in the fragments. The average heap size increases as the number of fragments increases when there is no materialization, or when the $C2B$ and CV are materialized, since the number of nodes in the boundary graph grows. When the CPV in the boundary graph is materialized, the average heap size decreases, because no A* are run in the boundary graph and the size of the fragments decreases.

The number of heap operations is a count of the number of inserts into and deletes from the heap data-structure and is shown in Figure 5(b). Materializing the hierarchical graph (F0) causes the number of heap operations to increase, as compared to running A* on the base graph due to running the K^2 A* in the boundary graph. Materializing the $C2B$ view (F1) decreases the number of heap operations slightly, but the K^2 factor still dominates. The hierarchical algorithm will have fewer heap operations than A* only when the CV is materialized (F2), since the K^2 boundary graph A* are replaced with lookups. Materializing the the CPV causes the heap operations to be minimized, since all computation is done in the fragment graphs. The number of fragments inversely affects the number of heap operations for two reasons. With a greater number of fragments, the number of boundary nodes per fragment decreases, lowering the K^2 term. Also, more fragments means that individual fragments will be smaller, thus requiring fewer heap operations to search them. In summary, the number of heap operations decreases as more views are materialized. since it is exploited by F1-F3 directly.

IO Cost Comparison: Clustering the views improves IO performance in the DRDB since the clustering is directly exploited by F1-F3. The *C2B* view is clustered around an internal node. Thus, one lookup to the clustered table retrieves the cost to all boundary nodes, as opposed the the K lookups (one for each boundary node in the fragment) the unclustered view would need. The *CV* is clustered around fragments – each boundary node stores the cost to all the boundary nodes in other fragments consecutively, resulting in K lookups instead of K^2 lookups.

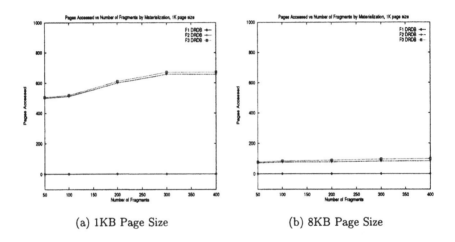

(a) 1KB Page Size   (b) 8KB Page Size

Fig. 6. IO Cost Per Path vs Number of Fragments

Logical Accesses We measured logical IOs to simplify interpretation and to avoid platform specific noise. The experiment setup counts the number of access to the views on disk and converts that number to the equivalent number of pages using the record page-size. We count each page to be a distinct IO. The experiment does not capture the effect of DRDB-buffering, page clustering, and related optimizations not apparent at the interface level.

Figure 6 shows how the number of accesses varies for the different hybrid materializations considered. F0 has no IO. In F1, the only IO comes from accessing the *C2B* view, which takes at most two accesses. F2 materializes the *CV* for the boundary graph and thus accesses more pages than F1. The difference in the amount of IO between F2 and F3 is the length of the path through the boundary graph. Increasing the number of fragments increases the number of boundary nodes, which has two effects on IO performance. Since the *CV* is clustered, an increase in the number of boundary nodes causes the size of each entry in that table to grow from 14K with 50 fragments to

47K with 400 fragments. The size of the entries is apparent in Figure 6(b) as the order of the materializations did not change as the page size grew. The second increase in IO comes from the increases length of the boundary path.

Storage Costs: Storage costs were attained by creating the necessary views for the candidate hybrid materializations. The graph was fragmented using a min-cut [10] technique with values of F ranging from 10 to 400. Notations are summarized in Table 2. All candidates store the base and fragment graphs, whose storage requirements are similar to those of the base graph. The storage for the $C2B$ view is $(N - B)K(S_c + S_{id})$ bytes. Each of the $(N - B)$ internal nodes have information for the cost to the border node for each of the K boundary nodes of their fragment. The CPV requires $B(B - 1)3S_{id}$ bytes of storage. Each of the B boundary nodes stores the boundary node identifier, and the source and destination identifiers for the $(B - 1)$ boundary nodes. Storing the CV takes $B(B - 1)(S_c + S_{id})$ bytes. Each of the B boundary nodes stored the cost to the $(B - 1)$ other boundary nodes. On our system, S_{id} was 4 bytes and S_c was 8 bytes. Storage was measured using the UNIX 'ls -l' command.

Figure 7(b) shows the storage requirements for the candidate hybrid materializations and for the views that compose those materializations. The storage required for the $C2B$ view decreases as the number of fragments increases, because there are fewer internal nodes to store costs from and fewer boundary nodes to store costs to. The amount of storage required for the CPV and CV increases as the number of fragments increases, due to the increased size of the boundary graph, i.e. there are more nodes for which to store costs/paths between. F0 requires nearly constant storage, with a slight increase due to a larger boundary graph. F1's storage requirement is dominated by the $C2B$ view and thus requires less storage as the number of fragments increases. The F2 and F3 materializations have their storage dominated by the CPV and CV, so their storage cost increases as the number of fragments increases. In summary, the storage costs increase with more materialization.

Summary - The Effect of Materializing Individual Views: Materialization of boundary graph views causes a decrease in the number of heap operations and a decrease in the average heap size. This materialization causes the number of IO operations to increase. IO was shown to not be sensitive to the page size used since the order of the candidate materializations did not change. It was also shown that this was caused by the size of the entries in the CV.

4.3 MMDB/DRDB Partitioning Candidates

Partitioning views in MMDB separates the data stored in the MMDB from the data stored in the DRDB. From Figures 6(a)(b) and 5(a)(b), the most

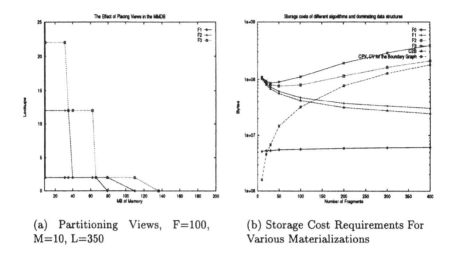

(a) Partitioning Views, F=100, M=10, L=350

(b) Storage Cost Requirements For Various Materializations

Fig. 7. View Partitioning and Storage Requirements

costly operation of the algorithm is accessing the boundary-graph cost view. If that view is virtually materialized, then the CPU cost becomes prohibitive, while if the view is fully materialized but in a DRDB, the IO is costly. Thus, the *CV* becomes the first candidate for placement in the MMDB. The next view to be placed in the MMDB is the *CPV*, since it is accessed M times, where M is the length of the boundary path. The last view to be placed in the MMDB should be the *C2B* table, since it is accessed twice at most.

Figure 7(a) shows the effect view-partitioning has on the number of disk lookups for each candidate hybrid materialization. This figure assume that the graph has 123,000 nodes, that the path length is 350, and that the boundary path length is 14 (variables characteristic a long path in the graph). For this experiment, a disk lookup is an access to the disk, not a count of the number of pages read. The first area of transitions occurs when enough memory becomes available to place the fragment graphs and boundary graphs in the MMDB. The next important transitions occur in the 30-40M range. At this time, enough memory is available for the *CV* to be in the MMDB. With around 70-90M of memory, the *C2B* can be placed in the MMDB, which eliminates IO for F1 and F2. IO for F3 is eliminated when enough memory becomes available for all three views, which occurs around 110M.

5   Conclusion

In this paper, we have addressed two important issues regarding the hierarchical routing algorithm. The first issue was to determine which views to materialize if storage is limited. We found that materializing the shortest-path-cost view for the boundary graph provides the greatest computational savings, followed by materializing the cost-to-boundary-nodes view for the fragments and then the shortest-path view for the boundary graph. Once the appropriate views are materialized, it is necessary to partition the views between the MMDB and the DRDB when the size of the MMDB is limited. We found that accessing the shortest-path-cost view incurs a high IO cost and should be placed in the MMDB first, followed by the shortest-path view for the boundary graph and then the cost-to-boundary-nodes view. In our future work, we plan on addressing clustering in DRDB, view maintenance and updates, graph fragmentation, and multi-level hierarchies. We also plan to implement a parallel version of the hierarchical routing algorithm.

References

1. R. Agrawal. "Alpha: An Extension of Relational Algebra to Express a Class of Recursive Queries". *IEEE Trans. on Software Engineering*, 14(7), 1988.
2. W.C. Collier and R.J. Weiland. "Smart Cars, Smart Highways". *IEEE Spectrum*, pages 27–33, April 1994.
3. Greg Frederickson. "Searching Among Intervals and Compact Routing Tables". *Algorithmica*, pages 448–466, 1996.
4. Hector Garcia-Molina and Kenneth Salem. Main Memory Database Systems: An Overview. *IEEE Transactions on Knowledge and Data Engineering*, 4(6):509–516, 1992.
5. R. H. Guting. GraphDB: Modeling and Querying Graphs in Databases. In *Proc. of Intl. Conference on Very Large Data Bases*, 1994.
6. Y. Ioannidis, R. Ramakrishnan, and L. Winger. "Transitive Closure Algorithms Based on Graph Traversal". *ACM Trans. on Database Systems*, 18(3), September 1993.
7. H.V. Jagadish, R. Agrawal, and L. Ness. "A Study of Transitive Closure As a Recursion Mechanism". In *Proc. of SIGMOD Intl. Conference on Management of Data*. ACM, 1987.
8. B. Jiang. "I/O Efficiency of Shortest Path Algorithms: An Analysis". In *Proc. of the Intl. Conference on Data Engineering*. IEEE, 1992.
9. Ning Jing, Yun-Wu Huang, and Elke Rudensteiner. "Hierarchical Optimization of Optimal Path Finding for Transportation Applications". In Proc. of ACM Conference on Information and Knowledge Management, 1996.
10. George Karypis and Vipin Kumar. "Analysis of Multilevel Graph Partitioning". Technical Report 95-037 see also http://www.cs.umn.edu/~karypis, University of Minnesota, Department of Computer Science, Minneapolis, MN 55455, August 1995.
11. Leonard Kleinrock and Farouk Kamoun. "Hierarchical Routing for Large Networks". *Computer Networks*, 1:154–174, 1977.
12. R. Kung, E. Hanson, and et. al. "Heuristic Search in Data Base Systems". In *Proc. Expert Database Systems*. Benjamin Cummings Publications, 1986.
13. Y. Kusumi, S. Nishio, and T. Hasegawa. "File Access Level Optimization Using Page Access Graph on Recursive Query Evaluation". In *Proc. Conference on Extending Database Technology*. EDTB, 1988.

14. H. Lu, K. Mikkilineni, and J.P. Richardson. "Design and Evaluation of Algorithms to Compute the Transitive Closure of a Database Relation". In *Proc. of Intl Conference on Data Engineering*. IEEE, 1987.

15. S. Shekhar, A. Kohli, and M. Coyle. Path Computation Algorithms for Advanced Traveler Information Systems. In *Proc. of the 9th Intl. Conference on Data Engineering*, 1993.

16. S. Shekhar and D. R. Liu. "A Connectivity-Clustered Access Method for Networks and Network Computation". In *Proc. Intl. Conf. on Data Engineering*, 1995. Complete version to appear in IEEE Transactions on Knowledge and Data Engineering.

17. Shashi Shekhar, Andrew Fetterer, and Brajesh Goyal. A Comparison of Hierarchical Algorithms for Shortest Path Computation in Advanced Travel Information Systems. Technical Report 96-046, University of Minnesota, 1996.

18. Gang Zhou, Richard Hull, Roger King, and Jean-Claude Franhitti. "Data Integration and Warehousing Using H20". *Bulletin of the Technical Committee of Data Engineering*, 18(2):29–40, June 1995.

Spatial Constraint Databases

Manipulating Spatial Data in Constraint Databases

Alberto Belussi[1]   Elisa Bertino[2]   Barbara Catania[2]

[1] Dipartimento di Elettronica e
Informazione
Politecnico di Milano
Piazza L. da Vinci 32
20133 Milano, Italy
e-mail: belussi@elet.polimi.it

[2] Dipartimento di Scienze
dell'Informazione
Università degli Studi di Milano
Via Comelico 39/41
20135 Milano, Italy
e-mail: {bertino,catania}@dsi.unimi.it

Abstract. Constraint databases have recently been proposed as a powerful framework to model and retrieve spatial data. In a constraint database, a spatial object is represented as a quantifier free conjunction of (usually linear) constraints, called *generalized tuple*. The set of solutions of such quantifier free formula represents the set of points belonging to the extension of the object. The relational algebra can be easily extended to deal with generalized relations. However, such algebra has some limitations when it is used for modeling spatial data. First of all, there is no explicit way to deal with the set of points representing a spatial object as a whole. Rather, only point-based computations can be performed using this algebra. Second, practical constraint database languages typically use linear constraints. This allows to use efficient algorithms but, at the same time, some interesting queries cannot be represented (for example, the distance between two objects cannot be computed). Finally, no update language for spatial constraint databases has been defined yet. The aim of this paper is to overcome some of the previous limitations. In particular, we extend the model and the algebra to directly deal with the set of points represented by a generalized tuple (a spatial object), retaining at the same time the ability of expressing all computations that can be expressed by other constraint database languages. Moreover, we discuss the introduction of external functions in the proposed algebra, in order to cover all the functionalities that cannot be expressed in the chosen logical theory. Finally, we propose an update language for spatial constraint databases, based on the same principles of the algebra.

1   Introduction

In the last few years, the area of spatial databases has attracted the interest of different application contexts, such as geographical information systems (GIS), VLSI design, geometric modeling in mechanical or building design (CAD). Issues related to the development of models able to represent spatial data, suitable for all these different environments, have been addressed in [2, 14, 26, 27, 28].

However, most of the proposed models are not general enough, often allowing only the representation of spatial domains of fixed dimensions.

Constraint databases have been recently proposed as a general framework for manipulating spatial data [16, 21, 22]. A constraint database uses constraints on a specific decidable logical theory both to model and retrieve data. At the data level, constraints are able to finitely represent possibly infinite sets of relational tuples. For example, the constraint $X^2 + Y^2 \leq 9 \wedge X > 0$ represents the positive semicircle with center in $(0,0)$ and with radius equal to 3. In general, in the constraint model a conjunction of constraints is called *generalized tuple*, a finite set of generalized tuples is called *generalized relation*, whereas the set of (multidimensional) points representing the solutions of a generalized tuple is called *extension* of such generalized tuple. With respect to data modeling, the main advantage of constraints is that they serve as a unifying data type for the (conceptual) representation of heterogeneous data. In particular, the benefit of this approach is emphasized when complex knowledge (for example, spatial or temporal data) has to be combined with some descriptive non-structured information (such as names or figures), or when several types of spatial objects with different dimensions (like points, lines, convex polygons and concave polygons) have to be represented [29].

At the query language level, constraints increase the expressive power of simple relational languages by allowing mathematical computations. The integration of constraints in existing query languages introduces several issues. Indeed, constraint query languages should preserve all the nice features of relational languages. For example, they should be closed and bottom-up evaluable [16]. From a spatial point of view, the main issue in defining constraint query languages is to determine which spatial queries can be expressed under a specific language, validating a given constraint language with respect to specific spatial requirements.

Different algorithms are in general applied to manipulate different types of spatial data [24]. On the contrary, constraint databases allow to represent heterogeneous data by using constraints on a certain theory. On this data, homogeneous computations can be applied. Different theories can be used to model and query different types of data. In spatial applications, the mostly used theory is that of linear polynomial constraints [16]. This choice is motivated by the fact that several efficient algorithms have already been defined for such class of constraints [19]. However, the use of constraints to model spatial data does not come for free. Indeed, spatial data is often represented in vector format, specifying the vertices of the spatial object to be represented. The migration from vector format to linear polynomial constraints has complexity $O(n \log n)$, where n is the number of points used to represent a given spatial object, since the cost of this operation is bound by the cost of decomposing the spatial object into convex polygons [24].

As for the relational model, the correct formalism to obtain both a formal specification of the language and a suitable basis for implementation is represented by a constraint relational algebra. In particular, the relational algebra can be extended to deal with constraint relations. The new algebra is called *generalized relational algebra* [15, 22]. However, if we consider this algebra from a spatial point of view, we can observe that it is not always the more appropriate language for manipulating spatial data. In particular:

1. The manipulation applied by constraint query languages to generalized relations is not always the more natural one for spatial databases. Constraint computations always see a constraint relation as a (possibly infinite) set of points and do not see it as a finite set of spatial objects. Therefore, common spatial queries have a complex representation in constraint languages, and therefore also in the generalized relational algebra [15].
2. The second shortcoming is related to the constraint language expressive power. Often, the chosen theory is not adequate to support all the functionalities needed by the specific application. For example, if we use the linear polynomial constraint theory, the distance between two points or the convex hull of n points cannot be computed [29], even if they can be modeled in the logical theory of (not necessarily linear) polynomial constraints. The simplest solution, i.e., the choice of a more powerful constraint theory, for example real polynomial constraints, is not always satisfactory, since all implementation advantages of linear constraints would be lost.
3. The third shortcoming is related to the lack of update operations for spatial constraint databases. The only approach to model updates in constraint databases we are aware of has been proposed by Revesz in [25], but it presents no concrete update primitives. Rather, it proposes a model-theoretic formalization of minimal changes, following the approaches proposed in [17]. Note that the definition of update languages is a relevant issue for spatial constraint databases, because spatial objects are often subject to transformations with respect to either their shape (for example, in rescaling or adding a new object component) or their position in the space (for example, in translation or rotation).

In this paper we extend the generalized relational algebra to overcome the above limitations. The main contributions of the paper can be summarized as follows:

1. In order to directly model typical spatial queries, we assign a new semantics to generalized relations. Indeed, each generalized relation can be seen as a set of (possibly) infinite sets, where each set represents the extension of a single generalized tuple (i.e., the extension of a single object), contained in the considered relation. Note that, under this meaning, generalized relations model a very simple kind of nesting [1]. Thus, the new semantics is called *nested semantics*. In this respect, the type of sets we are able to model by a single generalized tuple becomes important. Other logical connectives, and not only conjunction as proposed in [16], can be used. For example, to model concave sets inside a generalized tuple, disjunction must be used.

   Note that our approach does not aim at defining a new complex object model for constraint databases, as it has been done in [6, 7, 13]. Rather, it changes the semantics assigned to generalized relations in order to define a new algebra for constraint relational databases, more suitable from a spatial point of view. Thus, our main interest is in the definition of the language and not in the proposal of a new model.

2. We propose an algebra, based on the nested semantics, to manipulate generalized relations under two different points of view, either handling each generalized relation as a finite set of spatial objects or as a possibly infinite set of points (i.e., relational tuples). Indeed, the algebra we propose contains two classes of operators: one class handling generalized relations as an infinite set of points (*tuple operators*) and one class handling each generalized relation as a finite set of spatial objects (*set operators*). A preliminary version of this algebra has been presented in [3].

3. We introduce external functions in the proposed algebra and give examples of external functions that are useful from a spatial point of view. The use of external functions avoids the choice of a "complex" logical theory, with high computational complexity. Rather, it allows to adopt a "simple" logic, for example the linear polynomial inequality constraint theory, and to express specific functionalities by means of external functions.

   Note that while several approaches have been proposed to model aggregate functions inside constraint query languages [9, 18], as far as we know, no approach has been proposed to deal with arbitrary external functions.

4. An update language is proposed, based on the same principles on which the algebra is based. In particular a set and a tuple version for insert and delete operators are proposed. Then, we investigate how an update operator can be inserted in a constraint database manipulation language. We first present a general definition and we show under which hypothesis this operator collapses to delete and insert operators and under which other hypothesis it corresponds to some useful spatial operations, such as translation, rotation, etc.

The paper is organized as follows. Section 2 introduces the model and the generalized relational algebra, whereas Section 3 discusses the limitation of this algebra and proposes a new model and a new algebra, more suitable for spatial manipulations. Section 4 deals with the definition of the update language. Finally, Section 5 presents some concluding remarks, pointing out the issues related to the design of architectures supporting the proposed model.

2   The generalized relational model

The following definition formally introduces the constraint relational model, as defined in [16].

Definition 1 (Generalized relational model). [16] Let Φ be a decidable logical theory on a domain $\mathcal{D}$.

- A *generalized tuple* t over variables $x_1, ..., x_k$ in the logical theory Φ is a finite quantifier-free conjunction $\varphi_1 \wedge ... \wedge \varphi_N$, where each φ_i, $1 \leq i \leq N$, is a constraint in Φ. The variables in each φ_i are all free and among $x_1, ..., x_k$. We denote with $\alpha(t)$ the set $\{x_1, ..., x_k\}$ and with $ext(t)$ the set of relational tuples belonging to $\mathcal{D}^k$ which are represented by t.
- A *generalized relation* r *of arity* k in Φ is a finite set $r = \{t_1, ..., t_M\}$ where each t_i, $1 \leq i \leq M$, is a generalized tuple over variables $x_1, ..., x_k$ and in Φ. We denote with $\alpha(r)$ the set $\{x_1, ..., x_k\}$ and we call it the schema of r.
- A *generalized database* is a finite set of generalized relations. The schema of a generalized database is a set of relation names $R_1, ..., R_n$, each with the corresponding schema. □

In the generalized relational model, a generalized relation is interpreted as the finite representation of a possibly infinite set of relational tuples and this set represents its semantics.

Definition 2 (Relational semantics). Let $r = \{t_1, ..., t_n\}$ be a generalized relation. The relational semantics of r, denoted by $rel(r)$, is $ext(t_1) \cup ... \cup ext(t_n)$. Generalized relations r_1 and r_2 are r-equivalent (denoted by $r_1 \equiv_r r_2$) iff $rel(r_1) = rel(r_2)$. □

In order to represent geometry using constraints in the generalized relational model, a possible approach is to use a relation with n variables representing points of an n-dimensional space. Generalized tuples of this relation thus represent sets of points (i.e., spatial objects) embedded in that space.

Using different classical logical theories, different types of spatial objects can be defined within a large range of complexity. According to the definition of spatial data given in [14], linear polynomial constraints have the sufficient expressive power to describe the geometric component of spatial data in geographical applications. Linear polynomial constraints have the form "$p(X_1, ..., X_n) \; \theta \; 0$", where p is a linear polynomial with real coefficients in variables $X_1, ..., X_n$ and $\theta \in \{=, \neq, \leq, <, \geq, >\}$. This class of constraints is of particular interest, because it has been investigated in various fields (linear programming, computational geometry) and therefore several techniques have been developed to deal with them [19]. In the following, we denote with Φ_L the theory of linear polynomial constraints.

In order to show an example of spatial data for a real geographic database, we restrict our attention to the Euclidean Plane ($\mathcal{E}^2$). The types of point-sets of $\mathcal{E}^2$ which can be described using the formulas of Φ_L are shown in Table 1. The first three types, POINT, SEGMENT and CONVEX, correspond to conjunctions of formulas of the theory Φ_L (thus, they correspond to generalized tuples). The fourth type, COMPOSITE spatial type, corresponds to a disjunction of conjunctions of Φ_L. As generalized tuples represent conjunctions of constraints, for the representation of composite objects in the generalized relational model, several generalized tuples must be used, each representing a convex object belonging to the convex decomposition of the composite object. An identifier must

Graphical representation	Analytical representation	Representation in Φ_L
POINT (p) $\bullet p$	$p = (x, y)$	$C_p(p) = (X - x = 0) \wedge$ $(Y - y = 0)$
SEGMENT (s) P_2 r P_1	$s = \begin{cases} P_1 = (x_1, y_1) \\ P_2 = (x_2, y_1) \\ r : ax + by + c = 0 \end{cases}$	$C_s(s) =$ $(x_1 - X \leq 0) \wedge (X - x_2 \leq 0)$ $\wedge (aX + bY + c = 0)^a$
CONVEX (c) P_n r_i P_i r_n P_0 r_0	$c = \begin{cases} P_i = (x_i, y_i) \\ r_i : a_i x + b_i x + c_i = 0 \\ i = 0 \to n \end{cases}$	$C_{cx}(c) =$ $(sg(P_1, P_2)a_1 X + b_1 Y + c_1 \geq 0)$ $\wedge \dots \wedge$ $(sg(P_{n-1}, P_n)a_n X + b_n Y + c_n \geq 0)^b$
COMPOSITE (csp) $\bullet p_1$ s_3 s_2 s_1 c_2 c_1 $\bullet p_2$	$csp = (p_1 \cup \dots \cup p_n) \cup$ $(s_1 \cup \dots \cup s_m) \cup$ $(c_1 \cup \dots \cup c_l)$	$C_{ct}(csp) =$ $(C_p(p_1) \vee \dots \vee C_p(p_n)) \vee$ $(C_s(s_1) \vee \dots \vee C_s(s_m)) \vee$ $(C_{cx}(c_1) \vee \dots \vee C_{cx}(s_l))$

[a] One or both of the first two disjuncts of this formula can be removed if a semi straight line or a complete straight line has to be represented.

[b] The introduction of the function $sg()$ is necessary in order to take into account that the polygonal region represented by a simple polygon is always on the left side of the polygon itself. Thus, function $sg(P_1, P_2)$ returns 1 or -1 according to the direction of the line defined by P_1 and P_2.

Table 1. Representation of point-sets of the Euclidean Plane in Φ_L

be assigned to each generalized tuple, in order to identify all generalized tuples representing points of the same composite object.

The relational algebra can be extended to deal with generalized relations [15, 22]. The resulting algebra is called *Generalized Relational Algebra* (GRA(Φ))[1]. The operators of such algebra are shown in Table 2. The table presents for each

---

[1] Φ is a decidable logical theory. In order to guarantee the closure of GRA(Φ), only theories Φ admitting variable elimination and closed under complementation must be considered. A theory is closed under complementation iff, if c is a constraint of the theory, then $\neg c$ can also be represented in the theory [4].

Op. name	Syntax e	Restrictions	Semantics $r = \mu(e)(r_1, ..., r_n),$ $n \in \{1, 2\}$
selection	$\sigma_P(R_1)$	$\alpha(P) \subseteq \alpha(R_1)$ $\alpha(e) = \alpha(R_1)$	$rel(r) = \{t \mid t \in rel(r_1),$ $t \wedge P$ is satisfiable$\}$
renaming	$\varrho_{[A|B]}(R_1)$	$A \in \alpha(R_1), B \notin \alpha(R_1)$ $\alpha(e) = (\alpha(R_1) \setminus \{A\}) \cup \{B\}$	$rel(r) = \{t : t' \in rel(r_1),$ $t = t'[A \mid B]\}$
union	$R_1 \cup R_2$	$\alpha(e) = \alpha(R_1) = \alpha(R_2)$	$rel(r) = rel(r_1) \cup rel(r_2)$
projection	$\Pi_{[x_{i_1}, ..., x_{i_p}]}(R_1)$	$\alpha(R_1) = \{x_1, ..., x_m\}$ $\alpha(e) = \{x_{i_1}, ..., x_{i_p}\}$ $\alpha(e) \subseteq \alpha(R_1)$	$rel(r) = \{\pi_{x_{i_1}, ..., x_{i_p}}{}^a(t) :$ $t \in rel(r_1)\}$
natural join	$R_1 \bowtie R_2$	$\alpha(e) = \alpha(R_1) \cup \alpha(R_2)$	$rel(r) = \{t : t_1 \in rel(r_1),$ $t_2 \in rel(r_2), t = t_1 \bowtie {}^b t_2\}$
complement	$\neg R_1$	$\alpha(e) = \alpha(R_1)$	$rel(r) = \{t \mid t \notin rel(r_1)\}$

a This is the relational projection operator.
b This is the relational join operator.

Table 2. GRA(Φ) operators

operator of GRA(Φ) the *schema restriction* required by the argument relations and by the result relation. $R_1, ..., R_n$ are relation names, e represents the syntactic expression and μ is a function that takes an expression and returns the corresponding query function. The generalized relational algebra satisfies the following property, stating that GRA operators are a trivial extension of relational operators.

Proposition 1 (Relational homomorphism) *[15] Let Φ be a decidable logical theory, admitting variable elimination and closed under complementation. Let **OP** be a GRA(Φ) operator and let OP^{rel} be the corresponding relational algebra operator. Let r_i, $i = 1, ..., n$ be generalized relations on theory Φ. Then, $rel(\textbf{OP}(r_1, ..., r_n)) = OP^{rel}(rel(r_1), ..., rel(r_n))$.* □

3 A nested generalized relational model for spatial data

The generalized relational model is a simple and yet powerful model because of its ability of representing infinite sets of tuples. However, as the classical relational model, it has a number of shortcomings when dealing with more complex data objects, as spatial data. Indeed, spatial objects have the following two characteristics:

1. the *complex* (geometric) component of spatial data can be represented as an infinite set of points embedded in a reference space;
2. each object has a relevance not only as a set of values (points), but also as a single value (object).

From this consideration it follows that spatial data should be manipulated under two different points of view:

- For some manipulations, it is useful to see a set of spatial objects as a set of points in the reference space. For example, suppose to retrieve all object components contained in a certain region or to compute the intersection of two spatial objects. In both cases, the computations manipulate points belonging to the extension of the considered objects. Not all the points belonging to the extension of a given object are necessarily returned by the query. Rather, such extension is *modified* by the computation. This type of manipulation is called *point-based*.
- For other manipulations, it is useful to see each spatial object as a single value. For example, suppose to determine all spatial objects contained in a certain region or to compute all pairs of spatial objects that intersect. In both cases, the same computation must be applied to *all* the points belonging to the extension of a given object. In this case, the object itself, seen as a single value, must satisfy the query and not a part of it, as in the previous case. This type of manipulation is called *object-based*.

All computations that can be represented by GRA(Φ) are point-based. Note that this does not mean that object-based computations cannot be represented by GRA(Φ). However, typical object-based spatial manipulations have in GRA(Φ) a very complex representation, due to the fact that a point-based manipulation must be used to simulate them, as the following example shows.

Example 1. Consider the query selecting all spatial objects that are contained in the region of the space identified by a rectangle $rt \in \mathcal{E}^2$. This is a typical object-based manipulation. Assume that all spatial objects are represented inside a generalized relation R and that $C_{cx}(rt) = P$ (see Table 1). In order to represent such query in GRA(Φ), the schema of the generalized relation R must include a variable ID, representing the generalized tuple identifier. This identifier is needed to "glue" together all the points belonging to the extension of a given spatial object. The previous query can be expressed in GRA(Φ_L) as follows:

$$(\Pi_{[ID]}(R) \setminus (\Pi_{[ID]}(R \setminus \sigma_P(R)))) \bowtie R.$$

The previous expression has the following meaning:

- $\sigma_P(R)$ selects the points (X, Y) of R contained in P, together with the identifier of the object to which they belong.
- $R \setminus \sigma_P(R)$ selects the points (X, Y) that are not contained in P, together with the identifier of the object to which they belong.
- $\Pi_{[ID]}(R \setminus \sigma_P(R))$ selects the identifiers of the objects having at least one point not contained in P. Thus, all the retrieved identifiers correspond to objects not contained in P.
- $\Pi_{[ID]}(R) \setminus (\Pi_{[ID]}(R \setminus \sigma_P(R)))$ selects the identifiers of the objects contained in P.
- $(\Pi_{[ID]}(R) \setminus (\Pi_{[ID]}(R \setminus \sigma_P(R)))) \bowtie R$ selects the objects contained in P. $\diamond$

The previous expression is not very simple to write and to understand, even if the query is one of the most common in spatial applications. The problem is that the query deals with the extension of generalized tuples taken as single objects, whereas, in general, the algebra operators deal with single relational tuples, belonging to the extension of generalized tuples.

In order to be able to manipulate spatial objects as single values, we extend $GRA(\Phi)$ by providing two different classes of operators:

- *Set operators*: They apply a certain object-based computation to generalized relations. Each generalized relation must be interpreted as a set of spatial objects. For this reason, we call them *set operators*.
- *Tuple operators*: They apply a certain point-based computation to generalized relations. Each generalized relation must be *seen* as a possibly infinite set of points (i.e., of tuples). For this reason, we call them *tuple operators*.

The definition of set operators is possible only if the semantics assigned to generalized relations is changed, in order to *see* each generalized relation as a set of spatial objects. Moreover, tuple operators must be revised to deal with this new semantics. In the following, we first present the new semantics and then we present the algebra.

3.1 The nested generalized relational model

The relational semantics is not the only possible way to assign a meaning to generalized relations. Generalized relations can also be interpreted as *nested relations* [1]. A nested relation is a relation in which attributes may contain sets as values. If we interpret generalized relations as nested relations, each generalized tuple represents a possibly infinite set of relational tuples, implicitly represented by its extension, and a generalized relation is not any longer an (infinite) set of relational tuples (i.e., points of the embedding space) but is a *finite set of sets*, each representing a (possibly infinite) set of relational tuples (i.e., each set represents a spatial object). This consideration leads to the definition of the following semantics.

Definition 3 (Nested semantics). Let $r = \{t_1, ..., t_n\}$ be a generalized relation. The nested semantics of r, denoted by $nested(r)$, is the set[2]

$$\{ext(t_1), ..., ext(t_n)\} \setminus \{\{\}\}.$$

Two generalized relations r_1 and r_2 are n-equivalent (denoted by $r_1 \equiv_n r_2$) iff $nested(r_1) = nested(r_2)$. □

From now on, we call *nested generalized relational model* the generalized relational model in which the nested semantics is adopted.

---

[2] We assume to remove inconsistent generalized tuples, i.e., all generalized tuples t such that $ext(t) = \{\}$.

If a generalized relation is interpreted under the nested semantics, the set of logical connectives used inside generalized tuples becomes important, because it characterizes the types of (possibly) infinite sets that can be represented. As we have seen, when using only conjunction inside generalized tuples, only convex sets of points can be represented. To overcome this limitation, we extend generalized tuples to deal with disjunction.

Definition 1 (Disjunctive generalized tuples). Let Ψ be a decidable logical theory. A *disjunctive generalized tuple* (abbreviated as *d-generalized tuple*) over variables $x_1, ..., x_k$ in the logical theory Φ is a finite and quantifier-free disjunction of generalized tuples over variables $x_1, ..., x_k$. □

The concepts of generalized relation and generalized database do not change when replacing generalized tuples with disjunctive-generalized tuples. In the following, we always consider d-generalized tuples and for simplicity we simply refer to them as to generalized tuples. Notice that the use of disjunction allows the representation of all types of spatial data (see Table 1) inside a single generalized tuple.

Example 2. Figure 1 shows a possible geographic domain. The space is decomposed in four districts. Districts may contain towns, railway sections, and stations. Districts, towns, and railway sections are concave objects, whereas stations are convex objects. A possible representation of such domains in the nested generalized relational model is the following:

- Districts can be stored in a generalized relation D. Each d-generalized tuple represents a single district.
- Towns can be stored in a generalized relation T. Each d-generalized tuple represents a single town.
- Railway sections and stations can be stored in several ways. For example, each railway section, together with the stations located along the section, can be represented inside a single d-generalized tuple. We assume that railway sections and stations are represented in this way inside a generalized relation R.

We assume that the schema of generalized relations D, T, and R contains two variables X and Y, representing points belonging to the object extension, and a variable ID, representing the generalized tuple identifier. Notice that this identifier is not inserted to "glue" together the extension of different generalized tuples, as in the generalized relational model. Rather, it has been introduced to better identify the considered spatial objects in query expressions. ◇

3.2 The nested relational algebra

The algebra we propose is called *Nested Generalized Relational Algebra* and, when defined for a decidable logical theory Φ, admitting variable elimination

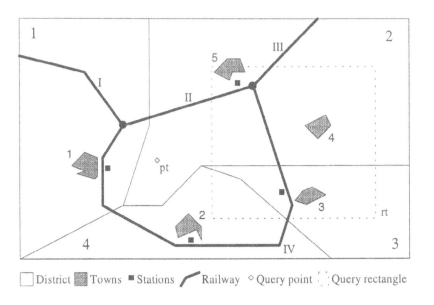

Fig. 1. *The map shows the content of the generalized relations D (representing districts), R (representing railways and stations) and T (representing towns)*

and closed under complementation, it is denoted by $\text{NGRA}(\Phi)$. The operators of $\text{NGRA}(\Phi)$, assuming that generalized relations are interpreted under the nested semantics, are the following:

- Tuple operators are exactly the operators introduced in Table 2. The only difference is that now they deal with a nested semantics. Thus, we force generalized relations obtained as query results to have a certain nested representation.
- Set operators are the following:
  1. *Set difference*: Given two generalized relations r_1 and r_2, this operator returns all generalized tuples contained in r_1 for which no equivalent generalized tuple exists in r_2. This is the usual difference operation in nested relational databases [1]
  2. *Set complement*: Given a generalized relation r, this operator returns a generalized relation containing a generalized tuple t' for each generalized tuple t contained in r, such that $t' \equiv_r \neg t$.
  3. *Set selection*: This operator selects from a generalized relation all the generalized tuples satisfying a certain *condition*. The condition has the form (Q_1, Q_2, θ), where $\theta \in \{\subseteq, (\bowtie \neq \emptyset)\}$ and Q_1 and Q_2 are either

Op. name	Syntax e	Restrictions	Semantics $r = \mu(e)(r_1,...,r_n),$ $n \in \{1,2\}$
		Tuple operators	
selection	$\sigma_P(R_1)$	$\alpha(P) \subseteq \alpha(R_1)$ $\alpha(e) = \alpha(R_1)$	$r = \{t : t_1 \in r_1, t = t_1 \wedge P\}$
renaming	$\varrho_{[A|B]}(R_1)$	$A \in \alpha(e), B \notin \alpha(e)$ $\alpha(e) = (\alpha(R_1) \setminus \{A\})$ $\cup\{B\}$	$r = \{t : t' \in r_1, t = t'[A \mid B]\}$
projection	$\Pi_{[x_{i_1},...,x_{i_p}]}(R_1)$	$\alpha(R_1) = \{x_1,...,x_m\}$ $\alpha(e) = \{x_{i_1},...,x_{i_p}\}$ $\alpha(e) \subseteq \alpha(R_1)$	$r = \{\pi_{x_{i_1},...,x_{i_p}}{}^a(t) : t \in r_1\}$
natural join	$R_1 \bowtie R_2$	$\alpha(e) = \alpha(R_1) \cup \alpha(R_2)$	$r = \{t : t_1 \in r_1,$ $t_2 \in r_2, t = t_1 \wedge t_2\}$
complement	$\neg R$	$\alpha(e) = \alpha(R)$	$r = \{\neg t_1 \wedge ... \wedge \neg t_n \mid$ $\{t_1,...,t_n\} = r\}$
		Set operators	
union	$R_1 \cup R_2$	$\alpha(R_1) = \alpha(R_2) = \alpha(e)$	$r = \{t : t \in r_1 \text{ or } t \in r_2\}$
set difference	$R_1 \setminus^s R_2$	$\alpha(R_1) = \alpha(R_2) = \alpha(e)$	$r = \{t : t \in r_1, \nexists t' \in r_2 :$ $ext(t) = ext(t')\}$
set complement	$\neg^s R_1$	$\alpha(e) = \alpha(R_1)$	$r = \{\neg t : t \in r_1\}$
set selection	$\sigma^s_{(Q_1,Q_2,\subseteq)}(R_1)$ $\sigma^s_{(Q_1,Q_2,\bowtie\neq\emptyset)}(R_1)$	$\alpha(Q_1) \subseteq \alpha(Q_2)$ $\alpha(e) = \alpha(R_1)$ $\alpha(Q_1) = \alpha(Q_2)$ $\alpha(e) = \alpha(R_1)$	$r = \{t : t \in r_1,$ $rel(t_1) \subseteq rel(\Pi_{\alpha(Q_1)}(t_2))\}$ $r = \{t : t \in r_1,$ $rel(t_1) \cap rel(t_2) \neq \emptyset\}$ $t_1 = Q_1(t), t_2 = Q_2(t)$

[a] This operator eliminates all variables not contained in $\{x_{i_1},...,x_{i_p}\}$, by applying a variable elimination algorithm [19].

Table 3. NGRA(Φ) operators

queries, whose unique argument is the tuple under analysis, seen as a singleton generalized relation and denoted by t, or constant functions returning a given generalized tuple P. In the last case, Q_1 (Q_2) is denoted by P. If Q_i, $i \in \{1,2\}$, is the identity query, i.e., it takes a tuple and returns the same tuple, it is simply denoted by t. Queries Q_i, $i = 1,2$, are constructed by recursively applying the subset of NGRA(Φ) operators given by $\{\sigma_P, \varrho_{A/B}, \Pi_{x_1,...,x_n}, \bowtie, \neg, \cap\}$, to the only argument t.

The set selection operator with condition (Q_1, Q_2, θ), applied to a generalized relation r, selects from r only the generalized tuples t for which there exists a relation θ between $Q_1(t)$ and $Q_2(t)$. In the following, given a NGRA(Φ) expression Q, $\alpha(Q)$ denotes the schema of the result of applying Q to a set of generalized relations. In a similar way, $\alpha((Q_1, Q_2, \theta)) = \alpha(Q_1) \cup \alpha(Q_2)$.

The possible meanings of θ operators are the following:

- $\theta \equiv \subseteq$: In this case, we require that $\alpha(Q_1) \subseteq \alpha(Q_2)$. It selects all generalized tuples t in r such that $rel(Q_1(t)) \subseteq rel(\Pi_{\alpha(Q_1)}(Q_2(t)))$.
- $\theta \equiv \bowtie \neq \emptyset$: In this case, we require that $\alpha(Q_1) = \alpha(Q_2)$. It selects all generalized tuples t in r such that $rel(Q_1(t)) \cap rel(Q_2(t)) \neq \emptyset$.

When a condition C is satisfied by a generalized tuple, we denote this fact with $C(t)$.

Table 3 presents set and tuple operators, according to the notation introduced in Section 2. Note that the union operator has been classified as a set operator, since it deals with the global extension of generalized tuples. The definition of additional operators derived from the proposed ones can be found in [4]. Among the possible derived operators we recall: set selections operators with conditions (Q_1, Q_2, θ), with $\theta \in \{\bowtie = \emptyset, \supseteq, =, \neq\}$; set selection operators using as condition a boolean combination of the conditions previously defined. The semantics of such operators directly follows from the semantics of the proposed set selection operator.

It can be proved that NGRA(Φ) operators can be seen as some nested-relational algebra computations [1], when extended to deal with constraints [4].

Proposition 2 (Nested homomorphism) *[15] Let Φ be a decidable logical theory, admitting variable elimination and closed under complementation. Let* **OP** *be a NGRA(Φ) operator. Let r_i, $i = 1, ..., n$ be generalized relations on theory Φ. There exists a nested relational algebra expression Q such that $nested(\mathbf{OP}(r_1, ..., r_n)) = Q(nested(r_1), ..., nested(r_n))$.* $\qquad\square$

In [4] it has been shown that NGRA(Φ) and GRA(Φ) are equivalent when specific generalized tuple identifiers are inserted in input generalized relations. Indeed, this is the only way to manipulate sets in the generalized relational model. However, NGRA(Φ) allows to represent some typical spatial queries in a simpler and much compact form. This result is similar to the equivalence result obtained for relational and nested relational databases [23]. Moreover, NGRA(Φ) has the same complexity of GRA(Φ), i.e., it is in NC.

Example 3. Consider the query selecting all spatial objects that are contained in the region of the space identified by a rectangle $rt \in \mathcal{E}^2$. Assuming that all spatial objects are represented in a generalized relation R and that $C_{cx}(rt) = P$ (see Table 1), the previous query can be expressed in NGRA(Φ_L) as $\sigma^s_{(t,P,\subseteq)}(R)$. Note that this expression is much more simpler than the one used to represent the same query in GRA(Φ_L) (see Example 1). $\qquad\diamond$

In spatial applications, the most interesting queries involve topological and metric properties of spatial objects [11, 14, 27, 28]. Topological properties are based on the definition of boundary and interior. These are the two parts in which a spatial object can be decomposed, according to the geometric application of the algebraic topology. A proposal for modeling the concept of classical

Type	Query	
	NGRA expression	**Conditions**
	Point interference queries	
POINT INTERFERENCE	select all districts in D that contain a given point $pt \in \mathcal{E}^2$ (see Figure 1 to locate pt)	
	$\sigma^s_{(t,P,\supset)}\,(D)$	$P \equiv^a C_p(pt)^b$
	Range queries	
RANGE INTERSECTION	select all districts in D that intersect the region of the space identified by the rectangle $rt \in \mathcal{E}^2$ (see Figure 1 to locate rt)	
	$\sigma^s_{(t,P,(\bowtie \neq \emptyset))}\,(D)$	$P \equiv C_{cx}(rt)$
RANGE CONTAINMENT	select all towns in T that are contained in the region of the space identified by the rectangle $rt \in \mathcal{E}^2$	
	$\sigma^s_{(t,P,\subseteq)}\,(T)$	$P \equiv C_{cx}(rt)$
RANGE CLIP	calculate all portions of districts obtained as intersection of each district in D with the rectangle $rt \in \mathcal{E}^2$	
	$\sigma_P\,(D)$	$P \equiv C_{cx}(rt)$
RANGE ERASE	calculate all portions of towns obtained as difference of towns in T with the rectangle $rt \in \mathcal{E}^2$	
	$T \setminus^t \sigma_P(T \cup \neg T)^c$	$P \equiv C_{cx}(rt)$

[a] $\equiv$ denotes syntactic equivalence.
[b] See Table 1 for the definition of functions $C_p()$, $C_{cx}()$.
[c] The symbol $\setminus^t$ is a short form for the derived operator: $T \setminus^t S = T \bowtie \neg S$.

Table 4. Examples of queries based on relationships with the embedding space

boundary [30], and therefore of topological relationship in the context of linear constraints, has been presented in [29]. The combinatorial boundary [20], which has a more complex definition than the classical one, can still be represented using linear constraints. Therefore, all spatial selections that involve topological relationships [10, 12] can be expressed in $\mathrm{GRA}(\Phi_L)$, and, in a more natural way, in $\mathrm{NGRA}(\Phi_L)$.

Tables 4 and 5 present some examples of spatial queries[3], including topological queries, expressed in NGRA(Φ_L). All the queries are assumed to be executed on the nested generalized database presented in Example 2. Each query in the tables is described by a textual description and by the mapping to NGRA(Φ_L). By contrast, metric queries cannot be expressed in NGRA(Φ_L) alone. Nest subsection shows how, by complementing NGRA(Φ_L) with external functions, metric queries can be expressed.

3.3 External functions

The introduction of external functions in database languages is an important topic. Functions increase the expressive power of database languages, relying on user defined procedures, without modifying the language definition. External functions can be considered as library functions, completing the knowledge of a certain application domain.

If we consider constraint query algebras, the introduction of external functions must preserve the closure of the language. The following definition introduces a class of functions for constraint databases, that satisfy this property. In the following, $DOM_{gentuple}(\Phi, S)$ is the set of all the possible disjunctive generalized tuples defined on theory Φ and having $\alpha(S)$ as schema, where $\alpha(S)$ denotes the set of variables in S and S is a tuple of variables (denoted by $[X_1, \ldots, X_n]$)[4].

Definition 5 (Admissible functions). Let Φ be a decidable logical theory. An admissible function f for Φ is a function from $DOM_{gentuple}(\Phi, S)$ to $DOM_{gentuple}(\Phi, S')$, where S and S' may be different. S is called the input schema of f and it is denoted by $is(f)$, whereas S' is called the output schema of f and it is denoted by $os(f)$. $\square$

When using external functions, two new operators, called *application dependent operators*[5] can be added to NGRA(Φ). The family of *Apply Transformation* operators allows to apply an admissible function to a generalized relation. Each operator of the family is specified by AT_f^{Sr}, where f is an admissible function and Sr is a tuple of variables. The result of the application of AT_f^{Sr} to a generalized relation r, whose schema contains $\alpha(Sr)$, is a new relation obtained from the previous one by replacing each generalized tuple t by a new tuple t'. The new tuple t' is obtained from t by modifying the set of values assigned to variables in Sr, according to the application of function f. The second operator (*Application dependent set selection*) is similar to the set selection of Table 3;

---

[3] All spatial queries presented in Tables 4, 5 and 7 must be considered as classical queries (as defined in [8]). We do not consider any definition of genericity for spatial queries with respect to geometric or topological transformations [22].

[4] The definition of S as a tuple simplifies the definition of application dependent operators (see Table 6).

[5] The term *application dependent operators* comes from the fact that functions reflect the application requirements.

Type	Query	
	NGRA expression	**Conditions**
	Topological queries	
ADJACENT QUERY	select all districts in D that are adjacent to the district with $ID = 1$	
	$\sigma^s_{c_1}(\sigma^s_{c_2}(D \bowtie D'))$ $D' \equiv \varrho_{[ID\|_{ID'},X\|_{X'},Y\|_{Y'}]}(\sigma_P(D))$	$P \equiv (ID = 1)$ $c_1 \equiv (Q_1(t), Q_3(t), (\bowtie = \emptyset))$, $c_2 \equiv (Q_2(t), Q_4(t), (\bowtie \neq \emptyset))$ $Q_1(t) \equiv Q_{INT}(\Pi_{[X,Y]}(t))$, $Q_2(t) \equiv Q_{BND}(\Pi_{[X,Y]}(t))^a$ $Q_3(t) \equiv \varrho_{[X'\|_X,Y'\|_Y]}(Q'(t))$ $Q'(t) \equiv Q_{INT}(\Pi_{[X',Y']}(t))$ $Q_4(t) \equiv \varrho_{[X'\|_X,Y'\|_Y]}(Q''(t))$ $Q''(t) \equiv Q_{BND}(\Pi_{[X',Y']}(t))$
CONTAINED QUERY	select all districts in D that contain the town identified by $ID = 4$	
	$\sigma^s_c(D \bowtie T')$ $T' \equiv \varrho_{[ID\|_{ID'},X\|_{X'},Y\|_{Y'}]}(\sigma_P(T))$	$P \equiv (ID = 4)$ $c \equiv (\Pi_{[X,Y]}(t), Q'(t), \supseteq)$ $Q'(t) \equiv \varrho_{[X'\|_X,Y'\|_Y]}(\Pi_{[X',Y']}(t))$
	Spatial Joins	
SPATIAL INTERSECTION	generate all portions of railway sections that are intersection between the district with $ID = 1$ and the railway network	
	$R \bowtie \sigma_P(D)$	$P \equiv (ID = 1)$
SPATIAL JOIN (intersection based)	generate all pairs (district d, railway section r) such that r reaches d	
	$\sigma^s_c(D \bowtie \varrho_{[X\|_{X'},Y\|_{Y'}]}(R))$	$c \equiv (Q_1(t), Q_2(t), (\bowtie \neq \emptyset))$ $Q_1(t) \equiv \Pi_{[X,Y]}(t)$, $Q_2(t) \equiv \varrho_{[X'\|_X,Y'\|_Y]}(\Pi_{[X',Y']}(t))$

[a] Q_{BND} and Q_{INT} represent a short form to indicate the queries that retrieve respectively the boundary and the interior of a spatial object [29].

Table 5. Examples of queries based on relationships with other objects in the same space

Op. name	Syntax e	Restrictions	Semantics $r = \mu(e)(r_1, \ldots, r_n)$ $n \in \{1, 2\}$
apply transformation	$AT_f^{Sr}(R_1)$	$f \in \mathcal{F}$ $is(f) = S, os(f) = S'$ $\alpha(Sr) \subseteq \alpha(R_1)$ $c(\alpha(Sr)) = c(\alpha(S))^b$	$r = \{\Pi_{[\alpha(t)\setminus\alpha(t')]}(t) \wedge t' : t \in r_1,$ $t' = \varrho_{[\alpha(S)|_{\alpha(Sr)}]}(f(t'')),$ $t'' = \varrho_{[\alpha(Sr)|_{\alpha(S)}]}(\Pi_{[\alpha(Sr)]}(t))\}^a$
set selection	$\sigma_{C_f}^s(R_1)$	$\alpha(e) = \alpha(R_1)$	$r = \{t : t \in r_1, \bar{C}_f(t)\}$

[a] Given two tuples of variables T and T', $\varrho_{[\alpha(T)|_{\alpha(T')}]}(R)$ replace in R the i-th variable in T with the i-th variable in T'.

[b] $c(s)$ returns the cardinality of the set s.

Table 6. NGRA$(\Phi, \mathcal{F})$ application dependent operators

the only difference is that now queries specified in the selection condition C_f may contain the operator AT_f.

Application dependent operators are summarized in Table 6. Given a decidable logical theory Φ, admitting quantifier elimination and closed under complementation, and a set of admissible functions $\mathcal{F}$, we denote with NGRA$(\Phi, \mathcal{F})$ the set of expressions obtained by composing application dependent operators presented in Table 6 and NGRA(Φ) operators. It can be shown that NGRA$(\Phi, \mathcal{F})$ is closed [4].

To show some examples of functions for spatial applications, we consider metric relationships. The metric relationships are based on the concept of distance referred to the reference space. The distance used in real applications is often the Euclidean one and a quadratic expression is needed to compute it. Thus, metric relationships can be represented in NGRA$(\Phi_L, \mathcal{F})$ only if proper external functions are introduced. Here we propose the following two functions defined as operations on the domain of all generalized tuples on Φ_L (we suppose that the reference space is $\mathcal{E}^2$):

- *Distance (Dis)*: Given a constraint c with four variables, representing two spatial objects, it generates a constraint $Dis(c)$ obtained from c by adding a variable D which represents the minimum distance between the two spatial objects. The following formula defines $Dis(c)$, supposing that c has variables (X, Y, X', Y'), where X, Y refer to the first spatial object and X', Y' refer to the second one (thus the input schema is $[X, Y, X', Y']$, whereas the output schema is $[X, Y, X', Y', D]$):

$$Dis(c) = c \wedge (D = min(\sqrt{(X' - X)^2 + (Y' - Y)^2} : c(X, X', Y, Y'))).$$

- *Buffer$_\delta$ (Buf$_\delta$)*: Given a constraint c, it generates the constraint $Buf_\delta(c)$ which represents all points that have a distance from c less than or equal to δ (thus the input and output schemas coincide and correspond to $[X, Y]$). It

Type	Query	
	NGRA expression	**Conditions**
	Metric queries	
DISTANCE QUERY	select all railway sections in R that are within 50 Km from the town identified by $ID = 4$	
	$\sigma_c^s(R \bowtie \varrho_{[X\mid_{X'},Y\mid_{Y'}]}(T'))$ $T' \equiv AT_{Buf_{50Km}}^{[\lambda,f]}(\sigma_P(T))$	$P \equiv (ID = 4)$ $c \equiv (\Pi_{[X,Y]}(t), \Pi_{[X',Y']}(t), \bowtie \neq \emptyset)$
SPATIAL JOIN (distance based)	generate all pairs (town t, railway section r) such that the distance between r and t is less than 40 Km, together with the real distance between r and t	
	$AT_{Dis}^{[X,Y,X',Y']}(\sigma_c^s(T \bowtie R'))$ $R' \equiv \varrho_{[ID\mid_{ID'},X\mid_{X'},Y\mid_{Y'}]}(R)$	$c \equiv (Q_1(t), Q_2(t), \bowtie \neq \emptyset)$ $Q_1(t) \equiv AT_{Buf_{40Km}}^{[X,Y]}(\Pi_{[X,Y]}(t))$ $Q_2(t) \equiv \varrho_{[X'\mid_X, Y'\mid_Y]}(\Pi_{[X',Y']}(t))$

Table 7. Examples of spatial queries using external functions

can be represented by the following formula:

$$Buf_\delta(c) = \exists X', Y' \ (c(X', Y') \wedge (X' - X)^2 + (Y' - Y)^2 \leq \delta^2).$$

Some relevant queries using external functions are reported in Table 7.

4 Update operators

Following the relational approach, at least three update operators must be defined for constraint databases: insertion, deletion, and modification of generalized tuples.

The distinction between point-based and object-based manipulation can be taken into account also in the definition of update operators. At this level, they have the following meaning:

- A point-based update modifies a set of spatial objects, seen as a possibly infinite set of points. Thus, it may add, delete, or modify some points, possibly changing the extension of the already existing objects.
- An object-based update modifies a set of spatial objects by inserting, deleting, or modifying a spatial object, seen as a single value.

Following the notation used in the definition of NGRA(Φ) operators, update operators applying a point-based manipulation are called *tuple operators*, whereas update operators applying an object-based manipulation are called *set*

Operator Name	Syntax e	Restrictions	Semantics $r_1 := \mu(e)(r_1)$
tuple insert	$Ins^t(R_1, C, u)$	$\alpha(u) \subseteq \alpha(R_1)$	$r_1 := \{t \vee u \mid t \in r_1 \wedge C(t)\} \cup$ $\{t \mid t \in r_1 \wedge \neg C(t)\}$
set insert	$Ins^s(R_1, u)$	$\alpha(u) \subseteq \alpha(R_1)$	$r_1 := r_1 \cup \{u\}$
set delete	$Del^s(R_1, C)$	$\alpha(C) \subseteq \alpha(R_1)$	$r_1 := \{t \mid t \in r_1 \wedge \neg C(t)\}$
tuple delete	$Del^t(R_1, C, u)$	$\alpha(u) \subseteq \alpha(R_1)$	$r_1 := \{t \wedge \neg u \mid t \in r_1 \wedge C(t)\} \cup$ $\{t \mid t \in r_1 \wedge \neg C(t)\}$
set update	$Upd^s(R_1, C, Q)$	$\alpha(Q) \subseteq \alpha(R_1)$ $\alpha(C) \subseteq \alpha(R_1)$	$r_1 := \{t \mid t \in r_1 \wedge \neg C(t)\} \cup$ $\{Q(t) \mid t \in r_1 \wedge C(t)\}$

Table 8. Update operators

operators. Note that the previous distinction has a counterpart only in the definition of insertion and deletion. Indeed, due to the nested semantics assigned to a generalized relation, point-based and object-based modify operations coincide. However, we choose to classify it as a set operator, since it always modifies a spatial object, represented by a generalized tuple.

In the following, we present insert, delete, and modify operators, pointing out the relationships existing among them. All the queries we consider in the definition of these operators are assumed to be expressed in NGRA($\Phi, \mathcal{F}$), for some decidable theory Φ, admitting variable elimination and closed under complementation, and set of admissible functions $\mathcal{F}$.

4.1 Insert operators

Specific application requirements lead to the definition of tuple and set insert operators. In particular:

- The definition of a *set insert* operator is motivated by the fact that a typical requirement is the insertion of a new spatial object in a generalized relation. The set insert satisfies this requirement by taking a generalized relation r and a generalized tuple t as input, and adding t to r, thus increasing the cardinality of r. Since r is a set, the set insert is a no-operation if t is already contained in r. This operation can be reduced to an equivalence test between generalized tuples. As generalized tuples are usually represented by using canonical forms [15], this test usually reduces to check whether two canonical forms are identical.
- Because a generalized relation contains sets of relational tuples, the user may be interested in inserting a relational tuple or a set of relational tuples into the existing sets of relational tuples. Note that this requirement is different from the previous one, since in this case we extend the extension of the already existing spatial objects, but we do not insert any new spatial object. Given a generalized relation r, a boolean condition C (see the definition of set selection at page 11) and a generalized tuple t, the *tuple insert* operator selects all generalized tuples of r that satisfy C and adds to them the

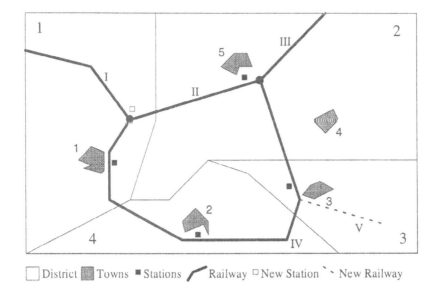

Fig. 2. *The map shows the content of the generalized relations D, R and T, together with the spatial objects to be inserted*

relational tuples contained in $ext(t)$. Notice that the tuple insert does not change the cardinality of the target generalized relation.

In Table 8, the syntax and the semantics of insert operations are presented following the style used in Tables 2 and 3. For update operations, μ is a function that takes an update expression and returns a function, representing the update semantics.

Example 4. Consider the map represented in Figure 2. The dashed line and the empty square represent spatial objects that have to be added to the database. In particular, the new railway section identified by $ID = 5$ is added to the R generalized relation and the new station is added to the railway sections identified by $ID \in \{1, 2, 4\}$, since it is an interchange node of the railway network. The first insertion is performed using the set insert operator, because a new spatial object has to be created. The second one uses the tuple insert operator, since only a modification of the extent of existing spatial objects has to be performed. Table 9 shows the expressions of the two insertions.   ◇

Description	NGRA update expression
insertion of a new railway section in R	$Ins^s(R, \langle ID = 5 \wedge 105 \leq X \leq 137^a \wedge Y = -\frac{9}{32}X + 50 \rangle)$
insertion of a new station belonging to the railway sections of R with $ID \in \{1, 2, 4\}$	$Ins^t(R, (t, (ID = 1 \vee ID = 2 \vee ID = 4), \bowtie \neq \emptyset),$ $\langle 53 \leq X \leq 55 \wedge 40 \leq Y \leq 42 \rangle)$
deletion of the town with $ID = 1$ contained in T	$Del^s(T, (t, (ID = 1), \bowtie \neq \emptyset))$
deletion of a specific station from the railway section in R with $ID = 2$	$Del^t(R, (t, (ID = 2), \bowtie \neq \emptyset),$ $\langle 65 \leq X \leq 67 \wedge 83 \leq Y \leq 85 \rangle)$

---

[a] $a \leq X \leq b$ is an abbreviation for $X \geq a \wedge X \leq b$.

Table 9. Examples of insertions and deletions using NGRA(Φ_L) update operators

4.2 Delete operation

For the delete operations the discussion is similar to the one presented for the insert operations. We therefore introduce two operators:

- The *set delete* operator, given a generalized relation r and a boolean condition C, deletes from r all generalized tuples that satisfy C.
- The *tuple delete* operator, given a generalized relation r, a boolean condition C, and a generalized tuple t, selects all the generalized tuples of r that satisfy C and removes from their extension the relational tuples contained in $ext(t)$.

In Table 8, the syntax and semantics of delete operations are presented. As an example, Table 9 shows the expressions to delete respectively the town with $ID = 1$ and the station of the railway section with $ID = 2$.

4.3 Modify operators

Traditional database systems provide a modify operation to deal with updates that are function of the old values of the tuples. In constraint database systems this case is really frequent since operations of this kind, like rescaling, translation or rotation, are often applied to spatial objects. Therefore the introduction of a modify operator (also called update operator) in a spatial oriented data model is a necessary action.

In a traditional DML (for example, SQL), the modify operation usually allows to compute the new value, to be assigned to the updated tuple, by a database query. Following the same approach we propose a *set update* operator with the following semantics. Given a generalized relation r, a boolean condition C, and a query $Q(t)$, the set update operator selects all tuples t of r that satisfy C and substitutes each t with $Q(t)$. The query $Q(t)$ acts on a single generalized tuple, denoted by t, at a time, as in the definition of the set selection operator. The generalized tuple t is considered as a generalized relation containing only one generalized tuple. This implies that all set operators of NGRA(Φ) are useless, since eventually they can only delete t. Note that, also the union operator cannot be used inside $Q(t)$, because it will necessary generate a relation with at least two generalized tuples. However, since an operator that generates the disjunction of two generalized tuples could be really useful to express some spatial transformations, we introduce a *tuple union* operator defined as

$$R_1 \cup^t R_2 = t_{1,1} \vee \ldots \vee t_{1,n} \vee t_{2,1} \vee \ldots \vee t_{2,m}$$

assuming that $R_1 = \{t_{1,1}, \ldots, t_{1,n}\}$ and $R_2 = \{t_{2,1}, \ldots, t_{2,n}\}$. Therefore, we restrict $Q(t)$ to be an expression of $(\text{NGRA}(\Phi, \mathcal{F}) \setminus \{\cup, \sigma_C^s, \setminus^s, \neg^s\}) \cup \{\cup^t\}$[6]. Table 8 presents the definition of the set update operation.

Notice that, depending on the operators used in the query $Q(t)$, a different modification of tuple t is obtained. In particular, the following proposition holds.

Proposition 3 *Given a generalized relation identifier R, a boolean condition C and a query Q, expressed using the operators of $NGRA(\Phi, \mathcal{F}) \setminus \{\varrho, \Pi, \cup, \sigma_C^s, \setminus^s, \neg^s\}$, a generalized tuple P exists such that:*

$$Upd^s(R, C, Q(t)) = Upd^s(R, C, \sigma_P(t)).$$

Proof. Since t is a single generalized tuple, $\sigma_P(t)$ is equal to $(t \wedge P)$. Thus, the thesis is proved if we show that a generalized tuple P always exists, such that $Q(t) = t \wedge P$. This is proved by induction on the structure of $Q(t)$:

- *Base step*: $Q(t) = t$. P representing the true formula on the schema of r satisfies the thesis.
- *Inductive step*:
  - $Q(t) = \sigma_{\overline{P}}(f(t))$.
    By inductive hypothesis, $f(t) = t \wedge P'$. Thus, from the definition of the selection operator, $Q(t) = t \wedge P' \wedge \overline{P}$, and therefore $Q(t) = t \wedge P$, where $P = P' \wedge \overline{P}$.
  - $Q(t) = f'(t) \setminus f''(t)$.
    By inductive hypothesis, $f'(t) = t \wedge P'$, $f''(t) = t \wedge P''$. Thus, from the definition of the difference operator, $Q(t) = t \wedge P' \wedge \neg(t \wedge P'')$, and therefore $Q(t) = t \wedge P' \wedge \neg P'' = t \wedge P$, where $P = P' \wedge \neg P''$.
  - $Q(t) = f'(t) \bowtie f''(t)$.
    By inductive hypothesis $f'(t) = t \wedge P'$, $f''(t) = t \wedge P''$. Thus, from the definition of the natural join operator, $Q(t) = t \wedge P' \wedge P''$, and therefore $Q(t) = t \wedge P$, where $P = P' \wedge P''$. □

---

[6] With $(\text{NGRA}(\Phi, \mathcal{F}) \setminus S) \cup S'$ we denote the expressions of the language obtained from $\text{NGRA}(\Phi, \mathcal{F})$ by not using operators in S but possibly using new operators contained in S'.

The previous result does not hold if the query $Q(t)$ contains a projection operator. Indeed, in this case, the query specified in the update operator may generate relational tuples that are not contained in $ext(t)$ (consider for example, the query $Q(t) = \Pi_{[X]}(t) \bowtie \Pi_{[Y]}(t)$ where $t \equiv (X = Y \wedge 1 < Y < 10)$). Thus, in general, a constraint P such that $Q(t) = \sigma_P(t)$ cannot be found. A similar consideration holds if $Q(t)$ contains the renaming operator.

Some examples of queries that can be used inside the set update operator in order to modify spatial data are shown in Table 10. For each query, together with the algebraic expression, an equivalent expression in the generalized relational calculus [16] is also proposed. Notice that, since the translation and the rotation of a spatial object of $\mathcal{E}^2$ can be expressed in NGRA(Φ_L), all the movements of a spatial object in $\mathcal{E}^2$ can be described in this language.

Some relationships exist between the proposed set update operator and the previously defined tuple operators, as stated by the following proposition. The proof trivially follows from the definition of the operators.

Proposition 4 (Tuple Operators Derivability) *Given a generalized relation identifier R, a boolean condition C, and a generalized tuple P, the tuple insert and the tuple delete operators can be expressed, respectively, as follows:*

$$Del^t(r, C, P) = Upd^s(r, C, \sigma_{\neg P}(t)) \quad Ins^t(r, C, P) = Upd^s(r, C, t \cup^t \sigma_P(t \cup^t \neg t)).^7$$

□

From the previous results it follows that the proposed set update operator is sufficient to model tuple insert and tuple delete operators, that therefore represent simpler syntactic forms to express data modifications.

5 Concluding remarks

Constraint databases are an important enabling technology for handling spatial data. In this paper we have pointed out some limitations of the generalized relational model and the generalized relational algebra when used to model spatial objects and we have proposed some solutions to overcome these limitations. The main contribution of this paper is the definition of a nested-based algebra. The algebra deals both with set operators, i.e., operators seeing a generalized relation as a finite set of (possibly) infinite sets (thus, a finite set of spatial objects), and tuple operators, i.e., operators seeing a generalized tuple as a (possibly) infinite set of tuples (thus, a possibly infinite set of points). We have shown through several examples that this algebra is more suitable than the generalized relational algebra for modeling spatial applications. The algebra has also been extended to deal with external functions and an update language has been proposed, based on the same principles of the algebra.

Some considerations are needed concerning the design of a constraint database system based on the proposed model, for modeling spatial data. Some relevant issues requiring further investigation are listed below.

---

[7] Notice, that P is generated by the query $\sigma_P(t \cup^t \neg t)$.

Description	
Calculus	**NGRA**
Projection $(Q^{prj}_{X_{i_1},...,X_{i_m}}(u))$ it projects the n-dimensional spatial value u onto the $X_{i_1},...,X_{i_m}$ $(m < n)$ coordinates	
$(\exists \{X_1,...,X_n\} \setminus \{X_{i_1},...,X_{i_m}\})(u)$	$\Pi_{[X_{i_1},...,X_{i_m}]}(u)$
Minimum Bounding Rectangle $(Q^{mbr}(u))$ it produces the Minimum Bounding Rectangle of the spatial value u, i.e., the rectangle of minimal area and with sides parallel to the axes, that contains the spatial value	
$(\exists Y')(u(X,Y')) \wedge (\exists X')(u(X',Y))$	$\Pi_{[X]}(u) \bowtie \Pi_{[Y]}(u)$
Translation $(Q^{tra}_{a,b}(u))$ it translates the spatial object u according to the vector $< a, b >$	
$(\exists X',Y')(u(X',Y') \wedge$ $X = X' + a \wedge Y = Y' + b)$	$\Pi_{[X,Y]}(\sigma_c(\varrho_{[X\|_{X'},Y\|_{Y'}]}(u) \bowtie (u \cup^t \neg u)))$ $c \equiv X = X' + a \wedge Y = Y' + b$
Rotation $(Q^{rot}_{\overline{X},\overline{Y},a_1,a_2,b_1,b_2}(u))$ it rotates the spatial value u according to the rotation coefficients $< a_1, a_2, b_1, b_2 >^a$	
$(\exists X',Y')(u(X',Y') \wedge$ $X = a_1(X' - \overline{X}) + a_2(Y' - \overline{Y}) \wedge$ $Y = b_1(X' - \overline{X}) + b_2(Y' - \overline{Y}))$	$\Pi_{[X,Y]}(\sigma_c(\varrho_{[X\|_{X'},Y\|_{Y'}]}(u) \bowtie (u \cup^t \neg u)))$ $c \equiv (X = a_1(X' - \overline{X}) + a_2(Y' - \overline{Y})) \wedge$ $(Y = b_1(X' - \overline{X}) + b_2(Y' - \overline{Y}))$

[a] The coefficients a_1, a_2, b_1, b_2 define the rotation, however only 2 of them are independent, i.e., $a_{1,1} = a$, $a_{1,2} = b$, $a_{2,1} = -b$ and $a_{2,2} = a$, where $a = \cos\alpha$, $b = \sin\alpha$, and α is the rotation angle.

Table 10. Examples of queries to be specified in the modify operator

- *Canonical forms.* In order to efficiently perform algebraic operations and reduce the redundancy of the representation, generalized tuples should be represented using some canonical form. A canonical form for dense-order constraints and its impact on the definition of generalized relational algebra have been discussed in [15]. Specific canonical forms for linear generalized tuples should also be developed.
- *Redundancy.* The use of the nested semantics changes the notion of redundancy typical of constraint databases. Indeed, under the relational semantics, a generalized tuple is redundant in a given generalized relation if its extension is already contained in the extension of the other generalized tuples

contained in the relation. Under the nested semantics, a generalized tuple is redundant iff there exists another generalized tuple in the same relation, with the same extension of the first one. Different redundancy tests must be applied in the two cases.

- *Logical optimization.* The introduction of new set operators leads to the definition of new equivalence relations for the expressions of NGRA(Φ). These new rules are important not only for applying logical optimization to NGRA(Φ) expressions, but, due to the equivalence between NGRA(Φ) and GRA(Φ) (modulo the use of generalized tuple identifiers) [4], these rewriting rules can also be used to apply in one step complex rewriting to GRA(Φ) expressions.
- *Indexing data structures.* The presence in NGRA(Φ) of set selection operators requires the definition of new index structures, able to deal with these new types of selection and having optimal worst-case complexity. For example, the definition of index structures to find all generalized tuples whose extension is contained in the extension of a given generalized tuple is an important topic. Some preliminary work on this topic has been presented in [5].
- *Cost-based optimization.* In relational databases, information on data structures should be used after the logical optimization step to determine the optimal execution plan. A similar situation arises in spatial databases. Techniques applied in both kind of systems should be integrated to define a cost-based optimizer for constraint databases. Preliminary results on this topic can be found in [7].
- *Visualization of spatial objects.* Issues related to the visualization of generalized tuples representing spatial objects must be addressed, developing efficient algorithms and evaluating the trade-off in using constraints instead of points.
- *Computational geometry algorithms.* The use of constraints might sometimes simplify the execution of some spatial queries. For example, the intersection-based spatial join can be computed on constraints by applying a satisfiability check, without using any computational geometry algorithm. This new approach to process spatial queries has to be compared with the classical one, based on the use of computational geometry algorithms.

References

1. S. Abiteboul and P. Kanellakis. Query Languages for Complex Object Databases. *SIGACT News*, 21(3):9–18, 1990.
2. W.G. Aref and H. Samet. Extending a Database with Spatial Operations. In *LNCS 525: Proc. of the Second Symposium on Advances in Spatial Databases*, pages 299-319, 1991.
3. A. Belussi, E. Bertino, M. Bertolotto, and B. Catania. Generalized Relational Algebra: Modeling Spatial Queries in Constraint Databases. In *LNCS 1034: Proc. of the First Int. CONTESSA Database Workshop, Constraint Databases and their Applications*, pages 40–67, 1995.

4. A. Belussi, E. Bertino, and B. Catania. An Extended Algebra for Constraint Databases. Technical report n. 179-96, University of Milano, 1996. Submitted for publication.

5. E. Bertino, B. Catania, and B. Shidlovsky. Towards Optimal Indexing for Segment Databases. Technical Report n.196-97, University of Milano, Italy. Submitted for publication.

6. E. Bertino, B. Catania, and L. Wong. Finitely Representable Nested Relations. Submitted for publication.

7. A. Brodsky and Y. Kornatzky. The $\mathcal{L}_{yri}\mathcal{C}$ Language: Querying Constraint Objects. In *Proc. of the ACM SIGMOD Int. Conf. on Management of Data*, 1995.

8. A.K. Chandra and D. Harel. Computable Queries for Relational Data Bases. *Journal of Computer and System Sciences*, 21:156–178, 1980.

9. J. Chomicki, D. Goldin, and G. Kuper. Variable Independence and Aggregation Closure. In *Proc. of the ACM SIGACT-SIGMOD-SIGART Symposium on Principles of Database Systems*, pages 40–48, 1996.

10. E. Clementini, P. Di Felice, and P. van Oosterom. A Small Set of Formal Topological Relationships Suitable for End-User Interaction. In *LNCS 692: Proc. of the Third Int. Symp. on Advances in Spatial Databases*, pages 277–295, 1993.

11. L. De Floriani, P. Marzano, and E. Puppo. Spatial Queries and Data Models. In *LNCS 716: Spatial Information Theory: a Theoretical Basis for GIS*, pages 123–138, 1993.

12. M.J. Egenhofer. Reasoning about Binary Topological Relations. In *LNCS 525: Proc. of the Second Int. Symp. on Advances in Spatial Databases*, pages 143-160, 1991.

13. S. Grumbach and J. Su. Dense-Order Constraint Databases. In *Proc. of the ACM SIGACT-SIGMOD-SIGART Symposium on Principles of Database Systems*, pages 66–77, 1995.

14. R.H. Güting and M. Schneider. Realm-Based Spatial Data Types: The ROSE Algebra. *VLDB Journal*, 4: 243–286, 1995.

15. P.C. Kanellakis and D.Q. Goldin. Constraint Programming and Database Query Languages. In *LNCS 789: Proc. of the Int. Symp. on Theoretical Aspects of Computer Software*, pages 96–120, 1994. See also Technical Report CS-94-31, Brown University, Providence, USA, 1994.

16. Paris Kanellakis, Gabriel Kuper, and Peter Revesz. Constraint query languages. *Journal of Computer and System Sciences*, 51:25–52, 1995.

17. H. Katsumo and A.O. Mendelzon. On the Difference Between Updating a Knowledge Base and Revising it. *Belief Revision*, Cambridge Tracts in Theoretical Computer Science, 1992. Cambridge University Press.

18. G.M. Kuper. Aggregation in Constraint Databases. In *Proc. of the First Int. Workshop on Principles and Practice of Constraint Programming*, 1993.

19. J.L. Lassez. Querying Constraints. In *Proc. of the ACM SIGACT-SIGMOD-SIGART Symposium on Principles of Database Systems*, pages 288–298, 1990.

20. E.E. Moise. *Geometric Topology in Dimension Two and Three*. Springer Verlag, 1977.

21. J. Paredaens. Spatial Databases, The Final Frontier. In *LNCS 893: Proc. of the Fifth Int. Conf. on Database Theory*, 1995.

22. J. Paredaens, J. Van den Bussche, and D. Van Gucht. Towards a Theory of Spatial Database Queries. In *Proc. of the ACM SIGACT-SIGMOD-SIGART Symposium on Principles of Database Systems*, pages 279–288, 1994.

23. J. Paredaens and D. Van Gucht. Possibilities and Limitations of Using Flat Operators in Nested Algebra Expressions. In *Proc. of the ACM SIGACT-SIGMOD-SIGART Symposium on Principles of Database Systems*, pages 29–38, 1988.

24. F. P. Preparata and M.I. Shamos. *Computational Geometry - an Introduction*, Springer Verlag, New York, 1985.

25. P.Z. Revesz. Model-Theoretic Minimal Change Operators for Constraint Databases. In *LNCS 1186: Proc. of the Sixth Int. Conf. on Database Theory*, 1997.

26. N. Roussopoulos, C. Faloutsos, and T. Sellis. An Efficient Pictorial Database System for PSQL. *IEEE Transaction on Software Engineering*, 14(5):639-650, 1988.

27. M. Scholl and A. Voisard. Thematic Map Modeling. In *Proc. of the Symp. on the Design and Implementation of Large Spatial Databases*, pages 167–190, 1989.

28. P. Svensson. GEO-SAL: a Query Language for Spatial Data Analysis. In *LNCS 525: Proc. of the Second Int. Symp. on Advances in Spatial Databases*, pages 119-140, 1991.

29. L. Vandeurzen, M. Gyssens, and D. Van Gucht On the Desirability and Limitations of Linear Spatial Database Models. In *LNCS 951: Proc. of the Fourth Int. Symp. on Advances in Spatial Databases*, pages 14–28, 1995.

30. A. Wallance. *An Introduction to Algebraic Topology*. Pergamon Press, 1967.

Constraint-Based Interoperability of Spatiotemporal Databases*

Jan Chomicki[1] and Peter Z. Revesz[2]

[1] Computer Science Department, Monmouth University,
West Long Branch, NJ 07764, USA, chomicki@moncol.monmouth.edu
[2] Department of Computer Science and Engineering
University of Nebraska–Lincoln, Lincoln, NE 68588, USA, revesz@cse.unl.edu

Abstract. We propose constraint databases as an intermediate level facilitating the interoperability of spatiotemporal data models. Constraint query languages are used to express translations between different data models. We illustrate our approach in the context of a number of temporal, spatial and spatiotemporal data models.

1 Introduction

Very large temporal and spatial databases are a common occurrence nowadays. Although they are usually created with a specific application in mind, they often contain data of potentially broader interest, e.g., historical records or geographical data. By *database interoperability* we mean the problem of making the data from one database usable to the users of another. Data sharing between different applications and different sites is often the preferable mode of interoperation [3]. But sharing of data (and application programs developed around it), facilitated by the advances in network technology, is hampered by the incompatibility of different data models and formats used at different sites. Semantically identical data may be structured in different ways. Also, the expressive power of some data models is limited.

A temporal database may have been built using one of the many temporal extensions of the relational data model (the recent book [TCG+93] describes at least 12 such extensions which are mutually incompatible), using a customized temporal data model, or simply using SQL (or some of *its* extensions). There may be many application programs and complex queries that have been developed for this database. The situation in the area of spatial databases is similar [Par95, Wor95], often with a considerable investment in software tools tuned to specific data models.

---

* The work of the first author was supported by NSF grant IRI-9632870. The work of the second author was supported by NSF grants IRI-9632871 and IRI-9625055, and by a Gallup Research Professorship.

[3] "Efficiency, security and availability all argue for shipping the data to the downstream database rather than providing integrated access to both systems."[RS94]

Temporal and spatial databases share a common characteristic: they contain *interpreted data*, associated with uninterpreted data in a systematic way. For example, a temporal database may contain the historical record of all the property deeds in a city. A spatial database may contain the information about property boundaries. Moreover, as this example shows, spatial and temporal data are often mixed in a single application.

In this research, we propose that *constraint databases* [KKR95] be used as a common language layer that makes the interoperability of different temporal, spatial and spatiotemporal databases possible. Constraint databases generalize the classical relational model of data by introducing *generalized tuples*: quantifier-free formulas in an appropriate constraint theory. For example, the formula $1950 \leq t \leq 1970$ describes the interval between 1950 and 1970, and the formula $0 \leq x \leq 2 \wedge 0 \leq y \leq 2$ describes the square area with corners $(0,0)$, $(0,2)$, $(2,2)$, and $(2,0)$. The constraint database technology makes it possible to finitely represent infinite sets of points, which are common in temporal and spatial database applications. We list below some further advantages of using the constraint database technology:

1. *Wide spectrum of data models.* By varying the constraint theory, one can accommodate a variety of different data models. By syntactically restricting constraints and generalized tuples, one can precisely capture the expressiveness of different models.

2. *Broad range of available query languages.* Relational algebra and calculus, Datalog and its extensions are all applicable to constraint databases. Those languages have well-studied formal semantics and computational properties, and are thus natural vehicles for expressing translations between different data models. Also, constraint query languages may be able to express queries inexpressible in the query languages of the interoperated data models, augmenting in this way the expressive power of the latter. (This is more a practical than a theoretical contribution. We simply mean that if, for instance, we have a TQuel database, then translation to a constraint database with dense order constraints allows querying by Datalog, a query language which is more expressive than TQuel. Similar comments apply to several other spatial and temporal data models in use.)

3. *Decomposability.* The problem of translating between two arbitrary data models, which is hard, is decomposed into a pair of simpler problems: translating one data model to a class C of constraint databases, and then translating C to the other data model. Also, by using a common constraint basis, we need to write only $2n$ instead of $n(n-1)/2$ number of translations for n different data models.

4. *Combination and interaction of spatial and temporal data within a single framework.* This is an issue of considerable recent interest ([FCF92], ESPRIT Chorochronos project).

In this paper we address the issue of *application-independent* interoperability of spatiotemporal databases. We show that the translations between different data models can be defined independently of any specific application that uses those models. We distinguish between *data* and *query* interoperability. For the former, it is the data that is translated to a different data model, while the latter concerns the translation of queries. The constraint database paradigm is helpful in both tasks. For data interoperability, constraint databases serve as a mediating layer and translations between different data models are expressed using constraint queries. For query interoperability, it is the constraint query languages themselves that serve as the intermediate layer. In an actual implementation, the presence of a mediating constraint layer may be completely hidden from the user. In this paper we study only *data interoperability*. Query interoperability is a topic of future research.

We show below two scenarios in which our approach may be useful in practice.

Scenario 1. The user of a data model Δ_2 wants to query a database D_1 developed under a data model Δ_1. He translates D_1 to a Δ_2-database D_2 (using constraint databases as an intermediate layer) that he can subsequently query using the query language of Δ_2. (As a practical matter, if a user is interested in a query Q2 in Δ_2, then only the part of the database that is relevant to the query needs to be translated.)

Scenario 2. The user of a data model Δ_1 wants to augment the power of the query language of Δ_1. For example, this language may be unable to express recursive queries. However, such queries can be formulated in an appropriate constraint query language. Thus whenever the user wants to run such a query on a database D_1, he first translates D_1 to a constraint database, runs the query in the constraint query language on it (using a constraint query engine), and translates the result back to Δ_1. (N.b., interoperating query results is an often neglected aspect of database interoperability.)

The plan of the paper is as follows. In Section 2 we define a very general notion of a data model and introduce a number of data models that will be studied in the rest of the paper: the TQuel data model for temporal databases, the 2-spaghetti model for spatial databases, and two spatiotemporal models (one of which is new). We believe that those models are representative of a large part of spatiotemporal data models that occur in practice. In Section 3 we characterize the expressiveness of the above data models using appropriately defined classes of constraint databases. In Section 4, which is the most technically involved part of the paper, we show that the bulk of the translations between the data models can be expressed using first-order constraint query languages. In particular, we show that the boundary, the corners and the edges of a convex polygon specified by linear arithmetic constraints can be defined using first-order queries with linear arithmetic constraints only. In Section 5 we discuss related work. In Section 6 we conclude the paper and point out directions for future work in this area.

2 Data Models

2.1 Database Interoperability

By *database interoperability* we mean the problem of making the data from one database usable to the users of another. There are many possible sources of mismatches between different databases [KCGS95]: they may use different data models, the schemas may not match, some data may be missing or inconsistent etc. In this paper we limit our attention to the differences in the data models and are thus concerned with *application-independent* interoperability.

2.2 Basic Notions

A *data model* Δ consists of a set of valid databases $I(\Delta)$ and a set of valid queries $L(\Delta)$. All valid databases are finite. We assume that for every valid database D in a data model Δ, the abstract semantics of D is given first-order structure $\theta_\Delta(D)$. (Often, the abstract semantics is not given in the published description of the data model but has to be inferred from it.) We will term D a *concrete* representation of the *unrestricted* database $\theta_\Delta(D)$. An unrestricted database may be infinite. Examples are given below.

Definition 2.1 Two databases $D_1 \in I(\Delta_1)$ and $D_2 \in I(\Delta_2)$ are *equivalent* if $\theta_{\Delta_1}(D_1) = \theta_{\Delta_2}(D_2)$.

Definition 2.2 [BNW91, BCW93] The *data expressiveness* of the data model Δ is the set $E_\Delta = \{\theta_\Delta(D) : D \in I(\Delta)\}$.

Currently used data models can substantially differ in terms of data expressiveness. For example, some can only represent finite unrestricted relations.

Given two data models Δ_1 and Δ_2, there are two fundamentally different ways to query databases from $I(\Delta_1)$ using queries from $L(\Delta_2)$. These two approaches are:

1. *Data interoperability:* for a given database $D_1 \in I(\Delta_1)$ an equivalent database $D_2 \in I(\Delta_2)$ is constructed. Then D_2 can be queried using queries in $L(\Delta_2)$. Data interoperability opens up the data of Δ_1 to the users of Δ_2, making direct data sharing possible.

2. *Query interoperability:* This means that for a given query $Q_2 \in L(\Delta_2)$ an equivalent query $Q_1 \in L(\Delta_1)$ is constructed. Then Q_1 can be evaluated over the given database $D_1 \in I(\Delta_1)$ and the result translated back to $I(\Delta_2)$. Query interoperability enables the users of Δ_2 to request the evaluation of queries within Δ_1. In this case data sharing is indirect.

Note that the above definitions characterize the *semantics* of data and query interoperability. In any *implementation* of data interoperability only a part of the database that is relevant for the given query will be translated. Note also that query interoperability relies on data interoperability to translate the query result. For some, e.g., Boolean, queries such translation will be trivial.

In this paper we concentrate on *data interoperability*. Note that for the data interoperability between Δ_1 and Δ_2 to be possible the data expressiveness E_{Δ_1} has to be contained in or equal to E_{Δ_2}. Otherwise, a database $D_2 \in I(\Delta_2)$ which is equivalent to a given database $D_1 \in I(\Delta_1)$ may not exist.

2.3 The TQuel Data Model

TQuel is a popular model for representing temporal data. (We chose TQuel over TSQL2 [Sno95] for the purpose of this presentation, because TQuel is simpler and TSQL2 is still in flux.) In the TQuel data model each relation contains two special attributes called *From* and *To* to represent *valid time*. The value of these temporal attributes must be integers or the special constants $-\infty$ or $+\infty$. The *From* and *To* values represent an interval. Such intervals in different tuples with identical nontemporal components have to be disjoint. (Another time dimension, transaction time, can also be present and is represented similarly to valid time. For simplicity we do not consider it here.)

Example 2.1 Table 1 is a TQuel representation of the unrestricted relation in Table 2.

Name	Company	From	To
Anderson	AT&T	1980	1993
Brown	IBM	1985	1996
Clark	Lotus	1990	1991

Table 1. TQuel DB researcher relation

The abstract semantics of a TQuel database is a relational database which has the same scheme as the TQuel database except the temporal attributes *To* and *From* are replaced with a single temporal attribute. For each TQuel tuple of the form $r(a_1, \ldots, a_k, b_1, b_2)$ the abstract model contains the tuples $r(a_1, \ldots, a_k, b_1), \ldots, r(a_1, \ldots, a_k, b_2)$. For example, the previous TQuel database has the abstract semantics:

2.4 The K-Spaghetti Data Model

The K-spaghetti data model [LT92] is a very popular model for representing spatial databases for CAD (Computer Aided Design) [KW87] and GIS (Geographic Information Systems)[Wor95] in K dimensions. In GIS applications typically $K = 2$ because the objects of interest are planar, while in CAD applications $K \geq 3$. The basic idea is to provide a general relational representation for geometric objects.

Name	Company	Year of Employment
Anderson	AT&T	1980
⋮	⋮	⋮
Anderson	AT&T	1993
Brown	IBM	1985
⋮	⋮	⋮
Brown	IBM	1996
Clark	Lotus	1990
Clark	Lotus	1991

Table 2. Unrestricted DB researcher relation

In this paper we concentrate on the 2-spaghetti (planar) data model. In this data model we can represent only objects that are composed of a finite set of closed polygons. As a matter of fact, each object can be represented as a set of triangles (some are degenerate triangles like line segments or points) where each triangle is represented by its three corners in a single relational database table. The issue of K-spaghetti for $K \geq 3$ is addressed in Section 6.

Example 2.2 Let us consider the figure shown below:

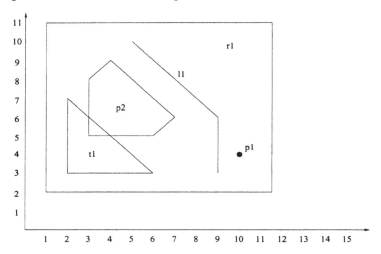

In the 2-spaghetti model the above figure is represented by the relation in Table 3.

Note that the rectangle is represented by two and the pentagon by three triangles. There are many good algorithms from computational geometry for triangulating polygons [PS85]. □

The abstract semantics of a 2-spaghetti data model is for each object the set of points that belong to the area of the plane that is within any of the triangles associated with that object. The spaghetti model comes with a set of built-in

ID	x	y	x'	y'	x''	y''
p1	10	4	10	4	10	4
l1	5	10	9	6	9	6
l1	9	6	9	3	9	3
t1	2	3	2	7	6	3
r1	1	2	1	11	11.5	11
r1	11.5	11	11.5	2	1	2
p2	0	0	0	0	4	9
p2	4	9	7	6	3	8
p2	3	5	7	6	3	8

Table 3. Triangular Representation

operators (e.g, overlap, intersection, etc.) that can be used in queries. These are of no concern to us here.

2.5 Worboys' Spatiotemporal Data Model

Worboys' spatiotemporal data model [Wor94] is a recent example of a data model that can represent both spatial and temporal information about objects in a database.

Example 2.3 Let us consider the following figure.

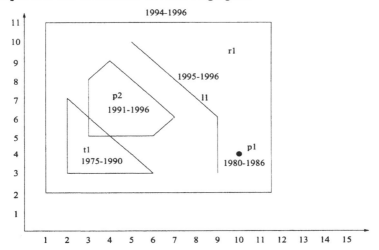

This is like the previous figure, but we added a temporal extent to each component of the figure. The temporal extent like in the TQuel data model tells us from which year to which year a component existed (valid time). For each component there could be several such intervals.

The above figure can be represented using a relation identical to the one in Table 3, except that two temporal attributes, *From* and *To*, are added to encode the appropriate intervals.

The abstract semantics of Worboys' data model is the cross product of the spatial semantics given for the 2-spaghetti data model and the temporal semantics given for the TQuel data model. An important feature of this model is that the temporal and the spatial arguments are independent. Hence we cannot describe for example the relationship that exists between time and the area covered by an incoming tide.

Worboys' model, like TQuel, allows one more dimension for time (transaction time) if necessary. Transaction time can be handled like valid time and we do not discuss it here.

2.6 Parametric 2-Spaghetti Data Model

This data model is a new model that we introduce in this paper. It generalizes Worboys' model by allowing an interaction between spatial and temporal attributes.

Example 2.4 Let us suppose that we have a rectangular area on a shore as shown in the figure below. A tide is coming in and the water level is shown as a line marked by the time the water rises to that line. The front edge of the tide water is a linear function of time.

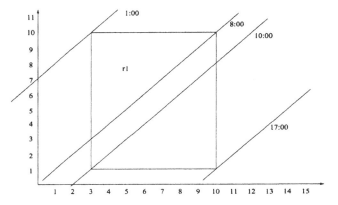

In this case the area flooded by the water will be a point at 1:00 am, a triangle at 8:00 am, a quadrangle at 10:00 am, and a pentagon at 1:00 pm. This data can be represented as in Table 4. Notice that this table is no longer a standard relation but rather a *parametric* one (with parameter t).

The abstract semantics of a parametric 2-spaghetti relation r is a possibly infinite set S, where each element of S is a relation consisting of all the instantiations of every tuple w in r by the values of the parameter t that fall between the *From* and *To* values in w and with the *From* and *To* columns dropped. Fixing a specific value for t yields a database which falls within the 2-spaghetti data model and has the same semantics.

ID	x	y	x'	y'	x''	y''	From	To
r1	3	10	3	10	3	10	1	$+\infty$
r1	3	10	3	11-t	2+t	10	1	8
r1	3	10	3	3	10	10	8	$+\infty$
r1	3	3	10	10	3	11-t	8	10
r1	10	10	3	11-t	10	18-t	8	10
r1	3	3	10	10	3	1	10	$+\infty$
r1	10	10	3	1	10	8	10	$+\infty$
r1	3	1	10	8	t-7	1	10	17
r1	t-7	1	10	18-t	10	8	10	17
r1	3	1	10	1	10	8	17	$+\infty$
r1	10	1	10	1	10	1	17	$+\infty$

Table 4. Parametric Triangular Representation

3 Data expressiveness

3.1 Constraint Databases

Definition 3.1 [KKR95] Let Φ be the set of atomic constraints of some constraint theory. A *generalized k-tuple* over variables $x_1, \ldots, x_k$ is of the form: $r(x_1, \ldots, x_k) :\!\!- \phi_1 \wedge \ldots \wedge \phi_n$ where r is a relation symbol, and $\phi_i \in \Phi$ for $1 \leq i \leq n$ and uses only the variables $x_1, \ldots, x_k$.
A *generalized relation r with arity k* is a finite set of generalized k-tuples with symbol r on left hand side.
A *generalized database* is a finite set of generalized relations. $\square$

Definition 3.2 [KKR95]. Let D be the domain over which variables are interpreted. Then the *model of a generalized k-tuple t* with variables $x_1, \ldots, x_k$ is the unrestricted k-ary relation $\{(a_1, \ldots, a_k) : (a_1, \ldots, a_k) \in D^k$ and the substitution of a_i for x_i satisfies the right hand side of t $\}$.
The *model of a generalized relation* is the union of the models of its generalized tuples.
The *model of a generalized database* is the set of the models of its generalized relations. $\square$
We consider the following classes of constraints:

- linear arithmetic constraints of the form $a_1 x_1 + \ldots a_k x_k \theta b$ where a_i and b are rational constants and x_i are variables on rational numbers and θ is one of $=, <, >, \leq, \geq$,

- order constraints $t_1 \theta t_2$ over integers where t_1 and t_2 are variables or constants (θ is one of $<, >, \leq, \geq$),

- equality constraints $t_1 = t_2$ over various domains where t_1 and t_2 are variables or constants.

We also consider restricted forms of constraints:

- unary order constraints of the form $x\theta c$ where x is a variable, c a constant and θ one of $<, >, \leq, \geq$),

- unary equality constraints of the form $x = c$ where x is a variable and c is a constant.

3.2 Constraint Basis

We use the constraint database framework to characterize data expressiveness of data models.

Definition 3.3 A *constraint basis* of a data model Δ is a class C of generalized databases such that every unrestricted database in $E_\Delta = \{\theta_\Delta(D) : D \in I(\Delta)\}$ is a model of some generalized database in C and vice versa.

In the following let $\alpha_1(x_1), \ldots, \alpha_n(x_n)$ be unary equality constraints over *thematic* (nonspatial and nontemporal) variables $x_1, \ldots, x_n$, $\phi(t)$ a conjunction of unary order constraints over a *temporal* variable t, and $\gamma(x, y)$ a conjunction of linear arithmetic constraints over *spatial* variables x and y. We assume that the planar object described by $\gamma(x, y)$ is closed and bounded.

Theorem 3.1 A constraint basis of the TQuel data model is a class of generalized databases with relations whose tuples are of the form

$$\alpha_1(x_1) \wedge \cdots \wedge \alpha_n(x_n) \wedge \phi(t).$$

Example 3.1 We can represent the relation in Table 2 as the following generalized relation:

$researcher(name, comp, t) :- name = "Anderson", comp = "AT\&T",$
$\qquad\qquad 1980 \leq t, t \leq 1993.$
$researcher(name, comp, t) :- name = "Brown", comp = "IBM",$
$\qquad\qquad 1985 \leq t, t \leq 1992.$
$researcher(name, comp, t) :- name = "Brown", comp = "IBM",$
$\qquad\qquad 1990 \leq t, t \leq 1996.$
$researcher(name, comp, t) :- name = "Clark", comp = "Lotus",$
$\qquad\qquad 1990 \leq t, t \leq 1991.$

Note that here two generalized tuples represent Brown's employment at IBM, while in the corresponding TQuel relation there was just one tuple with related information. In general the sets described by different generalized tuples need not be disjoint. For instance, different generalized tuples can represent information coming from different sources (this is common if multiple databases are interoperated).

Theorem 3.2 [VGVG95] A constraint basis of the 2-spaghetti data model consists of generalized databases with relations whose tuples are of the form

$$\alpha_1(x_1) \wedge \cdots \wedge \alpha_n(x_n) \wedge \gamma(x, y).$$

The planar object described by $\gamma(x, y)$ is supposed to be closed and bounded. Both conditions can be defined as first-order queries with linear arithmetic constraints, so the above constraint basis is effectively recognizable.

Example 3.2 For the 2-spaghetti model example, we can represent the abstract semantics by the following constraint database:

$object(p1, x, y) :- x = 10, y = 4.$
$object(l1, x, y) :- 5 \leq x, x \leq 0, y = x + 15.$
$object(l1, x, y) :- x = 9, 3 \leq y, y \leq 6.$
$object(t1, x, y) :- 2 \leq x, x \leq 6, 3 \leq y, y \leq 7, y \leq -x + 9.$
$object(r1, x, y) :- 1 \leq x, x \leq 11.5, 2 \leq y \leq 11.$
$object(p2, x, y) :- x \geq 3, y \geq 5, y \geq x - 1, y \leq x + 5, y \leq -x + 13.$

Theorem 3.3 A constraint basis of Worboys' spatiotemporal data model consists of generalized databases with relations whose tuples are of the form

$$\alpha_1(x_1) \wedge \cdots \wedge \alpha_n(x_n) \wedge \phi(t) \wedge \gamma(x, y).$$

(This means that in such relations the temporal attribute t and the spatial attributes x and y are syntactically independent in the sense of [CGK96].)

Example 3.3 We can represent the previous example of Worboys' model as the following generalized database.

$object(p1, x, y, t) :- x = 10, y = 4, 1980 \leq t, t \leq 1986.$
$object(l1, x, y, t) :- 5 \leq x, x \leq 9, y = -x + 15, 1995 \leq t, t \leq 1996.$
$object(l1, x, y, t) :- x = 9, 3 \leq y, y \leq 6, 1995 \leq t, t \leq 1996.$
$object(t1, x, y, t) :- 2 \leq x, x \leq 6, y \leq -x + 9, 3 \leq y, y \leq 7, 1975 \leq t, t \leq 1990.$
$object(r1, x, y, t) :- 1 \leq x, x \leq 11.5, 2 \leq y, y \leq 11, 1994 \leq t, t \leq 1996.$
$object(p2, x, y, t) :- x \geq 3, y \geq 5, y \geq x - 1, y \leq x + 5, y \leq -x + 13, 1991 \leq t,$
$\qquad t \leq 1996.$

For the parametric 2-spaghetti model we have only one-way containment.

Theorem 3.4 Every generalized database consisting of relations whose tuples are of the form

$$\alpha_1(x_1) \wedge \cdots \wedge \alpha_n(x_n) \wedge \psi(x, y, t)$$

where $\psi(x, y, t)$ is a conjunction of linear arithmetic constraints can be represented as a finite relation in the parametric 2-spaghetti model provided for each rational constant t_0, the formula $\psi(x, y, t_0)$ describes a closed bounded polygon[4].

The containment in the other direction does not hold, as shown by the following example.

Example 3.4 Consider the following parametric object (which is like the 4th tuple in Table 4).

---

[4] This condition can again be defined as a first-order query with linear arithmetic constraints.

ID	x	y	x'	y'	x''	y''	From	To
r1	3	3	10	10	3	11-t	8	10

This is a parametric triangle whose edge between the points $(3, 11 - t)$ and $(10, 10)$ has a slope that depends on t. This edge is contained in a line described by the equation:

$$7y = x(t - 1) + 10(8 - t).$$

The triangle cannot be specified using linear arithmetic constraints over x, y and t.

In many cases, however, a parametric object can be represented using linear arithmetic constraints.

Example 3.5 To represent Table 4, which is an instance of the parametric 2-spaghetti data model, note that flooded area is always described by the constraint $y \geq x + 8 - t$:

$$flooded(1, x, y, t) :- 1 \leq y, y \leq 10, 3 \leq x, x \leq 10, y \geq x + 8 - t.$$

We conjecture that as long as a relation in the parametric representation contains only objects whose boundaries involve only a finite number of slopes, there is a corresponding representation using linear arithmetic constraints. We think that the conjecture holds not only for triangular representations but also for representations involving rectangles, or higher degree convex polygons. In general, it is an open problem to precisely characterize relational representations of the class of generalized databases with relations whose tuples are of the form

$$\alpha_1(x_1) \wedge \cdots \wedge \alpha_n(x_n) \wedge \psi(x, y, t)$$

4 Data Translation Using Constraint Queries

4.1 Constraint wrapper

The notion of constraint basis defined in the previous section characterizes the *semantics* of a data model. Here we provide a constraint counterpart to the specific *syntax* of instances in this model.

Definition 4.1 A *constraint wrapper* for a data model Δ is a syntactically defined class C of generalized databases such that there is a simple correspondence between databases in $I(\Delta)$ and generalized relations in C.

By "simple correspondence" we mean that it is easy to construct the generalized relation in a wrapper of Δ from the instances in $I(\Delta)$ and vice versa. We are intentionally vague here, because we want to allow a broad class of wrappers. There may be more than one constraint wrapper for a data model. Unrestricted relations corresponding to a constraint wrapper of Δ may have a different arity than those corresponding to a constraint basis of Δ. For example, a constraint wrapper for TQuel consists of generalized relations whose elements are tuples over n data and two temporal variables t_1 and t_2. The temporal constraints in every such tuple are equality constraints of the form $t_i = c$ only.

Now the data translation between two data models Δ_1 and Δ_2 *such that the data expressiveness of Δ_1 is contained in or equal to the data expressiveness of Δ_2* is a composition of translations shown in Figure 1.

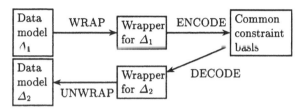

Fig. 1. Composition of translations

The WRAP/UNWRAP translations are outside the scope of the constraint database technology, because they are data-model-dependent. On the other hand, the ENCODE/DECODE translations are queries and may be expressible using constraint query languages. We expect many data models to share a common constraint basis. The interoperability of such data models is greatly simplified: instead of constructing $n(n-1)$ direct data translations between every pair of data models, it is enough to construct the WRAP/UNWRAP translations for every model, and ENCODE/DECODE for every constraint wrapper (at most $4n$ translations).

4.2 TQuel

A constraint wrapper for TQuel consists of constraint databases whose relations contain generalized tuples of the form

$$x_1 = c_1 \wedge \cdots \wedge x_n = c_n \wedge t_1 = a \wedge t_2 = b$$

where a and b are integers, $+\infty$ or $-\infty$. A constraint basis for TQuel consists of constraint databases whose relations contain generalized tuples of the form

$$x_1 = c_1 \wedge \cdots \wedge x_n = c_n \wedge \phi(t)$$

where $\phi(t)$ is a conjunction of order constraints over the integers.

Theorem 4.1 The ENCODE/DECODE translations for TQuel can be expressed as first-order constraint queries with order and equality constraints.

Proof: The ENCODE translation has to handle the difference of arities ($n+2$ versus. $n+1$) and eliminate $+\infty$ and $-\infty$. Let P be a generalized relation of the constraint wrapper. We define the corresponding generalized relation of the constraint basis using a relational calculus query $\gamma_P(\bar{x}, t)$ defined as

$$\exists t_1, t_2.(P(\bar{x}, t_1, t_2) \wedge \alpha(t_1, t_2, t))$$

where $\alpha(t_1, t_2, t)$ is

$$(t_1 \neq -\infty \ \wedge \ t_2 \neq +\infty \ \wedge \ t_1 \leq t \leq t_2 \vee t_1 \neq -\infty \ \wedge \ t_2 = +\infty \ \wedge \ t_1 \leq t \ \vee$$
$$t_1 = -\infty \ \wedge \ t_2 \neq +\infty \ \wedge \ t \leq t_2 \vee t_1 = -\infty \ \wedge \ t_2 = +\infty).$$

The DECODE translation is more complicated, as it involves coalescing tuples with the same nontemporal components (and nondisjoint intervals) and generating $+\infty$ and $-\infty$ where appropriate. Let R be a generalized relation of the constraint basis. We define the corresponding generalized relation of the constraint wrapper using a (domain) relational calculus query $\beta_R(\bar{x}, t_1, t_2)$ defined as

$$\beta_1(\bar{x}, t_1, t_2) \vee \beta_2(\bar{x}, t_2) \ \wedge \ t_1 = -\infty \vee \beta_3(\bar{x}, t_1) \ \wedge \ t_2 = +\infty \ \vee$$
$$\beta_4(\bar{x}) \ \wedge \ t_1 = -\infty \ \wedge \ t_2 = +\infty$$

where the query $\beta_1(\bar{x}, t_1, t_2)$ specifies tuples with bounded intervals:

$$\forall t.(t_1 \leq t \leq t_2 \Rightarrow R(\bar{x}, t) \ \wedge \ \exists t_3.(t_3 < t_1 \ \wedge \ \forall t_0.(t_3 \leq t_0 < t_1 \Rightarrow \neg R(\bar{x}, t_0))) \ \wedge$$
$$\exists t_4.(t_2 < t_4 \ \wedge \ \forall t_0.(t_2 < t_0 \leq t_4 \Rightarrow \neg R(\bar{x}, t_0)))).$$

The query $\beta_2(\bar{x}, t_2)$ specifies tuples with intervals unbounded to the left, $\beta_3(\bar{x}, t_1)$ tuples with intervals unbounded to the right, and $\beta_4(\bar{x})$ tuples unbounded on both sides. These queries can be defined similarly to $\beta_1(\bar{x}, t_1, t_2)$. $\square$

4.3 2-spaghetti

Lemma 4.1 Assume that closed convex polygons are represented as a generalized relation $object(i, x, y)$ with linear arithmetic constraints over x and y (i is the object identifier). The following relations:

- $boundary(i, x, y) \equiv$ the point (x, y) is on the boundary of a closed convex polygon i,

- $corner(i, x, y) \equiv$ the point (x, y) is a corner vertex of a closed convex polygon i,

- $edge(i, x, y, x', y') \equiv$ the points (x, y) and (x', y') form an edge of the boundary of a closed convex polygon i

can be defined as first-order queries with linear arithmetic constraints.

   Proof: (*Boundary.*) For a closed convex polygon each point on the boundary is a point of this polygon that is not an inside point of i. A point is an inside point iff it is inside a rectangle wholly contained within i:

$$inside(i, x, y) \Leftrightarrow object(i, x, y) \wedge \exists x', y', x'', y''.(x' < x < x'') \wedge (y' < y < y'') \wedge$$
$$\forall x''', y'''.((x' < x''' < x'') \wedge (y' < y''' < y'') \Rightarrow object(i, x''', y'''))$$

(*Corner and edge.*) We define two auxiliary relations (the rectangles are specified by their bottom-left and top-right corners):

- $iso(i, x, y, x', y', z, w, z', w') \equiv$ the piece of the planar object i contained in the rectangle (x, y, x', y') is the result of translating the piece of the same object contained in the rectangle (z, w, z', w') by the vector $(z - x, w - y)$.

$iso(i, x, y, x', y', z, w, z', w') \Leftrightarrow$

$$z' = x' + z - x \wedge w' = y' + w - y \wedge \forall s.\forall t.x \leq s \leq x' \wedge y \leq t \leq y' \Rightarrow$$
$$(object(i, s, t) \Leftrightarrow object(i, s + z - x, t + w - y))$$

- $nc(i, x, y) \equiv (x, y)$ is a boundary point of a closed convex polygon i but not a corner vertex of i.

$nc(i, x, y) \Leftrightarrow$

$$boundary(i, x, y) \wedge \exists c.\exists d.boundary(i, x + c, y + d) \wedge$$
$$iso(i, x - c, y - d, x, y, x, y, x + c, y + d)$$

This takes care of all the slopes except 0 and ∞ which are easily handled by separate formulas whose disjunction is then taken. Now

$$corner(i, x, y) \Leftrightarrow boundary(i, x, y) \wedge \neg nc(i, x, y).$$

Finally, $edge(i, x, y, x', y') \Leftrightarrow$

$$\forall c > 0.\forall d > 0.\exists s, s', t, t', z, z', w, w'.\ x \leq s \leq x + c \wedge x \leq s' \leq x + c \wedge s < s' \wedge$$
$$y \leq t \leq y + d \wedge y \leq t' \leq y + d \wedge t < t' \wedge$$
$$x' - c \leq z \leq x' \wedge x' - c \leq z' \leq x' \wedge z < z' \wedge$$
$$y' - d \leq w \leq y' \wedge y' - d \leq w' \leq y' \wedge$$
$$w < w' \wedge iso(i, s, t, s', t', z, z', w, w') \wedge$$
$$boundary(i, s, t) \wedge boundary(i, s', t') \wedge$$
$$boundary(i, z, w) \wedge boundary(i, z', w')$$

This handles any positive slope. Negative slopes are handled similarly. The slopes 0 and ∞ are again easy. $\Box$

A constraint wrapper for 2-spaghetti consists of databases with relations whose tuples are of the form

$$x_1 = c_1 \wedge \cdots \wedge x_n = c_n \wedge x = a \wedge y = b \wedge x' = a' \wedge y' = b' \wedge x'' = a'' \wedge y'' = b''.$$

Theorem 4.2 The DECODE translation for 2-Spaghetti can be expressed as a first-order constraint query with linear arithmetic constraints.

Proof. We assume that every generalized tuple has a distinct tuple identifier (this is to guarantee convexity). The construction in Lemma 4.1 gives the corner vertices and the edges of the corresponding polygon. The triangulation (necessary for our 2-spaghetti representation) can also be described in a first-order way by picking an arbitrary vertex v_0 (e.g., the least in the lexicographic ordering of all vertices) and constructing all nondegenerate triangles (v_0, v_1, v_2) such that (v_1, v_2) is an edge. $\Box$

Theorem 4.3 The ENCODE translation for 2-Spaghetti can be expressed as a first-order constraint query with polynomial inequality constraints.

Proof: The translation of points and line segments is straightforward. Let us look now at the translation of triangles. Let us assume that the corners of the triangle are $(x_1, y_1), (x_2, y_2)$ and (x_3, y_3). First note that $y_3 > \frac{y_2 - y_1}{x_2 - x_1}(x_3 - x_1) + y_1$ holds if and only if $y > \frac{y_2 - y_1}{x_2 - x_1}(x - x_1) + y_1$ holds for each point (x, y) within the triangle. This condition expresses that the points of the triangle are above (or below) the line segment defined by the points (x_1, y_1) and (x_2, y_2). Similar conditions hold for the other sides of the triangle. Hence let us define this as $find\_side(x, y, x_1, y_1, x_2, y_2, x_3, y_3) \Leftrightarrow$

$$((y_3 > \tfrac{y_2 - y_1}{x_2 - x_1}(x_3 - x_1) + y_1 \wedge y > \tfrac{y_2 - y_1}{x_2 - x_1}(x - x_1) + y_1) \vee$$
$$(y_3 < \tfrac{y_2 - y_1}{x_2 - x_1}(x_3 - x_1) + y_1 \wedge y < \tfrac{y_2 - y_1}{x_2 - x_1}(x - x_1) + y_1))$$

Now each triangle can be translated as

$$\exists x_1, y_1, x_2, y_2, x_3, y_3. \; triangle(o, x_1, y_1, x_2.y_2, x_3, y_3) \wedge$$
$$(x_1 \neq x_2 \vee y_1 \neq y_2) \wedge$$
$$(x_1 \neq x_3 \vee y_1 \neq y_3) \wedge (x_3 \neq x_2 \vee y_3 \neq y_2) \wedge$$
$$find\_side(x, y, x_1, y_1, x_2, y_2, x_3, y_3) \wedge$$
$$find\_side(x, y, x_2, y_2, x_3, y_3, x_1, y_1) \wedge$$
$$find\_side(x, y, x_3, y_3, x_1, y_1, x_2, y_2)$$

Note that the generalized relation resulting from translating any 2-spaghetti relation will contain only linear constraints. This is because even though the query involves quadratic constraints, the variables $x_1, x_2, x_3, y_1, y_2, y_3$ will be all replaced by constants. $\square$

Theorem 4.4 The DECODE translation for Worboys' spatiotemporal model can be expressed as a first-order constraint query with linear arithmetic constraints. The ENCODE translation for Worboys' spatiotemporal model can be expressed as a first-order constraint query with polynomial arithmetic constraints.

Proof: The DECODE construction is a combination of Theorems 4.1 and 4.2. Note that the independence of spatial and temporal attributes is crucial here. The ENCODE construction is a combination of Theorems 4.1 and 4.3. The independence of spatial and temporal attributes is again crucial here. $\square$

Theorem 4.5 The DECODE translation for the parametric 2-Spaghetti can be expressed as a first-order constraint query with linear arithmetic constraints.

Proof: A construction similar to Theorem 4.2 can be used here with the provision of an extra temporal argument in all predicates. $\square$

5 Related Work

Interoperability between a GIS database and application programs was studied in [SW93]. Interoperability in the sense of combining spatial and attribute data managers for the implementation of GIS systems was studied in [KNP93]. In this paper we are addressing a different issue, namely the data and query interoperability among various temporal and spatial data models.

Data interoperability of different temporal data models was studied in [JSS93]. The solution proposed there, the provision of a single unifying temporal data model to which other models could be mapped, is not sufficient. This still falls short of defining a *representation-independent* abstract semantics for temporal databases. In fact, the model of [JSS93] uses another notion of concrete temporal database, admittedly simpler and more general than others. This model is still limited in its data expressiveness as it is capable of representing only finite unrestricted databases. We also think that the model is unnecessarily complicated because it introduces a new notion of "bitemporal element". Moreover, translations between this model and other temporal data models are expressed using an ad-hoc procedural language.

Current work on the interoperability of temporal databases, e.g., Wang, Jajodia and Subrahmanian [WJS93, BWBJ95], addresses similar concerns as the present paper. However, the paper [WJS93] is limited to temporal databases and makes very strong assumptions about the concrete temporal databases that are to be interoperated. In particular, such databases have to provide a unified interface. This is not necessary in our approach. Moreover, the data expressiveness of the cited model is limited to sets of finite unrestricted databases. A follow-up work [BWBJ95] demonstrates a systematic approach of deriving implicit temporal information from the explicit information stored in a temporal database. Such derivations could very well be incorporated into our framework.

There is a quickly growing list of papers on various aspects of constraint databases. Among those, the papers that deal with specific constraint query languages and their properties [ACGK94, CK95, GS94, GS95, KG94, KKR95, KSW95, Kup94, Las90, PVdBVG94, Rev95, Sri93, VGVG95], those that address temporal [Cho94, KSW95, TCR94] and spatial [BJM93, BK95, BLLM95] database applications, as well as those that discuss implementation techniques [RS93, Tom95] are clearly relevant to the proposed work. However, none of the papers listed above deals with the issue of database interoperability and its specific problems.

6   Conclusions and Future Work

We have presented a novel way to apply the constraint database technology to temporal, spatial and spatiotemporal database applications. Constraint databases are used as a layer mediating between different data models. Constraint query languages are used to formulate the translations between the models.

Implementation. The research reported in this paper is just a first step in the direction of making spatiotemporal databases interoperable. We plan to implement the translations described here and address the practical issues involved in this task. One of these issues is *constraint simplification*: in Section 4 we didn't describe how the generalized databases resulting from a DECODE translation are simplified to yield databases in the appropriate constraint wrapper form.

Generalizations. Based on the framework proposed here, we plan to investigate a broader range of data models, as well as the issue of query interoperability. The spaghetti model can be extended to higher spatial dimensions [Wor95]. In higher dimensions each K-dimensional object is represented as a set of $K - 1$-dimensional facets. For representing this containment we need to have a separate table describing the object containment hierarchy. We believe that our techniques generalize to arbitrary fixed spatial (or temporal) dimensions.

Extensibility. Application-dependent interoperability issues of resolving semantic and representational mismatches and conflict detection and resolution have found a very elegant formulation using the language of first-order logic [QL94]. Thus, solutions to these problems can be seamlessly integrated with the techniques we propose for the translation between different data models. The problem of incomplete information can also be addressed in the same framework.

References

[ACGK94]   F. Afrati, S. Cosmadakis, S. Grumbach, and G. Kuper. Linear vs. Polynomial Constraints in Database Query Languages. In A. Borning, editor, *PPCP'94, Second International Workshop on Principles and Practice of Constraint Programming*, pages 181–192. Springer-Verlag, LNCS 874, 1994.

[AS91]   W.G. Aref and H. Samet. Extending a DBMS with Spatial Operations. In *International Symposium on Large Spatial Databases*, pages 299–318, 1991.

[BCW93]   M. Baudinet, J. Chomicki, and P. Wolper. Temporal Deductive Databases. In Tansel et al. [TCG+93], pages 294–320.

[BJM93]   A. Brodsky, J. Jaffar, and M.J. Maher. Towards Practical Constraint Databases. In *International Conference on Very Large Data Bases*, Dublin, Ireland, 1993.

[BK95]   A. Brodsky and Y. Kornatzky. The LyriC Language: Constraining Objects. In *ACM SIGMOD International Conference on Management of Data*, San Jose, California, 1995.

[BLLM95]   A. Brodsky, C. Lassez, J-L. Lassez, and M.J. Maher. Separability of Polyhedra for Optimal Filtering of Spatial and Constraint Data. In *ACM Symposium on Principles of Database Systems*, San Jose, California, 1995.

[BNW91]   M. Baudinet, M. Niézette, and P. Wolper. On the Representation of Infinite Temporal Data and Queries. In *ACM Symposium on Principles of Database Systems*, 1991.

[BWBJ95]   C. Bettini, X.S. Wang, E. Bertino, and S. Jajodia. Semantic Assumptions and Query Evaluation in Temporal Databases. In *ACM SIGMOD International Conference on Management of Data*, pages 257–268, San Jose, California, May 1995.

[CGK96]   J. Chomicki, D. Goldin, and G. Kuper. Variable Independence and Aggregation Closure. In *ACM Symposium on Principles of Database Systems*, Montreal, Canada, June 1996.

[Cho94]   J. Chomicki. Temporal Query Languages: A Survey. In D.M. Gabbay and H.J. Ohlbach, editors, *Temporal Logic, First International Conference*, pages 506–534. Springer-Verlag, LNAI 827, 1994.

[CK95]   J. Chomicki and G. Kuper. Measuring Infinite Relations. In *ACM Symposium on Principles of Database Systems*, pages 78–85, San Jose, California, 1995.

[FCF92]   A.U. Frank, I. Campari, and U. Formentini. *Theories and Methods of Spatio-Temporal Reasoning in Geographic Space*. Springer-Verlag, LNCS 639, 1992.

[GS94]   S. Grumbach and J. Su. Finitely Representable Databases. In *ACM Symposium on Principles of Database Systems*, pages 289–300, Minneapolis, Minnesota, May 1994.

[GS95]   S. Grumbach and J. Su. Dense-Order Constraint Databases. In *ACM Symposium on Principles of Database Systems*, pages 66–77, San Jose, California, May 1995.

[JSS93]   C.S. Jensen, M.D. Soo, and R.T. Snodgrass. Unification of Temporal Data Models. In *IEEE International Conference on Data Engineering*, 1993.

[KCGS95]   W. Kim, I. Choi, S. Gala, and M. Scheevel. On Resolving Semantic Heterogeneity in Multidatabase Systems. In W. Kim, editor, *Modern Database Systems*, pages 521–550. Addison-Wesley, 1995.

[KG94]   P.C. Kanellakis and D. Q. Goldin. Constraint Programming and Database Query Languages. In *Conference on Theoretical Aspects of Computer Software*. Springer-Verlag, 1994.

[KKR95]   P.C. Kanellakis, G.M. Kuper, and P.Z. Revesz. Constraint Query Languages. *Journal of Computer and System Sciences*, 51(1):26–52, August 1995.

[KSW95]   F. Kabanza, J-M. Stevenne, and P. Wolper. Handling Infinite Temporal Data. *Journal of Computer and System Sciences*, 51(1):3–17, August 1995.

[KNP93]   C.P.Kolovson, M-A. Neimat, and S. Potamianos. Interoperability of Spatial and Attribute Data Managers: A Case Study. *Proc. Symp. on Spatial Databases*, 239-264, Singapore, June 1993.

[Kup94]   G. Kuper. Aggregation in Constraint Databases. In *PPCP'93, First International Workshop on Principles and Practice of Constraint Programming*, pages 161–172. MIT Press, 1994.

[KW87]   A. Kemper and M. Wallrath. An analysis of geometric modeling in database systems. *ACM Computer Surveys*, 19(1), 1987.

[Las90]   J-L. Lassez. Querying Constraints. In *ACM Symposium on Principles of Database Systems*, pages 288–298, Nashville, Tennessee, April 1990.

[LT92]   R. Laurini and D. Thompson. *Fundamentals of Spatial Information Systems*. Academic Press, 1992.

[Par95]   J. Paredaens. Spatial Databases, The Final Frontier. In *International Conference on Database Theory*, pages 14–32, Prague, Czech Republic, January 1995. Springer-Verlag, LNCS 893.

[PS85]   F. Preparata and M. Shamos. *Computational Geometry: An Introduction*. Springer-Verlag, 1985.

[PVdBVG94] J. Paredaens, J. Van den Bussche, and D. Van Gucht. Towards a Theory of Spatial Database Queries. In *ACM Symposium on Principles of Database Systems*, pages 279–288, Minneapolis, Minnesota, 1994.

[QL94] X. Qian and T.F. Lunt. Semantic Interoperation: A Query Mediation Approach. Technical Report SRI-CSL-94-02, Computer Science Laboratory, SRI International, April 1994.

[Rev95] P. Z. Revesz. Datalog Queries of Set Constraint Databases. In *International Conference on Database Theory*, Prague, Czech Republic, January 1995. Springer-Verlag.

[RS93] R. Ramakrishnan and D. Srivastava. Pushing Constraint Selections. *Journal of Logic Programming*, 16(3&4):361–414, 1993.

[RS94] A. Rosenthal and L.J. Seligman. Data Integration in the Large: The Challenge of Reuse. In *International Conference on Very Large Data Bases*, pages 669–675, 1994.

[SW93] H.J. Sheck, and A. Wolf. From Extensible Databases to Interoperability between Multiple Databases and GIS Applications. *Proc. Symp. on Spatial Databases*, 207-238, Singapore, June 1993.

[Sno87] R. Snodgrass. The Temporal Query Language TQuel. *ACM Transactions on Database Systems*, 12(2):247–298, June 1987.

[Sno95] R.T. Snodgrass, editor. *The TSQL2 Temporal Query Language*. Kluwer Academic Publishers, 1995.

[Sri93] D. Srivastava. Subsumption and Indexing in Constraint Query Languages with Linear Arithmetic Constraints. *Annals of Mathematics and Artificial Intelligence*, 1993.

[TCG$^+$93] A. Tansel, J. Clifford, S. Gadia, S. Jajodia, A. Segev, and R. Snodgrass, editors. *Temporal Databases: Theory, Design, and Implementation*. Benjamin/Cummings, 1993.

[TCR94] D. Toman, J. Chomicki, and D.S. Rogers. Datalog with Integer Periodicity Constraints. In *International Logic Programming Symposium*, pages 189–203, Ithaca, New York, November 1994. MIT Press.

[Tom95] D. Toman. Top-Down Beats Bottom-Up for Constraint Based Extensions of Datalog. In *International Logic Programming Symposium*, Portland, Oregon, December 1995. MIT Press.

[VGVG95] L. Vandeurzen, M. Gyssens, and D. Van Gucht. On the Desirability and Limitations of Linear Spatial Database Models. In *International Symposium on Large Spatial Databases*, pages 14–28, 1995.

[WJS93] X.S. Wang, S. Jajodia, and V.S. Subrahmanian. Temporal Modules: An Approach Toward Federated Temporal Databases. In *ACM SIGMOD International Conference on Management of Data*, pages 227–236, 1993.

[Wor94] M.F. Worboys. A Unified Model for Spatial and Temporal Information. *Computer Journal*, 37(1):26–34, 1994.

[Wor95] M.F. Worboys. *GIS: A Computing Perspective*. Taylor&Francis, 1995.

Spatial Query Processing

Improving Spatial Intersect Joins Using Symbolic Intersect Detection

Yun-Wu Huang[1], Matthew Jones[1] and Elke A. Rundensteiner[2]

[1] University of Michigan, Ann Arbor MI 48109, USA
[2] Worcester Polytechnic Institute, Worcester, MA 01609, USA

Abstract. We introduce a novel technique to drastically reduce the computation required by the refinement step during spatial intersect join processing. This technique, called Symbolic Intersect Detection (SID), detects most of the true hits during a spatial intersect join by scrutinizing symbolic topological relationships between candidate polygon pairs. SID boosts performance by detecting true hits early during the refinement step, thus avoiding expensive polygon intersect computations that would otherwise be required to detect the true hits. Our experimental evaluation with real GIS map data demonstrates that SID can identify more than 80% of the true hits with only minimal overhead. Consequently, SID outperforms known techniques for resolving polygon intersection during the refinement step by more than 50%. Most state-of-the-art methods in spatial join processing can benefit from SID's performance gains because the SID approach integrates easily into the established two-phase spatial join process.

1 Introduction

The requirement for spatial data management in Geographic Information Systems (GIS), Cartography, image processing, and CAD/CAM has driven the fast-increasing market demand for spatial databases. Spatial join processing is critical in spatial databases because these applications require many classes of queries such as distance associated joins, orientation associated joins, and spatial intersect joins. Spatial intersect joins are used, for example, for *map overlap* [3, 9] which merges two map layers into a single layer consisting of the overlapping regions, and for *intersection detection* which returns all objects in one data set that intersect at least one object from a second data set. For example, "Find all road segments that intersect flooding areas" is a crucial spatial join operation used for path finding with spatial constraints (such as avoiding flooding areas) [7]. This paper focuses on a technique to improve the performance of queries requiring intersection detection[3].

To reduce cost, spatial join processing typically executes in two steps [12]:

- The *filter step* uses a conservative geometric approximation, such as the Minimum Bounding Rectangle (MBR), to represent each object. It then

---

[3] In the remainder of this paper, we use the term spatial joins to mean spatial joins requiring intersection detection.

processes a spatial join based only on the geometric approximations. The result is a set of object-pairs (called the candidate set) whose respective approximations intersect each other.

- The *refinement step* retrieves the full spatial data representation of the two objects in each candidate set tuple and runs a precise intersect computation algorithm to determine if the two objects really intersect. The most widely used algorithm for precise intersect computation is called the *plane sweep* algorithm [2, 14, 16] whose average time complexity is $O(n \times log(n))$ where n is the number of points representing the objects.

In summary, the two-step spatial join process benefits from two essential optimizations. One, the filter step reduces disk I/O by eliminating many object-pairs that *cannot* intersect before having to retrieve their storage-intensive spatial object representations. Two, the filter step attempts to reduce the CPU costs for the precise geometric intersection required in the refinement step.

Recent experiments conducted over GIS map data [13] reveal that the refinement step is more expensive (in terms of overall processing time) than the filter step, and most of the cost in processing the refinement step is on intersect computation. Spatial join processing therefore is a CPU-intensive process in which the intersect computation in the refinement step, based on current technology, is likely to be a bottleneck. The situation is even worse for fine-grained (high-resolution) map data[4] because while the filter step is independent of map resolution, the cost of the refinement step increases super-linearly as the resolution increases. To account for the emergence of map data with higher resolutions (e.g., images generated from fine-grained NASA measurement instruments), we are investigating improvements on spatial join processing that are scalable in handling such fine-grained data sets.

While most recent spatial join research [1, 6, 8, 13] presented improvements on the filter step, this paper focuses on the optimization of the refinement step. We propose a screen-test procedure to be executed before the *plane sweep* algorithm that substantially reduces the computation required during refinement. We call this procedure Symbolic Intersect Detection (SID). For each object-pair in the candidate set, SID uses the overlapping MBR (OMBR)[5] to clip all the segments of the two polygons which overlap the area bounded by the OMBR. During SID optimization, we represent each clipped segment in a compact symbolic form that identifies where the segments intersect the OMBR. SID generates two sets of such symbolic clipped segments, one for each object in the object-pair. Next, based on the symbolic representation of the clipped segments, SID can determine in constant time in many cases when any two clipped segments, one from each set, intersect each other. As a consequence, SID eliminates the expensive *plane sweep* computation for these detected true hits.

---

[4] A fine-grained map means its objects are represented by large numbers (hundreds or more) of points.

[5] The OMBR of an object-pair is the rectangle which is the overlap between the MBRs of the two objects.

We have evaluated SID using real, fine-grained GIS data sets and have observed that SID can identify a very high percentage ($>$ 80%) of true hits. Due to this high true hit detection, SID improves the performance of the refinement step by more than 50% over established techniques.

The outline of the remainder of this paper is as follows: In Section 2, we describe the background and related work on spatial join processing. Our proposed Symbolic Intersect Detection approach is introduced in Section 3. We present the experimental evaluation in Section 4 and conclusions in Section 5.

2 Background on Spatial Joins

2.1 Related Work

Recently, much research has focused on spatial join processing. In [11], the z-ordering technique is used to transform multi-dimensional data into the 1-dimensional domain. A spatial join is then conducted on the B^+-tree structures that store z-ordering values of the spatial data. In [15], spatial join indexes are computed using Grid files [10] to index the spatial data. In [4], a model of the generalization tree is proposed to compare the tree-based spatial joins with the alternative approaches using cost estimation. Spatial joins based on depth-first traversal of R-trees [5] were proposed in [1]. In [6], we developed an alternative R-tree join technique based on breadth-first traversal which was shown to have a better overall performance than the depth-first approach.

Recent spatial join research has focused on joining spatial data when the associated spatial indexes do not exist for both data sets. In [8], a spatial hash join is proposed that uses spatial partitioning as the hash function. Parallel to the work of [8], a similar partition-based spatial-merge join is proposed in [13].

All above mentioned research focuses mainly on developing a more efficient filtering step (i.e., to optimize false hit reduction). To our knowledge, the only related work that achieves a significant percentage of true hit reduction is the progressive approximation techniques presented in [2]. Their approach achieves roughly 35% true hit detection in the filter step. However, unlike SID, their approach incurs additional pre-computation and storage costs in order to create and store the additional approximation for each object.

Work in [2] proposed another approach using trapezoid decomposition as the spatial representation for polygons. Although very efficient in spatial join processing, such an approach incurs even more extensive pre-computation and storage costs. Furthermore, it requires creating a spatial representation for polygons that is incompatible with the vector format most commonly used in spatial data applications.

In contrast, our SID approach not only achieves an impressively higher true hit detection rate ($>$ 80%, see Section 4), but it is also independent of the access structures deployed in the filter step and is based on the vector representation of the polygons. Therefore, SID is compatible with the popular spatial data (index and storage) structures and can be used in conjunction with almost all spatial join optimizations of the filter step, particularly those techniques that focus on early detection of false hits [1, 2, 6, 8, 13].

Our focus in this paper is on the computationally more expensive refinement step. Therefore, for the purpose of this paper, we assume a *candidate set* has already been computed (by any of the popular techniques) and is available as input to the refinement step. The remainder of this section gives background of the refinement step processing adopted by most recent spatial join techniques [2, 14, 10].

2.2 The State-of-the-Art Refinement Processing

Upon retrieving from the candidate set a polygon-pair and their vector representations, the state-of-the-art refinement approach [2, 13] determines the *intersect* relation between the two polygons by processing the *restricting* phase and the *sweeping* phase (see Figure 1a).

The Restricting Phase. If a polygon-pair in the *candidate set* is a true hit (its two polygons intersect), the area of intersection must be enclosed by their OMBR. Therefore, any line segments of one polygon that do not overlap[6] the area bounded by the OMBR will not intersect any line segments in the other polygon. Therefore, the *restricting* phase eliminates the line segments that do not overlap the OMBR. Only the ones that overlap the OMBR are retained as input to the next phase: the *sweeping phase*. If no line segment in either polygon is marked, the two polygons do not intersect, and the candidate pair is eliminated from further computation.

The Sweeping Phase. During the *sweeping phase*, the query processor runs the *plane sweep* algorithm [2, 14, 16] to determine whether or not the candidate polygons intersect. The *plane sweep* algorithm sorts the marked line segments from both vectors by the x-coordinate values of their left-most vertex. Next, the algorithm "sweeps" a vertical line across all sorted line segments from left to right. The sweep process evaluates the sorted line segments by stopping at the left-most vertex of each segment it encounters. At each of these vertices, the algorithm conducts an *intersect* test between the encountered line segment and all line segments from the other polygon whose MBRs intersect that of this line segment. If at any time an intersection is found, the sweep process terminates and the candidate satisfies the *intersect* relation. If the sweep process exhausts all line segments and does not find an intersection, the candidate is discarded as the two candidate polygons cannot intersect.

Although the number of iterations for the inner loop of the *plane sweep* algorithm can be large in the worst case, our experimental results show that the average number of iterations is a small constant (close to 3 with our GIS map data sets). The average time complexity to perform the *sweeping phase* on real map data therefore is bounded by the cost to sort the line segments $(O(n \times log(n)))$ [2] received from the *restricting* phase.

---

[6] We use *overlap* to mean *intersects* or *contains* or *is contained by*.

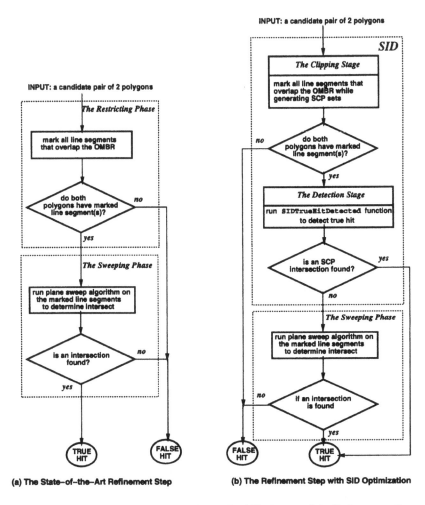

Fig. 1. The State-of-the-Art Approach vs. the SID Approach in Refinement Step.

3 The Symbolic Intersect Detection (SID) Optimization

We now introduce a novel approach, called Symbolic Intersect Detection (SID), which improves the performance of the refinement step. While our discussion focuses on the *boundary intersect* computation[7], SID can be easily extended to handle generalized intersect predicates such as *area intersect*.

---

[7] The set of object-pairs that satisfy the *area intersect* predicate are the union of those that satisfy the *boundary intersect* predicate and those that satisfy the *contain* predicate. The most common technique in resolving the *contain* predicate is called the point test [14] which has a linear time complexity in the length of the vector representation. Therefore, the *contain* test can be much less expensive than the *intersect* test which dominates the cost in *area intersect* joins.

3.1 Overview

Figure 1 contrasts the state-of-the-art refinement step (henceforth called the state-of-the-art approach) with the refinement that incorporates our SID optimization (the SID approach) presented in the remainder of this section. As shown in Figure 1, the SID approach improves the state-of-the-art approach by directly replacing the *restricting* phase with our proposed SID optimization.

The basis for SID stems from our observation that clipping the two candidate polygons using their OMBR as a window creates two sets of polygon segments, one for each polygon. By examining the topological relationship between these polygon segments and the OMBR, in many cases, one can detect without ambiguity that two polygons intersect. This technique improves performance during the refinement step by reducing the number of polygons that must be passed on to the compute-intensive *plane sweep* algorithm. Our SID optimization is comprised of two stages, the *clipping* stage and the *detection* stage.

3.2 The Clipping Stage of SID

Clipped Polygon Segments. By clipping a candidate polygon with the OMBR, we create a set of polygon segments (in the shape of an *arc*) enclosed by the OMBR, that are parts of the boundaries of this polygon. Each polygon segment is entirely enclosed by the OMBR with its two end points on the OMBR boundary. For example, in Figure 2, during the *clipping* stage while evaluating the candidate polygons r and s, three clipped polygon segments are created. They are x and y of polygon r, and z of polygon s.

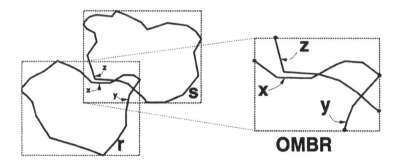

Fig. 2. Clipping the Polygon Segments Using OMBR.

Classification of Clipped Polygon Segments. Based on the topological relationship between the OMBR and the two end points of the clipped polygon segments, we categorize these clipped polygon segments into 10 classes. These 10 classes are **Vertical, Horizontal, Left, Up, Right, Down, UpperLeft, Up-perRight, LowerRight**, and **LowerLeft**. Figure 3 illustrates these ten classes. This classification is complete in the sense that all possible cases of clipped polygon segments are covered by this set.

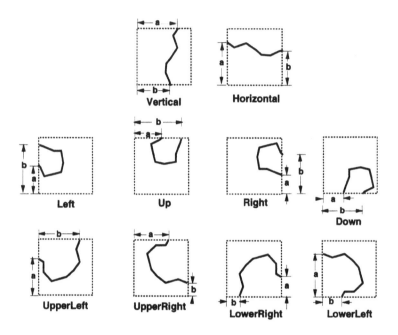

Fig. 3. The Ten Classes of Clipped Polygon Segments.

Symbolic Clipped Primitives (SCP). In SID, each clipped polygon segment is modeled by a symbolic representation we call a Symbolic Clipped Primitive (SCP). Each SCP is represented by a 3-tuple $< class, a, b >$. The first item in the tuple *class* is one of the ten classes (shown in Figure 3) to which the polygon segment modeled by this SCP belongs. The values a and b in $< class, a, b >$ are offsets from the corners of the OMBR that identify the location of the segment endpoints. For each SCP, the locations of the corners, from which the offsets a and b are measured from, are determined by the class of the SCP. The relevant corners and offsets for each SCP class are illustrated in Figure 3.

The Clipping Stage Creates Two SCP Sets. During the *clipping* stage, an SCP is generated for each clipped polygon segment. All the SCPs for a candidate polygon are then stored in a set which we call the SCP set. The output of the *clipping* stage therefore corresponds to two SCP sets, one for each candidate polygon.

172

3.3 The Detection Stage of SID

Detecting Intersection Using SCPs. After the two SCP sets are created for the candidate polygons during the *clipping* stage, SID performs true hit detection by a simple comparison of SCPs from the two sets. To illustrate how true hits can be detected by comparing SCPs we use the examples depicted in Figure 4. Let r and s be the two polygons in a candidate. Then SCP(r) represents a 3-tuple $< class_r, a_r, b_r >$ in the SCP set of r whereas SCP(s) represents a 3-tuple $< class_s, a_s, b_s >$ in the SCP set of s.

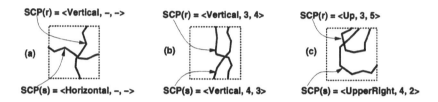

Fig. 4. Three Examples of Intersection Detected by SID.

Figure 4a shows that if SCP(r) = <**Vertical**, -, -> and SCP(s) = <**Horizontal**, -, -> ("-, -" means don't-care), then the polygon segments represented by SCP(r) and SCP(s) must intersect. As shown in Figure 4a, SID requires in this case only knowledge of the classes of the two SCPs to determine unambiguously that r and s intersect.

In Figure 4b, SCP(r) = <**Vertical**, 3, 4> and SCP(s) = <**Vertical**, 4, 3>. Because $a_r = 3 < 4 = a_s$ and $b_r = 4 > 3 = b_s$, the SCPs must intersect. Therefore, r and s intersect in this case. More generally, if both classes are **Vertical** and we have ($a_r \leq a_s$ and $b_r \geq b_s$) or ($a_r \geq a_s$ and $b_r \leq b_s$), then r and s intersect. Figure 4c is similar to 4b except that the intersection is assured because of the condition $a_r = 3 < a_s = 4 < b_r = 5$. The offset b_s is irrelevant for the intersect detection.

In contrast, Figure 5 illustrates some of the cases where SID cannot detect an intersection. Note that an intersection does occur in Figures 5b and 5c. However the symbolic information kept by the SCPs is insufficient to draw a conclusive decision, i.e., it is non-resolvable. For instance, Figures 5a and 5b have the same SCPs, but Figure 5b corresponds to an intersection whereas Figure 5a does not.

The examples in Figure 4 are a small subset of cases for which SID can determine polygon intersection. We present all possible cases using a two-dimensional table (Symbol Intersect Resolution Table, or SIRT) in Table 1.

Pair-Wise Evaluation of the Two SCP Sets. The processing of the *detection* stage corresponds to the **SIDTrueHitDetected** function shown in Figure 6 (also depicted in Figure 1). The **SIDTrueHitDetected** function nested-loops over the two SCP sets and, for each SCP pair, it calls the **SymbolIntersect** function

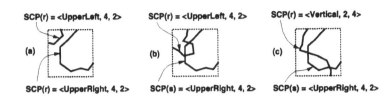

Fig. 5. Three Examples of Non-Resolvable Cases.

$$SCP(r) = < class_r, a_r, b_r > (\text{row})$$
$$SCP(s) = < class_s, a_s, b_s > (\text{column})$$

class	Vert	Horiz	Left	Up	Right	Down	UpL	UpR	LowR	LowL
Vert	*	T	F	$a_r \leq a_s$ $b_r \geq a_s$	F	$a_r \leq b_s$ $b_r \geq b_s$	$b_r \geq a_s$	$a_r \leq a_s$	$b_r \leq b_s$	$b_r \geq b_s$
Horiz	T	*	$a_r \leq a_s$ $b_r \geq a_s$	F	$a_r \leq b_s$ $b_r \geq b_s$	F	$a_r \leq a_s$	$b_r \leq b_s$	$a_r \geq b_s$	$a_r \geq a_s$
Left	F	$a_r \leq b_s$ $a_r \geq a_s$	*	F	F	F	$a_r \leq b_s$ $a_r \geq a_s$	F	F	$a_r \leq b_s$ $a_r \geq a_s$
Up	$a_r \leq b_s$ $a_r \geq a_s$	F	F	*	F	F	$b_r \leq b_s$ $b_r \geq a_s$	$a_r \leq a_s$ $a_r \geq a_s$	F	F
Right	F	$b_r \leq b_s$ $b_r \geq a_s$	F	F	*	F	F	$b_r \leq b_s$ $b_r \geq a_s$	$a_r \leq b_s$ $a_r \geq a_s$	F
Down	$b_r \leq b_s$ $b_r \geq a_s$	F	F	F	F	*	F	F	$b_r \leq b_s$ $b_r \geq a_s$	$b_r \leq b_s$ $b_r \geq a_s$
UpL	$a_r \leq b_s$	$a_r \geq a_s$	$a_r \leq a_s$ $b_r \geq a_s$	$a_r \leq b_s$ $b_r \geq b_s$	F	F	*	$a_r \leq b_s$	F	$a_r \geq a_s$
UpR	$a_r \geq a_s$	$b_r \geq b_s$	F	$a_r \leq a_s$ $b_r \geq a_s$	$a_r \leq b_s$ $b_r \geq b_s$	F	$b_r \geq a_s$	*	$a_r \geq b_s$	F
LowR	$b_r \geq b_s$	$b_r \leq a_s$	F	F	$a_r \leq a_s$ $b_r \geq a_s$	$a_r \leq b_s$ $b_r \geq b_s$	F	$b_r \leq a_s$	*	$b_r \geq b_s$
LowL	$b_r \leq b_s$	$a_r \leq a_s$	$a_r \leq a_s$ $b_r \geq a_s$	F	F	$a_r \leq b_s$ $b_r \geq b_s$	$a_r \leq a_s$	F	$b_r \leq b_s$	*

Key:
1. "T" for TRUE means intersect detected without conditions.
2. "F" for FALSE means intersect cannot be determined.
3. " * " $= ((a_r \leq a_s) \wedge (b_r \geq b_s)) \vee ((a_r \geq a_s) \wedge (b_r \leq b_s))$.
4. An entry is TRUE if the conjunction of all its clauses evaluates to TRUE.

Table 1. Symbol Intersect Resolution Table (SIRT).

```
    Boolean SIDTrueHitDetected(SCP set: x, SCP set: y)
    {
1       for each SCP r in x do
2           for each SCP s in y do
3               if (SymbolIntersect(r, s)) then
4                   return TRUE; // polygons intersect
                end if;
            end do;
        end do;
5       return FALSE; // pass this candidate-pair to plane sweep
    }
```

Fig. 6. Function for Early True Hit Detection using SCPs.

(line 3 in Figure 6) to detect a true hit. The SymbolIntersect function uses the two SCP classes to index into the SIRT, evaluates the predicate based on the offsets of the two SCPs, and returns a boolean value. If SymbolIntersect returns TRUE, the two SCPs intersect, whereas a FALSE means that SID cannot determine if the two SCPs intersect.

Once an intersection between two SCPs is detected (line 4 in Figure 6), a candidate is known to be a true hit. If SID exhausts all SCP pairs without finding an intersection (line 5 in Figure 6), it then passes this candidate to the *sweeping* phase to perform the traditional *intersect* testing using *plane sweep* (as done by the state-of-the-art-approach).

The SID approach improves refinement processing by deploying the techniques in the *clipping* and *detection* stages as a replacement for the *restricting* phase in the state-of-the-art approach (Figure 1).

4 Experimental Evaluation

In this section, we present an experimental evaluation on real maps to measure how SID improves on the state-of-the-art in refinement processing.

4.1 Experimental Design

Testbed. All experiments were conducted on a SUN Sparc-20 workstation running the SunOS 4.1.3 operating system. The implementation of the filtering step in the spatial join is based on techniques proposed in [1]. For the refinement step, we implemented both the traditional refinement step and the refinement step that incorporates the SID optimization technique presented in this paper. All programs were written in C++.

Maps. The map data used in our experiments were extracted from the polygon data sets in the Sequoia 2000 Storage Benchmark [17]. In particular, they are regions of various land use and land cover classifications in the State of California and Nevada. The number of line segments per polygon ranges from tens to thousands. Because our SID optimization is targeted to handle complex spatial data, we extracted several sets of polygons and created eight classes of data with the minimum number of line segments set to 100, 200, 300, 400, 500, 600, 700, and 800. This resulted in eight groups of data sets with an average number of line segments of 850, 950, 1100, 1300, 1450, 1650, 1750, and 1850, respectively (see Table 2).

map class number	1	2	3	4	5	6	7	8
Min. no. of line segments per polygon	100	200	300	400	500	600	700	800
Avg. no. of line segments per polygon	850	950	1100	1300	1450	1650	1750	1850

Table 2. Map Classifications.

4.2 Comparing Performance of the Refinement Steps

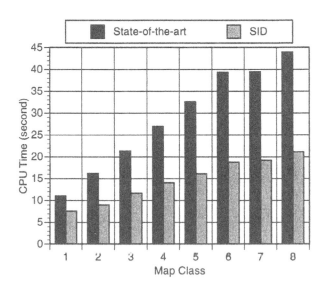

Fig. 7. CPU Usage Time: SID vs. State-of-the-art.

In Figure 7, we compare the CPU usage times between SID and the state-of-the-art approach for all map classes (1-8). The results show that SID achieves a significant performance improvement for all map classes. In Figure 8, we show

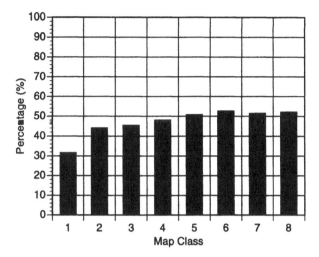

Fig. 8. Percentage of the CPU Time Achieved by SID vs That Achieved by the State-of-the-art Approach.

the percentage of the CPU costs of SID over that of the state-of-the-art approach (X-axis) for all map classes (Y-axis). The results show that the performance gain by SID map classes 2 ($n = 950$) to 4 ($n = 1300$) is above 40% whereas an over 50% improvement is achieved for map classes higher than 4 ($n > 1300$).

5 Conclusion

In this paper, we consider the optimization of the most computation-expensive component in spatial join processing: the refinement step. To this end, we present the Symbolic Intersect Detection (SID) technique that significantly improves the efficiency of the refinement step. For each candidate polygon pair, SID performs efficient true hit detection using the abstract information characterizing the topological relation between the two polygons and their overlapping minimum bounding rectangle. This constant-time true hit detection improves performance because further *intersect* computation such as the computationally intensive *plane sweep* algorithm is not needed for the true hit candidates detected by SID.

We present an experimental evaluation of SID by comparing it with the state-of-the-art approach based on real GIS maps with complex polygons that are taken from the Sequoia 2000 benchmark [17]. Our results show an impressive percentage (greater than 80%) of the true hits are detected by SID with only negligible overhead. Consequently, with the SID optimization, the time to resolve polygon intersection in the refinement step is improved by over 50%. Because SID optimizes the refinement step, it is complementary to many state-of-the-art spatial join techniques [1, 6, 8, 13] that have focused on improving the filter step. The combination of an impressive performance gain achievable by

SID and its compatibility with other approaches therefore promises a significant improvement for existing solutions to spatial join processing.

References

1. Brinkhoff, T., Kriegel, H., and Seeger, B.: Efficient Processing of Spatial Joins Using R-trees. Proc. ACM SIGMOD Int. Conf. on Management of Data. (1993) 237–246
2. Brinkhoff, T., Kriegel, H., Schneider, R., and Seeger, B.: Multi-Step Processing of Spatial Joins. Proc. ACM SIGMOD Int. Conf. on Management of Data. (1994) 197–208
3. Burrough, P.A.: Principles of Geographic Information Systems for Land Resources Assessment. Oxford University Press (1986)
4. Gunther, O.: Efficient Computation of Spatial Joins. Proc. IEEE 9th Int. Conf. on Data Engineering. (1993) 50–59
5. Guttman, A.: R-tree: a dynamic index structure for spatial searching. Proc. ACM SIGMOD Int. Conf. on Management of Data. (1984) 45–57
6. Huang, Y.W., Jing, N., and Rundensteiner, E.A.: Breadth-First R-tree Joins with Global Optimizations. EECS Technical Report, University of Michigan (1996)
7. Huang, Y.W., Jing, N., and Rundensteiner, E.A.: Integrated Query Processing Strategies for Spatial Path Queries. Proc. IEEE 13th Int. Conf. on Data Engineering. (1997)
8. Lo, M.L. and Ravishankar, C.V.: Spatial Hash-Joins. Proc. ACM SIGMOD Int. Conf. on Management of Data. (1996) 247–258
9. Maguire, D.J., Goodchild, M.F., and Rhind, D.W.: Geographic Information Systems, volume 1. John Wiley & Sons, Inc. (1991)
10. Nievergelt, J. and Hinterberger, H.: The Grid File: An Adaptable, Symmetric Multikey File Structure. ACM Transactions on Database Systems. $9(1)$ (1984) 39–71
11. Orenstein, J.A.: Spatial Query Processing in an Object-Oriented Database System. Proc. ACM SIGMOD Int. Conf. on Management of Data. (1986)
12. Orenstein, J.A.: A Comparison of Spatial Query Processing Techniques for Native and Parameter Spaces. Proc. ACM SIGMOD Int. Conf. on Management of Data. (1990) 343–352
13. Patel, J.M. and DeWitt, D.J.: Partition Based Spatial-Merge Join. Proc. ACM SIGMOD Int. Conf. on Management of Data. (1996) 259–270
14. Preparata, F.P. and Shamos, M.I.: Computational Geometry. Springer (1985).
15. Rotem, D.: Spatial Join Indices. Proc. IEEE 7th Int. Conf. on Data Engineering. (1991) 500–509
16. Shamos, M.I. and Hoey, D.J.: Geometric Intersection Problems. Proc. Conf. on Foundations of Computer Science. (1976) 208–215
17. Stonebraker, M., Frew, J., Gardels, K., and Meredith, J.: The SEQUOIA 2000 Storage Benchmark. Proc. ACM SIGMOD Int. Conf. on Management of Data. (1993)

Data Partitioning for Parallel Spatial Join Processing

Xiaofang Zhou[†], David J. Abel[†] and David Truffet[‡]

[†]CSIRO Mathematical and Information Sciences
GPO Box 664, Canberra, ACT 2601, Australia
{xiaofang.zhou, dave.abel}@cmis.csiro.au
[‡]Department of Computer Science
University of Queensland, QLD 4072, Australia
david@it.uq.oz.au

Abstract. The cost of spatial join processing can be very high because of the large sizes of spatial objects and the computation-intensive spatial operations. While parallel processing seems a natural solution to this problem, it is not clear how spatial data can be partitioned for this purpose. Various spatial data partitioning methods are examined in this paper. A framework combining the data-partitioning techniques used by most parallel join algorithms in relational databases and the filter-and-refine strategy for spatial operation processing is proposed for parallel spatial join processing. Object duplication caused by multi-assignment in spatial data partitioning can result in extra CPU cost as well as extra communication cost. We find that the key to overcome this problem is to preserve spatial locality in task decomposition. We show in this paper that a near-optimal speedup can be achieved for parallel spatial join processing using our new algorithms.

1 Introduction

Two of the most successful database research areas in the last decade are the parallel databases which can improve database performance drastically via parallel processing [6, 18], and the spatial databases which can support spatial data types such as points, lines and polygons[9]. Most major database vendors have already provided parallel processing services for their relational database products, and more and more database vendors are starting to integrate spatial data types into their mainstream products. The combination of these two trends defines the new and challenging research area of parallel spatial databases.

Spatial joins, which combine two spatial data sets by their spatial relationship such as intersection and enclosure, are among the most important operations in spatial databases. Their processing costs are intrinsically high because of three factors. First, spatial data are often large in size and therefore expensive to transfer from disk. Second, the spatial operations, often implemented using computational geometry algorithms, are CPU intensive. Third, most join algorithms efficient for one-dimensional data, such as the sort-merge and the hash join algorithms, cannot be applied to multi-dimensional spatial data. Efficient spatial join algorithms have long been the focus of spatial database research[4, 8, 16, 17, 19, 20]. However, parallel spatial join processing has not been studied until very recently[5, 11].

In this paper we consider parallel processing of spatial joins with particular attention to examining whether the techniques which are efficient in parallel relational databases, such as the data partitioning methods and load-balancing algorithms, can be applied to spatial joins. Data partitioning parallelism has been successfully explored in parallel relational databases[6]. The basic idea to join relations R and S (denoted as $R \bowtie S$) in parallel is to partition R and S, by hashing or sorting on their join columns, into buckets $R_1, R_2, \ldots R_n$ and $S_1, S_2, \ldots S_n$ such that for all $1 \le i, j \le n$, $R_i \cap R_j = \emptyset$ and $S_i \cap S_j = \emptyset$ if $i \ne j$, and $R \bowtie S \equiv \cup_{i=1}^{n}(R_i \bowtie S_i)$. Such a data partitioning method has the *single assignment, single join* (SASJ) property as one tuple can be mapped into one and only one bucket, and one R bucket only needs to join with one S bucket. Each pair of R and S buckets, called a *task*, can then be assigned to a processor for parallel execution[15]. For joining spatial data which have extents, unfortunately, there does not exist a data-independent

SASJ partitioning method. This can be seen by considering an example where two polygons A and B are mapped into different buckets using an SASJ method and then a polygon C, which overlaps with both A and B, is added into the data set. Therefore, a spatial data partitioning method is either *multi-assignment, single-join* (MASJ), or *single-assignment, multi-join* (SAMJ)[17]. A typical partitioning method for spatial data is based on *space decomposition*. That is, the space is decomposed into regions such that spatial objects can be partitioned according to their spatial relationship, such as enclosure and overlapping, with these regions. The R-tree approach ([10]) is SAMJ in the sense that an object is only assigned to one R-tree node but in order to find objects in a given area it is possibly necessary to search more than one R-tree node[4]. On the other hand, the R^+-tree approach ([22]) and the Z value based approach ([3, 19]) can be regarded as MASJ. The former keeps regions of tree nodes non-overlapping but assigns an object to all the nodes whose regions overlap with the object; and the latter can associate an object with multiple regions which are possibly at different resolution levels. While both MASJ and SAMJ data partitioning methods are popular as spatial indexing mechanisms[9], we argue in this paper that the MASJ type of data partitioning method is preferable for the purpose of parallel spatial join processing. In comparison with the task decomposition methods for one dimensional data, the multi-assignment spatial data partitioning methods have a distinct property - *object duplication*. That is, there can be duplicated objects among both the operand buckets and the results of different tasks (i.e., processors). Object duplication can impact negatively on the performance of parallel spatial join algorithms by the extra communication cost for sending one object to many processors and by the extra workload for processing the same pair of objects at different processors. Therefore, a data partitioning algorithm for parallel spatial join processing should not only balance the workload among tasks as most parallel relational join algorithms do, but also minimise the total workload which can vary depending on the partitioning methods.

Binary polygon intersection joins are considered in this paper, though our results can be extended easily to many other types of spatial joins. We concentrate on some fundamental issues which are general to data partitioning parallel spatial join algorithms instead of focusing on a specific parallel architecture. Our aim is to maximise CPU parallelism and minimise communication cost. While most previous parallel spatial join algorithms have largely evolved from sequential spatial join algorithms with the focus on utilisation of certain types of spatial indices, we adopt a different approach based on simple space partitioning. The load-balancing issue with non-uniform data distribution in the join columns, known as the *data skew* problem [14], is approached through a low-cost bottom-up merging process instead of the top-down recursive decomposition methods used by previous research. Combining the data partitioning techniques widely adopted by parallel relational databases and the filter-and-refine strategy for spatial operation processing, we present a seven-step framework for parallel spatial join processing. We identify a set of optimisation objectives by examining each step of the framework. Several novel algorithms are proposed for each step to achieve these objectives. The central idea, not surprisingly, is to preserve spatial locality during task decomposition. Using real spatial data, a performance analysis of the parallel join algorithm based on our framework shows a near-optimal speedup for CPU cost and only a moderate increase of total amount of data exchanged among processors when the number of processors increases. Our main contribution in this paper is to show that the parallel join algorithms which have proven successful for relational databases can also be applied to spatial joins with certain modifications.

In Section 2 we give a brief review of related work. A framework for parallel spatial join algorithms is given in Section 3, together with a set of optimisation objectives regarding to the CPU costs and communication costs. Various design alternatives for each step of the framework are discussed in Section 4. In Section 5, the performance of a complete parallel spatial join algorithm based on the framework using the algorithms identified as most efficient for each step is analysed with real spatial data. Our conclusions are presented in Section 6.

2 Related Work

2.1 The Filter-And-Refine Strategy

A common strategy to reduce the CPU and I/O costs for spatial join processing is the *filter-and-refine* approach which employs a filtering step followed by a refinement step. In the filtering step, a weaker condition for the spatial operation is applied to approximations of spatial objects to produce a list of *candidates* which is a superset of the final join results. These candidates are then checked by applying the spatial operation on the full descriptions of the spatial objects to eliminate the "false drops". The join cost can be reduced because the weaker condition is usually computationally less expensive to evaluate and the approximations are smaller in size than the full geometry of spatial objects. A common weaker condition for many spatial operations is the intersection of *minimum bonding rectangles* (MBRs) of spatial objects. Although the MBR intersection operation can be implemented as a pure aspatial operation using the coordinates of MBRs, the filtering cost can still be very high because those efficient relational join algorithms cannot be applied the MBR intersection joins. Spatial indices can be used to reduce the filtering cost. It is proposed to sort the objects by their Z values and to merge them sequentially just like in the conventional sort-merge join algorithm except that a controlled traversal is required for the objects whose Z values are represented at higher than the resolution level[19]. Brinkhoff, Kriegel and Seeger propose to use R-trees for the filtering operation[4]. They identify overlapping MBRs at two levels using the plane-sweep algorithm. First, the MBRs for the R-tree leave nodes are examined to identify all R-tree node pairs, each from one operand table, with overlapping regions. Then, the objects within a pair of R-tree nodes are joined to produce candidates using the similar algorithm. The idea to organise spatial objects into a tree structure for efficient spatial join processing is generalised by Günther[8]. Lo and Ravishankar give an algorithm to build spatial indices to perform spatial join operations without spatial indices[16]. As the join operations are usually postponed to after selection operations by the query optimiser in a database system, the operand tables of a spatial join operation usually do not have spatial indices because they can be interim results of other operations. Some recent studies find that the data partitioning based spatial join algorithms using techniques similar to those used for relational data partitioning outperform the spatial join algorithms which build spatial indices on-the-fly [17, 20].

Previous research on spatial join algorithms has largely focused on the filtering step only. Partly because of this and partly because of the intrinsically high costs of the operations involved, the refinement cost becomes dominant for the queries with high join selectivity factors. Using the same weak condition, all filtering algorithms will produce the same set of candidates. However, different filtering algorithms may or may not produce duplicated candidates; and they may produce candidates in different orders. Such differences can influence significantly on the refinement cost. This problem has been recently studied by Abel *et al* [1].

2.2 Parallel Join Processing

A query is usually executed as a sequence of operations (e.g., fetch, sort and merge). While different queries and independent operations within a query can be executed in parallel, some consecutive operations in a query can also be executed concurrently by overlapping the execution (i.e., *pipelining parallelism*). Additional parallelism, known as *data partitioning parallelism*, can be obtained for set-oriented operations by partitioning the input data and applying the operator on part of data. Data partitioning parallelism is well positioned to explore the benefits promised by modern parallel computers [6]. Assuming the data are distributed among processors before join processing starts, a typical data-partitioning parallel join algorithm for relational data has three phases:

1. *the task decomposition phase*: the join operand data are partitioned in parallel into buckets, and bucket pairs are grouped into tasks. It is critical to maintain load-balancing among tasks such that all processors can finish at the same time in the third phase.

2. *the data redistribution phase*: the operand data are redistributed among processors such that the processors can execute their tasks in phase 3 independently.

3. *the parallel task execution phase*: the processors execute the tasks assigned to them in parallel. No communication is required unless phase 1 fails to balance the workload among processors. In that case, workload need to be re-balanced dynamically during execution.

Many task decomposition algorithms are proposed for relational joins (see [24] for a survey). When the data distribution expected by the data partitioning method differs from that in the join operand data, some tasks may contain more data than other and thus take longer time to process[14]. A data partitioning method is said to be *balanced* if it does not result in fluctuation of workload among tasks even when data skew is present. The algorithms to handle data skew can be classified into two categories: those using sampling techniques to find a tailor-made data partitioning method for a given join operation [7], and those by decomposing data into much smaller units and merging them into larger but balanced tasks [13, 14]. Dynamic load-balancing inevitably interrupts parallel task execution. Its cost can be high because of dynamic workload monitoring and additional work for data redistribution. It is not used unless the join workload is unpredictable due to a lack of sufficient statistics data about the operation.

Spatial data skew is not uncommon. For instances, most land parcels and utility lines are around large cities. Therefore, it is important for a spatial data partitioning method to be data skew robust. In order to prevent load-imbalancing, two approaches can be used for partitioning spatial data:

1. *the top-down approach*: the space is partitioned into a number of regions, each corresponding to a bucket. Once a bucket is found to have more than expected workload (e.g., in terms of number of objects), the region for the bucket is split and the objects inside are re-assigned to the new buckets according to the new region definition. Some rules are required to reduce the possibility of further splitting, or to ensure the smaller regions contain the same amount of objects. In order to end up with a required number of buckets, an algorithm to merge buckets after splitting may be necessary. Most spatial index mechanisms, such as the R-tree and R$^+$-tree, decompose the space in the top-down manner recursively. Hoel and Samet adopt this approach to balance amount of spatial data by building a PMR-quadtree using recursive object decomposition [11]. The process of building the PMR-quadtree is parallelised by using a special parallel platform.

2. *the bottom-up approach*: the space is initially decomposed into N cells where $N >> n$, n is the number of processors. After the objects are assigned into N small buckets, these small buckets are merged into n larger but balanced buckets (usually applying heuristic rules to the *bin-packing* problem [12]). This approach is widely used to handle the data skew problem in parallel relational join algorithms [13, 14]. It is also used by Patel and DeWitt to partition spatial data for sequential join processing [20].

We consider the bottom-up approach in this paper because it is simple and each object is assigned to one or more buckets once only. In addition to finding a proper spatial data partitioning method, other new issues related to the design of a load-balanced bottom-up task decomposition algorithm for parallel spatial join processing include estimating spatial join workload and handling object duplication.

The algorithm in [5] is also based on data partitioning. However, it is completely different from our work on three grounds: 1) their task decomposition is based on pre-existing R-tree indices while we do not assume any type of spatial indices on the join columns; 2) they solve the load-imbalancing problem by rebalancing workload dynamically using global virtual memory while we prevent the data skew problem in the task decomposition stage; and 3) they do not consider refinement operation (by assuming the candidates are refined at where they are produced) while we consider the complete process of spatial join processing. Our work is similar to [11] in two aspects: both explore CPU parallelism in data parallel environments and neither relies on pre-existing spatial indices. Their work uses top-down object decomposition to balance the join workload. This requires constant data exchange during the partitioning stage, and so requires a specialised parallel platform. Our work, however, is based on bottom-up data partitioning and does not decompose spatial objects. Our algorithms do not rely on specific parallel architectures.

3 Parallel Spatial Join Algorithms

3.1 Assumptions

Let the scheme of the operand relations be

$$(id : \text{number}, mbr : \text{rectangle}, size : \text{number}, boundary : \text{polygon})$$

where id is the identifier of the spatial object; mbr is represented by the lower left corner (x_{low}, y_{low}) and the upper right corner (x_{high}, y_{high}) of the MBR of polygon; and $boundary$, represented as a sequence of vertices, describes the geometry of the polygon. The number of vertices, $size$, is introduced for the purpose of workload estimation. Columns (id, mbr, $size$) are called the key-pointer data. The join we consider is of the form $\pi_{R.id, S.id}(R \bowtie_{R.boundary\ intersect\ S.boundary} S)$.

We assume a shared-nothing architecture in this paper[6]. It consists a set of n identical parallel processors $p_1, \ldots p_n$ interconnected by a communication network, through which and only through which different processors can exchange data. We also assume that the cost of sending the same amount of data between any two processors are identical. Each processor has its own memory and can execute tasks independently (i.e., MIMD parallelism). As in [4], there is a processor p_0 designated as the interface processor which can communicate with every other processor directly and simultaneously. The parallel computer is connected to the outside by p_0 (i.e., to receive data from the database server where spatial data are stored). We assume the initial data distribution among the parallel processors is *even*. That is, before the parallel join processing starts we have $R = \cup_{i=1}^{n} R_i$, $S = \cup_{i=1}^{n} S_i$, R_i, S_i reside on p_i, $1 \leq i \leq n$, and $|R_1| = \ldots = |R_n|$ and $|S_1| = \ldots = |S_n|$ where $|R|$ denotes the number of objects in R. Further, we assume all the data can be stored in the aggregate memory of the parallel processors. Although it is known that a clustered spatial data distribution can reduce the join cost, this is not always possible as the join operand tables are often the intermediate results from other database operations. Instead of considering the operation of data clustering (which can be time-consuming) as a pre-processing step, we consider it as part of the parallel join processing by assuming the initial distribution is *random* (i.e., the spatial objects in R and S are distributed into R_i and S_i, $1 \leq i \leq n$, in a round-robin way). Our assumptions about initial data distribution model the cases where the join operand data are the results of some previous operations, or where data can be bulk loaded into parallel processors from outside (e.g., a parallel file system or a database server). Final join result collection is not considered. These assumptions are common for parallel join research[11, 13, 14].

3.2 A Framework

The filter-and-refine strategy in multiprocessor can, in addition to reducing CPU and I/O costs, reduce the amount of data to be transferred among processors (this has been demonstrated in a distributed spatial database environment by Abel *et al* [2]). It is natural for a parallel spatial join algorithm to adopt this two-step approach. In light of the paradigm of parallel relational join algorithm and the filter-and-refine spatial operation processing strategy, we propose a framework for parallel spatial join processing. The framework consists of a *parallel filtering phase* (steps F1 - F4) and a *parallel refinement phase* (steps F5 - F7). Only the key-pointer data are used in the filtering phase. The candidates in the refinement phase determine which spatial objects are required for this phase and where they are to be used. Each phase has three steps (i.e., task decomposition, data redistribution and task execution). The filtering phase has an additional step to redistribute its results (i.e., candidates). The seven-step framework for n processors is as follows:

1. **F1** (*filtering task decomposition*): each p_i, $1 \leq i \leq n$, partitions the key-pointer data of R_i and S_i into R_{ij}, S_{ij}, where $1 \leq j \leq N$ and $N >> n$, using a **spatial data partitioning** algorithm which defines a mapping between objects and regions (called *cells*). Then, p_0 collects the sizes of small buckets R'_i and S'_i, $1 \leq i \leq N$, where $R'_i = \cup_{j=1}^n R_{ji}$ and $S'_i = \cup_{j=1}^n S_{ji}$, and merges the small buckets into n large buckets R''_i and S''_i, $1 \leq i \leq n$, using a **bucket merger** algorithm. We denote the indices of the large R and S buckets which the ith small R and S bucket is merged into as $f(i)$ and $g(i)$ respectively (unlike in parallel relational join algorithms, one small bucket can be merged into multiple large buckets here).

2. **F2** (*key-pointer data redistribution*): each p_i sends the key-pointer data for the objects in R_{ij} and S_{ij} to processors $p_{f(j)}$ and $p_{g(j)}$, $1 \leq j \leq N$, and puts the data received into R''_i and S''_i with the processors from which the data come from recorded.

3. **F3** (*parallel filtering*): each p_i performs its filtering task $R''_i \bowtie_{MBR\ intersect} S''_i$ to produce candidates $\{(r.id, s.id, cost, cell)\}$ using an **MBR intersection join** algorithm, where *cost* is the estimated refinement cost of the candidates (see Section 4.5), and *cell* is the cell number of object $r.id$ (if there is more than one cell for the object, choose the smallest number). The duplicate candidates produced locally, if any, are removed at each processor independently.

4. **F4** (*candidates distribution*): each p_i sends the *full candidate* ($r.id$, $s.id$, $cost$, $cell$) to the processor which has object $r.id$ (we call it the *home processor* of object $r.id$), and sends the *data-request candidate* ($s.id$, $cell$) to the home processor of object $s.id$ if the home processors of $r.id$ and $s.id$ are different.

5. **F5** (*refinement task decomposition*): each p_i processes the full candidates it receives to remove the duplicate candidates produced by different processors (which cannot be removed in step F3). The refinement cost for a cell is calculated in parallel by adding up the costs of all full candidates a processor receives with the cell number. A cell-to-task mapping is determined at p_0 using a **workload re-balancing** algorithm which takes as input the estimated refinement workload of each cell.

6. **F6** (*full object redistribution*): each p_i sends full objects used by either full candidates or data-request candidates to their destination processors according to the cell-to-task mapping determined at F5. The full candidates (but not the data-request candidates) themselves are also sent to the destination processors.

7. **F7** (*parallel refinement*): each p_i performs its refinement task using the full candidates and the full objects received using a **polygon intersection check** algorithm.

In this framework, most duplicate candidates are removed at where they are produced before they are sent back to their home processors. The remaining duplicate candidates (produced by different processors) are removed at home processors. As all the candidates referencing to the same R object have the same home processor, there are no duplicate candidates after these two steps. Note that data-request candidates do not participate in duplicate removal or refinement cost estimation; they are only used for requesting S objects.

Clearly, the performance of a parallel spatial join algorithm based on the above framework depends on the algorithms it uses for each step. Before we discuss these algorithms in the next section, we analyse the costs of these steps to identify a set of optimisation objectives.

3.3 Optimisation Objectives

The complex problem of modelling the cost of a parallel spatial join algorithm can be simplified by assuming synchronised steps (i.e., no processor can proceed to the next step until all processors finish the current step). Therefore, the total cost is the sum of the costs for the component steps. Now we analyse the communication costs and the CPU costs for these steps.

Data exchange in step F1 occurs between p_0 and the parallel processors: the parallel processors send their numbers of local objects in each cell to p_0 and p_0 broadcasts back cell-to-filtering-task mapping information. Clearly, the communication cost of this step is solely determined by the number of cells. As each object is processed only once in step F1 because of the bottom-up data partitioning approach, the CPU cost of step F1 is simply the cost of a linear scan for local R and S objects to map them into small buckets (experiments show that the CPU cost difference between multi-assignment and single-assignment using MBRs is negligible). Bucket merging is done at p_0. Under the assumption of even initial distribution, the CPU cost for this step is uniform across all processors and independent of the partitioning algorithms used.

In step F2, the key-pointer data are re-organised before being transferred to other processors (according to the cell-to-filtering-task mapping), and the objects duplicated in different small buckets sending to the same destination are removed. The CPU costs for removing duplicates are uniform among processors because of even data distribution. However the number of data sent by a processor can vary depending on the cell-to-filtering-task mapping as one object can be sent to several processors if the object is assigned to multiple filtering tasks. The communication cost of this step depends directly on the bucket merger algorithm which defines the filtering tasks, and indirectly on the data partitioning algorithm that determines the input to the bucket merger algorithm. Under the assumption of even and random initial data distribution, each processor has an equal share of the extra amount of data exchanged, which is equivalent to the total number of objects duplicated in the filtering tasks. Therefore, we need, as the first optimisation objective, to *minimise object duplication among filtering tasks*.

Step F3 incurs no communication cost. The CPU cost is determined by the filtering task which requires the largest number of MBR comparisons. Thus the merger algorithm must *minimise the number of MBR comparisons for filtering tasks*. In step F4, candidates are sent back to home processors for requesting objects for refinement. The communication cost of this step, clearly, depends on the number of candidates duplicated among processors. Therefore, the merger algorithm must also *minimise the number of candidates duplicated among filtering tasks*. Another factor affecting the communication cost of this step is whether some processors produce significantly more candidates than others (this is known as the *join product skew* problem [24]). Prevention of this problem requires estimation of the MBR join selectivity for filtering tasks; this is beyond the scope of this paper.

The cost of step F5 is similar to that of step F1 except for an additional operation used for removing duplicate candidates. This cost can be regarded as uniform among processors because of even and random initial data distribution. Refinement task creation is a centralised operation at p_0, and data exchange occurs only between p_0 and the parallel processors. Again, the amount of data communication is determined only be the number of cells (i.e., sending the estimated refinement cost for each cell to p_0 and receiving candidates-to-refinement-task mapping from p_0). Full geometry descriptions of the spatial objects need to be transferred in step F6. A local processing is performed to determine which objects to send where and to remove the redundant objects sending to the same processor. Note that this cost is independent of the filtering algorithm after the duplicated candidates are removed. Under the assumption of even and random data distribution, the objective of minimising the communication cost of this step can be stated as to *minimise the number of objects duplicated among refinement tasks*, which means not sending one object to multiple processors. Step F7 incurs only computational costs. The optimisation objective for this step is obvious: *balancing the workload among refinement tasks*.

4 Design Issues

In this section we discuss the algorithms to be used for each step in the parallel join framework to achieve the optimisation goals identified in Section 3.

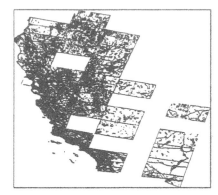

Fig. 1. Sequoia island polygons.

Fig. 2. Sequoia land use polygons.

4.1 Data Sets

The algorithms discussed in this section are compared using the Sequoia data set[23], which are publicly available and have been used by several published spatial join performance analysis [4, 20]. It consists of two sets of polygons: the "land use" polygons (denoted as R hereafter) that describe land uses in California, and the "island" polygons (denoted as S) which are the holes in land use polygons (so all polygons are simple polygons). Data distribution of the data set is illustrated in Fig 1 and 2. There are 37 land use types with from a few to more than 20000 polygons in each type. The data set contains 79385 polygons ($|R| = 58411$, $|S| = 20974$) with 3.65 million vertices in total (see Table 1 for details). A vertex is represented as a pair of integers. The total amount of data, including the key-pointer data, is 29.67 MBytes. The ratio of the area of the spatial extent over the average MBR size is about 248×248, 282×282 for R and S data respectively. Each object is assigned with a unique identifier by concatenating the land use code and the sequence number of the object in the land use type. Note that the sequence numbers bear a co-relationship with spatial locality (i.e., for the objects of the same land use type, the objects with "close" sequence numbers are close in the space). The query used in this paper is, as in[20], to find all the holes in polygons. The MBR filtering phase returns 84862 distinct candidates, which reference to 54015 different objects; and the final result returns 47533 pairs of polygons.

Num of points	1-10	10-30	30-50	50-100	100-500	500-1000	1000+
Num of polygons	18564	36158	11027	7873	5009	444	310

Table 1. Description of the Sequoia polygon data

4.2 Local Join Algorithms

As does any filter-and-refine based spatial join algorithm, our framework uses two join algorithms: an MBR intersection join algorithm to find candidates and an polygon intersection join algorithm for refinement. The MBR intersection join algorithms have been extensively investigated in the literature. The algorithms based on spatial indices are not attractive to our parallel spatial join framework as spatial indices are not available in our parallel environment. Because a large bucket is merged from a set of small buckets, it is natural for us to use the partitioning-based MBR join

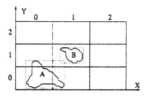

Fig. 3. The space is divided into 3 × 3 cells.

algorithm [17, 20]. That is, a pair of large buckets is joined by joining each pair of the component small buckets which correspond to the same cell; and a pair of small buckets are joined using the nested-loops method (i.e., to check each object in one small bucket with all objects in another bucket) because the sizes of small buckets are small enough to yield satisfactory performance.

The polygon intersection algorithm as a computational geometry problem has been well studied. It is a consensus that the most efficient algorithm is the plane-sweep algorithm, which requires $O((m + n) \log(m + n))$ time to check polygons with m and n vertices respectively[21]. As we assume that all the objects can be held in the distributed memory of parallel processors, candidate sequencing ([1]) is not considered here. In other words, the polygon intersection join algorithm used in this paper is to process sequentially the candidates assigned to a processor; and for each pair of polygons, the plane sweep algorithm is used to check if the polygons have cutting lines after whether a vertex of one polygon is inside another are checked.

4.3 Spatial Data Partitioning

As mentioned before, a spatial data partitioning method can be either MASJ or SAMJ. Partitioning the *spatial extent* of the join operand data (i.e., the MBR for all the polygons in both tables) into rectangles (i.e., *cells*) of the same size (see Fig 3), an MASJ partitioning method maps all objects overlapping with a cell into the bucket for the cell; and an SAMJ method groups objects according their geometric centers. For example, object A in Fig 3 is mapped to cell 00 using the SAMJ method, and to cells 00, 01, 10, 11 under the MASJ mapping. Object B is mapped into cell 11 for both methods. A spatial data partitioning method can affect the number of MBR comparisons required for the MBR intersection join, the number of duplicated objects among large buckets and the number of duplicated candidates produced by filtering tasks. We compare these two partitioning methods by these three factors.

Number of MBR Comparisons Fig 4 illustrate how the number of MBR comparisons can be reduced under different data partitioning methods. The workload of $R \bowtie_{MBR\ intersect} S$ is represented by the shaded area in Fig 4. The outer unshaded rectangle is the MBR join workload

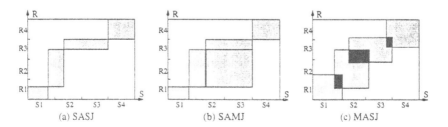

Fig. 4. Workload reduction by data partitioning.

187

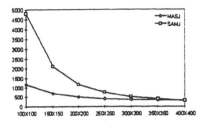 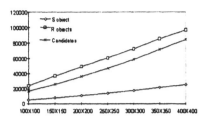

Fig. 5. Number of MBR comparisons (×1000) against the number of cells.

Fig. 6. Number of duplicated objects against the number of cells.

when the nested-loops algorithm is used. Obviously, an SASJ method does not result in overlapping objects between different buckets, thus no redundant join workload presents under that method. This method is, however, not applicable for spatial joins as we mentioned before. Both MASJ and SAMJ methods incur some redundant join workload. In Fig 4 (b), bucket R_2 needs to join with both S_2 and S_3 (i.e., multi-join). In Fig 4 (c), there can have duplicated objects between neighbouring buckets (i.e., the same pair of objects can be duplicated in several filtering tasks).

Now we show that the SAMJ method usually requires a larger number of MBR comparisons than the MASJ method. First, we consider the simple case where both operand tables have uniform data distribution. Let the space be partitioned into n cells. For the SAMJ method, we have $R = \cup_{i=1}^{n} R_i$ and $S = \cup_{i=1}^{n} S_i$, and for any $1 \leq i,j \leq n$, $R_i \cap R_j \equiv \emptyset$ and $S_i \cap S_j \equiv \emptyset$; for the MASJ method, we have $R = \cup_{i=1}^{n} R_i'$ and $S = \cup_{i=1}^{n} S_i'$ where for some $1 \leq i,j \leq n$, $i \neq j$, $R_i' \cap R_j' \neq \emptyset$ and $S_i' \cap S_j' \neq \emptyset$. Further, we assume all bucket sizes are about the same after partitioning. Combining these assumptions, we have $|R_i| = |R|/n$, $|S_i| = |S|/n$, $|R_i'| = (1+\alpha)|R|/n$, and $|S_i'| = (1+\beta)|S|/n$, where α and β are the percentage of object duplication caused by multi-assignment in R and S buckets respectively. Therefore, the total number of MBR comparisons for MASJ and SAMJ mapping methods can be expressed as

$$C_{MASJ} = \sum_{i=1}^{n} |R_i'||S_i'| = \frac{(1+\alpha)(1+\beta)}{n}|R||S|, \qquad C_{SAMJ} = \sum_{i=1}^{n} n|R_i||S_i| = \frac{k}{n}|R||S|$$

where k is the average number of S buckets for a R bucket to join with under the SAMJ mapping. Thus, $C_{MASJ} \leq C_{SAMJ}$ if $(1+\alpha)(1+\beta) \leq k$. When $\beta = 0$ (i.e., object duplication occurs only in one table), $C_{MASJ} \leq C_{SAMJ}$ if $\alpha \leq k-1$; and when $\alpha = \beta$ (i.e., object duplication occurs for both tables and the two tables have similar data distribution in the space), $\alpha \leq \sqrt{k}-1$. Under the SAMJ mapping, one R bucket can join with up to 9 neighbouring S buckets. Assume $k = 3$, then the MASJ method has a smaller search space if the average percentage of overlapping objects among buckets are less than 200% and 73% for the above two cases. Moreover, when n is large, another join algorithm (e.g., the plane-sweep algorithm in[4]) is needed to find which pair of buckets need to be joined when the SAMJ method is used.

The total number of MBR comparisons required by the SAMJ and MASJ partitioning methods for the Sequoia data is shown in Fig 5. The data partitioning costs for the two methods are practically identical. When n increases from 100×100 to 400×400, k is reduced from 4.74 to 3.90 for the SAMJ methods, and the total comparison numbers for the two methods are getting closer. However, the average number of objects in each bucket reduces to 1 when $n > 300 \times 300$, and the bucket management cost (e.g., to determine which buckets are to be joined) becomes dominant. When the number of cells is not too large (e.g., no more than 200×200 cells for the Sequoia data), the benefit of the MASJ method is clear.

Object Duplication The benefit of not having duplicated objects in small buckets for an SAMJ method does not imply that there is no object duplication among the filtering tasks as a small bucket from one table often needs to be duplicated among several filtering tasks because of multi-join. In addition to a higher CPU cost as seen above, this also means a higher communication cost. For example, if there are some objects overlap with both region i and j, all R_i objects (including those not overlapping with region j) need to be sent to where S_j resides unless these objects can be distinguished from each other (e.g., using the algorithm in[4]). This will require additional computation and communication costs. An SAMJ method can have a particularly poor performance when there are some large objects spanning over many cells.

Fig 6 shows the trend of object duplication against numbers of cells in the filtering step under the cell-overlapping MASJ mapping. While the objects duplicated in both operand tables and in the resultant candidate set rise steadily with the increase of cell numbers, the number of MBR comparisons does not drop at the same rate (see Fig 5). Recalling the ratio of the spatial extent over object sizes given in section 4.1, it is clear that the cell size should not be smaller than the average size of MBRs.

One advantage of an SAMJ method is that it does not produce duplicate candidates across processors. However we found that the cost of removing duplicate candidates negligible in comparison to the cost caused by a significantly larger search space resulted from an SAMJ method as well as caused by bucket management. An MASJ method is clearly the preferred data partitioning method for parallel spatial join processing. We use the cell-overlapping MASJ partitioning method in the rest of this paper. Next we show that object duplication associate with the MASJ method can be greatly reduced by using a proper bucket merger algorithm.

4.4 Merging Buckets

The bucket merger algorithm merges N small buckets into n filtering tasks to balance the workload and minimise duplication of the key-pointer data among filtering tasks. Note that the problem of merging small buckets into balanced large buckets alone is an NP-hard problem. We consider three heuristic-based algorithms here:

1. *the round-robin algorithm* (RR): use the round-robin method to merge small buckets into large buckets. This method, originally proposed to handle data skew in parallel relational join processing by Kitsuregawa[15, 14], is used by Patel and DeWitt for spatial data partitioning [20].

2. *the Z Order greedy algorithm* (ZOrder): associate each small bucket with the Z value of the corresponding cells, and sort the small buckets by their Z values. Then, as many as consecutive small buckets (in the sorted order) are merged into a large bucket until a pre-set large bucket capacity C is reached. If there are any small buckets left after n large buckets are formed, the remaining small buckets are assigned, one by one, into the large bucket with the smallest estimated load. This algorithm minimises object duplication by merging together neighbouring cells (which are close to each other after sorting by the Z values), and balances the workload using the greedy heuristic. We set $C = \sum_{i=1}^{n} |R_i|/n$ and increase the workload of the ith large bucket C_i after small bucket R_j is merged into it using formula $C_i = C_i + |R_j|$.

3. *the improved Z Order greedy algorithm* (ImpZOrder): the ZOrder algorithm does not consider object duplication in large bucket size estimate. In order to improve the accuracy of large bucket size estimation, the number of duplicated objects needs to be recorded in the data partitioning algorithm by recording the number of objects overlapping between neighbouring cells. Let s_{ij} and e_{ij} be the numbers of duplicated objects between bucket R_{ij} and its southern and eastern neighbours respectively (note that when an object overlaps with both the southern and eastern cells, we record this duplication only once to avoid over-estimate the number of duplicate objects). We set $C = (1 + \gamma)|R|/n$ where γ is an allowance for duplicate objects in the merged buckets (we set $\gamma = 0.02$ in our tests), and $C_i = C_i + |R_j| - s_j - e_j$ after R_j is merged in.

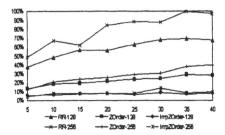

Fig. 7. Variation of large bucket sizes against number of filtering tasks.

The RR method is used to test if this commonly used method for parallel relational join processing is suitable for parallel spatial joins. The other two are used to test the influences of spatial locality and object duplication respectively. We ran two sets of experiments using the land use polygons (i.e., R) for $N = 128 \times 128$ and 256×256. In Fig 7, the bucket size *variation* is defined as $(B_{max} - B_{opt})/B_{opt}$ where B_{max} is the number of objects in the largest merged bucket, and B_{opt} is the optimal size which is the number of total distinct polygons in the data set divided by the number of buckets (i.e., $|R|/n$). Fig 7 demonstrates clearly that algorithm RR is not suitable for spatial data partitioning. It can increase bucket sizes by at least 40% because of object duplication. This overhead increases with the number of cells as well as with the number of filtering tasks. Algorithm ZOrder can reduce this overhead significantly. However, it fails to balance the load among buckets because of inaccurate estimation of object numbers after merging. By considering the number of objects duplicated among small buckets, algorithm ImpZOrder can reduce size variation to less than 10%. Note that this method is also robust to the number of buckets and the number of cells. Fig 8 gives the actual number of distinct objects in the largest merged bucket merged from 256×256 cells.

4.5 Re-Balancing Workload for Refinement

A refinement task is defined as a set of candidates. Two criteria for the refinement task creation step are to minimise the communication cost by minimising the number of objects duplicated among refinement tasks and to minimise the response time by balancing refinement workload. In this subsection we examine the impact of possible join product skew in the filtering phase before giving an algorithm to overcome this problem.

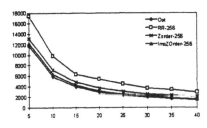

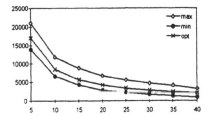

Fig. 8. Maximum large bucket sizes of merging 256×256 cells.

Fig. 9. Numbers of candidates produced by filtering tasks (product skew).

MBR Join Selectivity As we will further discuss later, the refinement workload is determined by not only the number of candidates but also the number of vertices of the objects referenced by the candidates. The spatial data partitioning algorithm and the bucket merger algorithm in the filtering phase cannot balance the refinement workload because the MBR join selectivity are unknown at that stage. MBR join selectivity estimation requires the statistical data about the operand tables such as the object sizes and spatial distribution of the objects. Even for the bucket merger algorithm which tries to balance the number of objects in the filtering tasks, Fig 9 shows that there still exists a significant amount of product skew caused by non-uniform MBR join selectivity in the filtering tasks, which are created using the MASJ partitioning method merged from 256 × 256 cells using using algorithm ImpZOrder. One can see that some filtering tasks can produce as many as 50% more candidates than the optimal case (which is the total number of distinct candidates divided by the number of processors). Fig 9 also shows that the proportion of the difference between the maximum and minimum numbers of candidates among filtering tasks becomes larger with the increase of the number of filtering tasks. This suggests that an algorithm of processing candidates where they are produced cannot lead to load-balanced refinement.

Object Duplication in Refinement Tasks Now we consider how to reduce object duplication among refinement tasks. For two candidates referencing to the same object, the object is duplicated if the candidates are assigned to different refinement tasks. Object duplication here will increase the communication cost for sending the full geometry of spatial objects. One straightforward method is to sort the candidates by their identifiers (say, of R then S) and assign the candidates sharing the same R object to the same refinement task. This method, however, cannot at the same time avoid duplicating objects from another table. In order to minimise duplicated objects for both tables, we apply the same idea as used for bucket merging. That is, we sort the candidates by the Z values of their cells and group candidates into refinement tasks in the sorted order using certain load-balancing criteria (to be discussed later). The reason of doing so is that the candidates referencing to the objects which are spatially close to each other are more likely to share common objects for both tables.

Fig 10 shows the empirical results about duplication of full objects in refinement tasks, which are defined by merging 256 × 256 cells with the objective of balancing the number of candidates in refinement tasks using various methods (i.e., sorting by R identifiers, by S identifiers and by the Z values). It is interesting to notice the effect of relationship between object identifiers and their locations. As mentioned in Section 4.1, there exists a co-relationship between the identifiers of the Sequoia objects of the same land use type and the spatial locations of the objects. Therefore, sorting by the S identifiers effectively groups objects spatially close to each other together. Note that the S objects are regarded as one type of land use - "island". This is not true for R objects which are from 37 types of different land uses. As Fig 10 illustrates, sorting by the Z values of

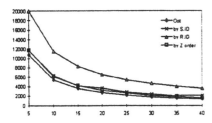

Fig. 10. Maximum number of distinct objects in varying number of refinement tasks.

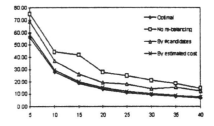

Fig. 11. Response time (seconds) of parallel refinement using different re-balancing algorithms against the number of processors.

the cells is slightly better than sorting by the S identifiers, while both of them are significantly better than sorting by R identifiers and both are close to the optimal values (which is the total number of distinct objects used by the candidates divided by the number of tasks). Thus, grouping candidates by the Z order can greatly eliminate object duplication among refinement tasks. This is particularly useful when object identifiers bear no relationship with spatial locations or such a relationship is unknown to the query optimiser. It should be pointed out that the R identifiers here are not completely random as they are still spatially clustered for the objects of the same land use (thus the results in Fig 10 can be better than the worst case with completely random identifiers).

In our refinement load-rebalancing algorithm, the candidates in a cell are grouped as a unit with the sum of costs of the member candidates as the unit cost. In comparison to considering each candidate as a unit for refinement task creation, our method can reduce the communication cost (from the parallel processors to p_0) as well as the task scheduling cost at p_0. It makes these costs dependent only on the number of cells, not the number of candidates.

Refinement Workload Estimation As one can see from Table 1, the number of vertices in a polygon ranges from a few to thousands. Thus, it is inadequate to regard the refinement costs for different candidates as the same. It is difficult to give an accurate cost model for polygon intersection as the cost depends on whether two polygons intersect or not as well as how they intersect. Let m and n be the number of vertices in polygon r and s. For the best case the polygon join algorithm described in Section 4.2 takes only $O(n)$ time when the point of polygon r used for testing point-in-polygon is inside polygon s; and at the worst takes $O(m) + O(n) + O((m+n)\log(m+n))$ when the two polygons do not intersect. Here we use $m + n$ as the formula to estimate the polygon intersection cost. Fig 11 shows the refinement response time for the following three methods (the optimal response time is calculated by dividing the single-processor refinement time by the number of processors):

1. *no re-balancing*: refines candidates at where they are produced. Both local and inter-processor duplicate candidates are removed.

2. *re-balancing by candidate numbers*: the candidates are sorted by their Z values and grouped by that order into refinement tasks such that each task has the same number of candidates.

3. *re-balancing by estimated cost*: this method is similar to the above but balances the workload using the cost estimate function $m + n$. Costs are estimated when candidates are produced, and the cost for the candidates in a cell is calculated in parallel in step F5.

As shown in Fig 9, different processors can produce significantly different numbers of candidates. When the refinement workload is not re-balanced, as Fig 11 shows, the processors with larger numbers of candidates jeopardise the response time for parallel refinement. As the cost to check polygon intersection depends on the number of vertices in the polygons, refinement load-balancing cannot be achieved by only balancing the number of candidates among processors. Fig 11 shows that our cost function can reasonably reflect the real refinement cost, and a near-optimal speedup can be achieved for parallel refinement after workload re-balancing.

5 Performance

We have discussed the algorithms required by the parallel spatial join framework individually. In this section we examine the overall performance of the complete parallel spatial join algorithm based on the framework using the most efficient algorithm for each step (i.e., the cell-overlapping based MASJ algorithm with 128×128 cells for data partitioning, algorithm ImpZOrder for bucket merging and the Z order based greedy algorithm with $(m + n)$ as cost estimate for refinement workload re-balancing).

5.1 CPU Costs

First we look at the CPU costs for the operations involved in the parallel spatial join algorithm. We summarise these operations here. Each processor first partitions the key-pointer data of its local objects into buckets. The bucket sizes, together with the information about object duplication between neighbouring cells are sent to p_0 which merges the cells into filtering tasks. Each processor then reorganises the local key-pointer data according to the filtering task definition such that no redundant data are sent to the same processor. After the parallel MBR join processing using the cell-based nested-loops algorithm, each processor removes local duplicate candidates before sending them to the home processors to request full objects. The inter-processor duplicate candidates are removed after each processor receives all candidates from others. Candidates are grouped by cells in parallel, and mapped into refinement tasks at p_0. Finally, the full descriptions of the objects required by the candidates are packed at each processor (with duplicate objects for the same processor removed) and are sent out for parallel refinement.

Num of procs	Partition		Create filtering tasks	Reorganise key-pointer data		MBR join		Remove duplicate candidates		Create refinement tasks	Pack full objects		Refine	
	Total	Max		Total	Max	Total	Max	Total	Max		Total	Max	Total	Max
5	1.00	0.23	0.09	2.04	0.42	5.75	1.29	3.10	0.63	0.02	4.20	0.93	276.7	61.56
10	1.21	0.16	0.09	2.09	0.22	6.17	0.79	3.03	0.32	0.02	4.10	0.42	277.3	31.56
15	1.49	0.14	0.10	2.39	0.18	6.39	0.58	2.89	0.21	0.02	3.97	0.31	277.3	21.40
20	1.92	0.15	0.12	2.56	0.16	6.68	0.52	2.89	0.15	0.02	3.87	0.21	277.6	16.24
25	1.99	0.14	0.12	2.78	0.13	6.87	0.41	2.86	0.15	0.02	3.67	0.22	277.8	13.23
30	1.89	0.11	0.13	2.97	0.13	6.79	0.32	2.82	0.11	0.02	3.39	0.13	277.9	10.99
35	2.00	0.12	0.14	3.13	0.11	6.94	0.30	2.78	0.10	0.02	3.20	0.11	278.3	9.44
40	1.87	0.11	0.16	3.31	0.10	6.92	0.25	2.79	0.09	0.02	3.06	0.09	278.5	8.50

Table 2. CPU costs (seconds) for operations in the parallel spatial join algorithm.

Table 2 reports the total time and the time required by the slowest processor to perform the operations on the Sequoia polygons. These figures are obtained by running the algorithms on a SunSPARC 10 workstation. Two facts can be learnt from Table 2. Firstly, the centralised operations for creating filtering and refinement tasks take negligible time. Secondly, the spatial join cost is clearly dominated by the refinement cost, with the filtering cost takes less than 5% of the total CPU time. Thus, there is no doubt that the overall performance is critically dependent on load-balancing in the refinement phase for those spatial joins with relatively large number of candidates (recall that for the Sequoia join $|R| = 58411$, $|S| = 20974$ and the number of candidates is 84862). Although the load-balancing issue for the filtering phase becomes less important because of its very small total cost, the cell-based partitioning and bucket merger algorithms are still necessary as they not only provide the foundation for the whole framework by associating objects with cells, but also reduce significantly the key-pointer data duplicated among filtering tasks (as shown in Fig 10). In addition, they also reduce the number of candidates duplicated among filtering tasks to minimum (there are only 544 to 4193 inter-processor duplicate candidates for from 5 to 40 filtering tasks).

Fig 12 shows the CPU costs for the parallel spatial join algorithm against varying number of processors, where the costs for all pre-refinement operations are added up together. As Fig 12 illustrates, a near-linear CPU speedup has been achieved. This confirms the satisfactory performance of our cell-wise cost based refinement load re-balancing algorithm.

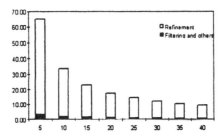

Fig. 12. CPU cost (seconds) of parallel join processing with varying number of processors.

5.2   Communication Costs

Communication cost is measured by the total and the maximum per-processor amount of data exchange (object sizes are listed in Table 3). There are two types of data exchange: that between p_0 and the parallel processors for creating filtering and refinement tasks, and that among the parallel processors for redistributing key-pointer data, candidates and full objects. For the first type of data exchange, the amount of data is solely determined by the number of cells regardless the operand tables. For $N \times N$ cells, each processor sends no more than $N \times N$ numbers twice to p_0 for reporting non-empty small bucket sizes and for the estimated refinement cost of the full candidates in each cell; and p_0 broadcasts $N \times N$ numbers twice for the cell-to-task mapping for the filtering and refinement tasks.

Object Name	Object Structure	Size (bytes)
bucket size/cell refinement cost	one number	4
key-pointer data	id, mbr, number of vertices	$6 \times 4 = 24$
full candidates	id1, id2, cell number, estimated cost	$4 \times 4 = 16$
data-request candidate	id, cell number	$2 \times 4 = 8$
full objects	id, n vertices as per object	$(n + 1) \times 4$

Table 3. Sizes of the objects to be exchanged for parallel spatial join.

Fig 13 shows the amount of the second type of data exchange for the Sequoia join with varying number of processors. Fig 13 (a) gives the total number of key-pointer data redistributed. Note that the total number of objects for the join is 79385 and for n processors the floor number of data exchange (when one object is assigned to one task) is $79385 \times (n - 1)/n$ under the assumption of even and random initial data distribution (i.e., a minimum of 63508 and 77400 objects need to be exchanged for $n = 5$ and 40 respectively). Fig 13 (b) is the number of full and data-request candidates redistributed (local duplicate candidates are removed, but not the inter-processor candidates). The number of data-request candidates is close to the number of full candidates because of the random data distribution across processors. The number of inter-processor duplicate candidates increases, as expected, with the number of processors though it has been kept to very small by the bucket merger algorithm. Fig 13 (c) and (d) show the numbers and sizes (in Kbytes) of full objects transferred. As clearly illustrated in Fig 13, the amount of data exchange for all the three of the second type data redistribution operations increases with the number of processors but at a much slower rate, in contrast to fast decrease of CPU cost as shown in Fig 12. The benefit of minimising object duplication for the filtering phase can be seen by comparing Fig 13 with Fig 6,

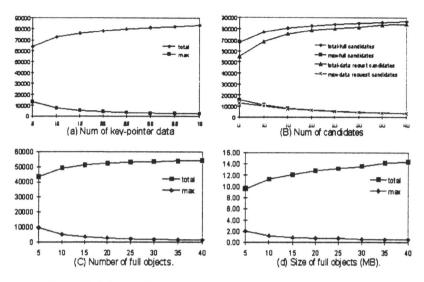

Fig. 13. Full object redistribution costs with varying number of processors.

while the impact of our algorithm on the duplication of full objects has already been shown in Fig 10.

Fig 14 compares the amount for the first type and the three second type data redistribution operations. Although the amount of the first type data exchange increases linearly with the number of processors, its communication cost can be a constant because for many parallel computers the parallel processors can communicate with the interface processor p_0 directly and simultaneously. The total amount of data exchange for redistributing key-pointer data and candidates is very small in comparison to that for full object redistribution. It is clear that the major communication cost for parallel spatial join processing is for full object redistribution.

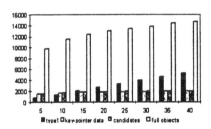

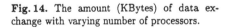

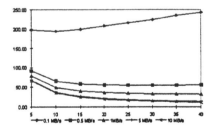

Fig. 14. The amount (KBytes) of data exchange with varying number of processors.

Fig. 15. Response time (seconds) of the parallel spatial join algorithm with varying number of processors and different effective communication bandwidth.

5.3 Response Time

Now let us consider communication bandwidth. Communication bandwidth of typical parallel database systems ranges from 10-100 MBytes per second [18]. The *effective* bandwidth can be lower because of the overhead associated with communication protocols and network traffic. Assuming the parallel processors are connected to a bus (thus the communication time is determined by the total amount of data exchange, not the maximum per processor amount of data exchange as for the parallel computers with parallel communication channels). All communication operations in our parallel join framework have been separated from local processing, thus the communication and local processing operations do not interfere with each other. Note that the CPU costs for pre-processing before data exchange, such as packing objects and removing duplicates, have been considered already in Section 5.1. Fig 15 shows the response time of parallel spatial join processing with varying number of processors and varying effective communication bandwidth from 0.1-10 MBytes/s. The response time here is the sum of the maximum per-processor CPU time for the operations in Table 2 plus the estimated communication time for transferring the total amount of data shown in Fig 14. (Note that the possible overlapping of communication time among processors is not considered here.)

When effective communication bandwidth is higher than 5 MBytes/s, parallel spatial join processing becomes CPU-bound, thus it can be effectively parallelised to achieve the speedup close to that in Fig 12. When effective bandwidth is lower than 1 MB/s, our join algorithm is communication-bound with little response time reduction when the number of processors increases to more than 5; when effective bandwidth is as low as 0.1 MB/s, Fig 15 shows that the response time actually increases with the number of processors. This means that for a given spatial join operation a minimum communication bandwidth is required for efficient parallel processing. Note that the communication bandwidth requirements are mainly raised by redistribution of full objects. For most types of modern parallel computers, it is safe to conclude, at least for the join operations of sizes comparable to that considered in this paper, that the spatial join problem is CPU-bound, and that a near-linear speedup can be achieved. This conclusion holds for symmetric multiprocessor computers, most parallel database systems and even workstation clusters with Ethernet connections.

6 Conclusions

In this paper we have proposed a framework for parallel spatial join processing, with algorithms given for each step of the framework to minimise both CPU- and communication-costs. Our framework based on the filter-and-refine strategy adopts the popular structure for data partitioning parallel relational join algorithms. We found that the MASJ data partitioning method is more suitable than the SAMJ method for parallel spatial join processing because it is simple, easy to be parallelised, and has a smaller number of MBR comparisons. Our method is very similar to conventional parallel join algorithms. It differs primarily in including new mechanisms overcome the data skew problem caused by object duplication in both filter and refinement tasks.

Object duplication clearly imposes higher CPU and communication costs. Preservation of spatial locality is the key to minimising object duplication. This has been accomplished through a cell approach in both the filer and refinement steps.

The analysis of individual steps as well as the complete parallel spatial join process using the Sequoia data set has demonstrated a significant performance gain over other algorithms. We have shown that for the joins with a relatively high MBR join selectivity the response time is dominated by the refinement CPU cost. Using our algorithms, the spatial join operations can be effectively parallelised to achieve a near-optimal speedup for large joins in parallel environments.

Acknowledgment: The authors would like to thank Robert Power for many helpful discussions.

References

1. D. J. Abel, V. Gaede, R. A. Power, and X. Zhou. Resequencing and clustering to improve the performance of spatial join. Technical report, CSIRO Mathematical and Information Sciences, Australia, 1997.

2. D. J. Abel, B. C. Ooi, K.-L. Tan, R. Power, and J. X. Yu. Spatial join strategies in distributed spatial dbms. In *LNCS 951: Proceedings of 4th Int. Symp. on Large Spatial Databases (SSD'95)*, pages 346 – 367. Springer-Verlag, 1995.

3. D. J. Abel and J. L. Smith. A data structure and algorithm based on a linear key for a rectangle retrieval problem. *Computer Vision, Graphics and Image Processing*, 24(1):1–13, 1983.

4. T. Brinkhoff, H. P. Kriegel, and B. Seeger. Efficient processing of spatial joins using R-trees. In *Proc. ACM SIGMOD Int. Conf. on Management of Data*, pages 237–246, 1993.

5. T. Brinkhoff, H. P. Kriegel, and B. Seeger. Parallel processing of spatial joins using R-trees. In *Proceedings of 12th International Conference on Data Engineering*, 1996.

6. D. J. DeWitt and J. Gray. Parallel database systems: the future of database processing. *C. ACM*, 35(6):85–98, 1992.

7. D. J. DeWitt et al. Practical skew handling in parallel join. In *Proc. 18th Int. Conf. on Very Large Data Bases*, pages 27–40, Vancouver, Canada, 1992.

8. O. Günther. Efficient computation of spatial joins. In *Proceedings of 9th International Conference on Data Engineering*, pages 50–59, Vienna, Austria, 1993.

9. R. H. Güting. An introduction to spatial database systems. *VLDB Journal*, 3(4):357–399, 1994.

10. A. Guttman. R-trees: A dynamic index structure for spatial searching. In *Proc. ACM SIGMOD Int. Conf. on Management of Data*, pages 47–54, 1984.

11. E. G. Hoel and H. Samet. Performance of data-parallel spatial operations. In *Proc. 20th Int. Conf. on Very Large Data Bases*, pages 156–167, 1995.

12. E. Horowitz and S. Sahni. *Fundamentals of Computer Algorithms*. Computer Science Press, 1978.

13. K. A. Hua and C. Lee. Handling data skew in multiprocessor database computers using partition tuning. In *Proceedings of 17th International Conference on Very Large Data Bases*, pages 523–535, Barcelona, 1991.

14. M. Kitsuregawa and Y. Ogawa. Bucket spreading parallel hash: A new, robust, parallel hash join method for data skew in the super database computer (SDC). In *Proc. 16th Int. Conf. on Very Large Data Bases*, pages 210–221, 1990.

15. M. Kitsuregawa, H. Tanaka, and T. Motooka. Application of hash to database machine and its architecture. *New Generation Computing*, 1(1):66–74, 1983.

16. M. L. Lo and C. V. Ravishankar. Spatial joins using seeded trees. In *Proc. ACM SIGMOD Int. Conf. on Management of Data*, pages 209–220, 1994.

17. M. L. Lo and C. V. Ravishankar. Spatial hash-join. In *Proc. ACM SIGMOD Int. Conf. on Management of Data*, pages 247–258, Montreal, Canada, 1996.

18. H. J. Lu, B. C. Ooi, and K. L. Tan. *Query Processing in Parallel Relational Database Systems*. IEEE Computer Society Press, 1994.

19. J. Orenstein and F. A. Manola. Probe spatial data modeling and query processing in an image database application. *IEEE Trans. Software Eng.*, 14(5):611–629, 1988.

20. J. M. Patel and D. J. DeWitt. Partition based spatial-merge join. In *Proc. ACM SIGMOD Int. Conf. on Management of Data*, pages 259 – 270, Montreal, Canada, 1996.

21. F. P. Preparata and M. I. Shamos. *Computational Geometry: an introduction*. Springer-Verlag, 1985.

22. T. Sellis, N. Roussopoulos, and C. Faloutsos. The R^+-tree: a dynamic index for multi-dimensional objects. In *Proc. 13th Int. Conf. on Very Large Data Bases*, pages 3–11, 1987.

23. M. Stonebraker, J. Frew, K. Gardels, and J. Meredith. The SEQUOIA 2000 storage benchmark. In *Proceedings of ACM SIGMOD Int. Conf. on Management of Data*, pages 2–11, Washington, DC, 1993.

24. X. Zhou. *Parallel Processing in Relational Database Systems*. PhD thesis, University of Queensland, 1994.

Orthogonal Polygons as Bounding Structures in Filter-Refine Query Processing Strategies

Claudio Esperança[1] * and Hanan Samet[2] **

[1] COPPE, Prog. Eng. Sistemas
Universidade Federal do Rio de Janeiro
Cidade Universitária, C.T., Sala H319
Rio de Janeiro, RJ, 21945-970, Brazil
E-mail: esperanc@lcg.ufrj.br
[2] Computer Science Department and
Center for Automation Research and
Institute for Advanced Computer Science
University of Maryland at College Park
College Park, Maryland 20742
E-mail: hjs@umiacs.umd.edu

Abstract. The use of bounding structures in the form of orthogonal polygons (also known as rectilinear polygons) with a varying number of vertices in contrast with a minimum bounding rectangle (an orthogonal polygon with just 4 vertices in two dimensions) as an object approximation method is presented. Orthogonal polygons can be used to improve the performance of the refine step in the filter-refine query processing strategy employed in spatial databases. The orthogonal polygons are represented using the vertex representation implemented as a vertex list. The advantage of the vertex representation implemented as a vertex list is that it can be used to represent orthogonal polygons in arbitrary dimensions using just their vertices. This is in contrast to conventional methods such as the chain code which only work in two dimensions and cannot be extended to deal with higher dimensional data. Algorithms are given for varying the number of vertices used to represent the objects. It is shown that the use of non-trivial orthogonal polygons (i.e., with more than four vertices) is of benefit when a spatial index is used in the filter step for processing spatial queries such as point-in-object and windowing. If no spatial index is used, then all objects must be examined. In this case, many of the objects are small thereby not benefiting from the variation in the number of vertices that they have as the simple bounding box is adequate.

1   Introduction

The efficient processing of queries is an important issue in spatial databases [10]. This is facilitated, in part, through the use of spatial indexing

---

* The support of Conselho Nacional de Desenvolvimento Científico e Tecnológico (CNPq), Projeto GEOTEC / PROTEM II is gratefully acknowledged.
** The support of the National Science Foundation under Grant IRI-92-16970 is gratefully acknowledged.

methods (e.g.,[22,23]) coupled with approximations. These techniques serve as the basis of a query processing strategy known as *filter-and-refine* [3]. In this case, the spatial indices serve to restrict the search to a subset of the data (termed the *filter* step) which is usually a subspace of the data. Once the subset has been defined, the queries are often executed with approximations of the underlying data as well as the query objects (termed the *refine* step). The rationale is to avoid excessive computation where the data is likely not to satisfy the query.

For example, determining the intersection point of two straight line segments is a surprisingly complex process due to issues of limited precision [15,17]. If we can determine that the segments do not intersect, then we can avoid the need to compute the intersection point thereby saving a considerable amount of effort. This can be done by approximating the two segments with some objects in which they are guaranteed to lie such as their enclosing rectangles, and then checking if the two enclosing objects overlap. This is a a relatively easy process especially if the rectangles are axes-aligned (i.e., their sides are parallel to the coordinate axes). This test can be made even more selective by ensuring that the rectangles fit the lines as closely as possible (termed a *minimum bounding rectangle (MBR)*).

We can generalize the line intersection problem to arbitrary curves in which case the selectivity of the refine step can be increased by loosening the requirement that the sides of the enclosing rectangles are parallel to the coordinate axes, thereby obtaining an enclosing object which is truly an MBR. Unfortunately, we are now confronted with our original problem of determining whether two arbitrary straight lines in space intersect since we must check if a pair of sides (one from each of the MBRs) intersect. Thus the only alternative is to loosen the restriction that the enclosing objects are rectangles and permit them to be orthogonal polygons[1] (also known as rectilinear polygons [5] having a varying number of vertices (i.e., more than four).

Interestingly, orthogonal polygons are not always the result of the approximation of the data. Instead, at times the original data is in the form of an orthogonal polygon due to the manner in which it was acquired (e.g., using raster conversion). For example, in a GIS application, orthogonal polygons can be used to indicate in what parts of a country a given crop is grown or the boundaries of lakes and water bodies.

These orthogonal polygons can be represented using the vertex representation [6,24]. The use of vertex representation as the representation of orthogonal polygons is the subject of this paper. The advantage of these methods is that they work for data of arbitrary dimensionality while still only recording the vertices of the orthogonal polygon. This is quite powerful as most higher-dimensional object representations make use of several of the

---

[1] In this paper we use the term *orthogonal polygon* in its most general sense to refer to data of arbitrary dimensionality.

lower dimensional components to represent the object. For example, in three dimensions, we may use the winged-edge representation [1,23] (also known as a BRep or a boundary model [16]) which represents the boundary of the object in terms of its vertices, edges, and faces.

In contrast, the vertex representation enables the representation of the object just using its vertices and the implicit assumption of orthogonal boundary elements. Notice that such implicit orthogonality assumptions about the boundary connectivity are also made when using chain codes [7] to represent two-dimensional data. However, it is impossible to generalize chain codes to higher dimensions as unlike the vertices (really edges) in two dimensions which are ordered in sequence of connectivity since each boundary element can be adjacent to just two elements, for $d > 2$ there is no such order associated with $(d - 1)$-dimensional hyperplanes which are the boundary elements in d-dimensions.

The rest of this paper is organized as follows. Section 2 reviews the concept of a bounding structure and gives the key properties that make it useful in the processing of queries in spatial databases. Section 3 describes the use of the vertex representation of orthogonal polygons as bounding structures. This includes a definition of the vertex list implementation of the vertex representation. Section 4 discusses the actual process of dynamically varying the number of vertices in the bounding structures (termed *coarsening*) and gives an algorithm for doing so. The results of some experiments using the vertex representation as a bounding structure are given in Sect. 5 while concluding remarks are drawn in Sect. 6.

2 Properties of Bounding Structures

As mentioned in Sect. 1 in some situations it is advantageous to trade an exact geometric representation for an approximate one, provided it can be processed more efficiently. This is the key observation behind what we call the *bounding structure* technique. The most common type of bounding structure is the axes-aligned rectangle, mainly because of its simple geometry. Minimum bounding rectangles are used extensively in spatial databases, mainly in the design of spatial access methods. The R-tree [11] and its variations [2,25], for instance, rely heavily on the bounding properties of rectangles. Polygons, on the other hand, are in general too complex to be suitable as bounding structures, although they have been investigated and have been found promising in certain applications [4,13]. An alternative to the bounding structure as an object approximation is a decomposition of the space spanned by the object into a collection of cells (possibly, but not necessarily, disjoint). Such representations include the region quadtree [12,14] as well as the R+-tree [25]. The variation in the number of cells as an approximation technique has been studied [8,18]. We do not discuss this approach further in this paper.

One of the main uses of bounding structures is to help in the evaluation of spatial predicates. Consider a spatial object S, where S may be a line, a polygon, a collection of line segments or any other object that can be regarded as a set of points. We may use set notation to denote operations and predicates about spatial objects. For example, we may use predicate $S \cup R = \emptyset$ to say that spatial objects S and R do not overlap in space. Depending on the nature of S and R, the evaluation of such a predicate may vary in complexity from trivial (e.g., when both S and R are points) to very hard (e.g., when both S and R are sets of polygons). On the other hand, bounding structures are usually very simple geometric objects, so that evaluating common spatial predicates is never too costly.

One property of bounding structures that enables us to evaluate predicates on their corresponding bounded objects is the *subset property*. If $B(S)$ is a bounding structure for S, then S must be a subset of $B(S)$ (written $S \subseteq B(S)$). For example, if two objects have a non-empty intersection, then their bounding structures also intersect, i.e.:

$$S \cap R \neq \emptyset \Longrightarrow B(S) \cap B(R) \neq \emptyset.$$

An equivalent assertion is to say that if two bounding structures *do not* intersect, then their bounded objects cannot intersect either, i.e.:

$$B(S) \cap B(R) = \emptyset \Longrightarrow S \cap R = \emptyset.$$

Another frequent use of bounding structures is to help estimate distances between spatial objects. Let $\delta(a, b)$ denote the distance between two points a and b according to some metric. Then, the distance between two sets S and R, written $\delta(S, R)$, is defined as the minimum distance between two points belonging to S and R, respectively. Formally,

$$\delta(S, R) = \min_{s \in S, r \in R} \delta(s, r).$$

One could also consider the *maximum distance* (written $\Delta(S, R)$) as a way of defining the distance between two sets, i.e.,

$$\Delta(S, R) = \max_{s \in S, r \in R} \delta(s, r).$$

Bounding structures are of little use if the exact distance between two objects is to be computed. However, in many applications, distance computations are used in predicates to measure closeness between two objects, e.g., "$\delta(S, R) < 10$" or "$\delta(S, R) > 5$". In these cases, bounding structures can be used to give estimates which are sometimes sufficient to avoid computing the actual distance between the corresponding bounded objects. In particular, it is easy to see that the minimum and maximum distances between bounding structures provide bounds for the (minimum) distance between the bounded objects, i.e.,

$$\delta(B(S), B(R)) \leq \delta(S, R) \leq \Delta(B(S), B(R))$$

Depending on the nature of the bounding structure $B(S)$, it is possible to estimate a better (tighter) maximum bound for $\delta(S, R)$. As an example, consider a minimum bounding rectangle (MBR). If S is an object defined in $\mathcal{R}^d$, then its MBR has $2 \cdot d$ faces and, by construction, each face is guaranteed to contain at least one point of S; this is called the *MBR face property* [21]. If this property holds for bounding structure $B(S)$, then an upper bound for $\delta(S, R)$ can be established by computing the maximum distance between each pair of faces of $B(S)$ and $B(R)$ and taking the minimum of these. Let us call this measure $\Delta_\partial(B(S), B(R))$. Then, using the notation $\partial(B(S), i)$ to refer to each of the $2 \cdot d$ faces of $B(S)$, we can define

$$\Delta_\partial(B(R), B(S)) = \min_{i=1\ldots2\cdot d, j=1\ldots2\cdot d} \Delta(\partial(B(S), i), \partial(B(R), j))$$

Actually, we may think of $\partial(B(S), i)$ as any subset of $B(S)$ which is guaranteed to contain a point of S. Notice that $\Delta_\partial(B(R), B(S))$ is often[2] less than and can never be greater than $\Delta(B(R), B(S))$ (see Fig. 1).

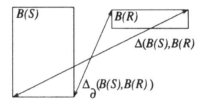

Fig. 1. Comparison between $\Delta(B(S), B(R))$ and $\Delta_\partial(B(S), B(R))$ for two sample objects.

We end this section by summarizing properties of a bounding structure $B(S)$ with respect to its corresponding bounded object S.

1. $S \subseteq B(S)$.
2. $B(S) \cap B(R) = \emptyset \implies S \cap R = \emptyset$.
3. $\delta(B(S), B(R)) \leq \delta(S, R) \leq \Delta(B(S), B(R))$.
4. If we can extract from $B(S)$ (and $B(R)$) subsets $\partial(B(S), i)$ such that each of them contains at least one point of S, then $\delta(S, R) \leq \Delta_\partial(B(S), B(R)) \leq \Delta(B(S), B(R))$.

3 Orthogonal Polygons as Bounding Structures

An orthogonal polygon is a polygon whose sides are perpendicular to the coordinate axes. Axes-aligned rectangles are the most common instances of

---

[2] In fact, it can be shown that $\Delta_\partial(B(R), B(S))$ is always smaller than $\Delta(B(R), B(S))$, unless the boundary of $B(S)$ (and $B(R)$) is equal to $B(S)$ itself.

orthogonal polygons and the fact that they can be represented and processed efficiently has made them popular as bounding structures. Simple non-orthogonal polygons can be represented economically as a circular list of its endpoints and can be tuned to approximate a given region in two-dimensional space. However, certain key bounding structure operations are relatively hard to compute on polygons, e.g. testing whether two polygons intersect or computing the distance between them. Moreover, although polygons may easily be defined in multi-dimensional spaces, the effort to represent and process them in more than two dimensions makes them unattractive as bounding structures.

Recently, a new representation for orthogonal objects based on a data structure called the *vertex representation* has been presented [6,24]. This representation can be implemented in a manner that makes it possible to store an orthogonal polygon in any (integer) number of dimensions as a list of its vertices (termed a *vertex list*)[3]. Many common operations on orthogonal polygons represented by vertex lists may be efficiently computed by using algorithms based on the *sweeping plane* [20] approach. In particular, set operations such as those required for the processing of bounding structures have expected $O(d \cdot N)$ time complexity[6], where d is the number of dimensions and N is the total number of vertices in the vertex list(s). Our contribution is in the use of orthogonal polygons as bounding structures as well as the computation of morphological operations such as coarsening which is described in Sect. 4.

3.1 Vertex Representation

A polygon in two dimensions is usually represented as a circular list of its endpoints. This scheme cannot be extended to three or more dimensions since there is no obvious way in which vertices can be enumerated sequentially. The vertex list implementation solves his problem by using the order in which a hyperplane (e.g., a line in 2D or a plane in 3D) would encounter the vertices while sweeping the space from minus- to plus-infinity along one of the coordinate axes. Vertices that lie on the same hyperplane are similarly ordered by imagining a hyperplane of lower dimension sweeping that space. In addition to the coordinate values of its position in space, each vertex is accompanied by a scalar value termed the vertex *weight*[4]. Fig. 2 depicts an orthogonal polygon represented as a vertex list.

---

[3] Note that a polygon in three dimensions is called a *polyhedron* or, in general, a *polytope*. In this paper we will mostly deal with examples in two dimensions, and thus we opted to use the less precise term "polygon", although the described techniques can be applied to orthogonal polytopes of any number of dimensions

[4] Henceforth, we will use the term "vertex" to refer to weighted points, and the term "vertex list" to refer to a list of such weighted points. The term "vertex list" is not to be confused with the common way of representing two-dimensional polygons as a circular list of its endpoints.

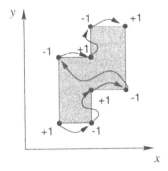

Fig. 2. An orthogonal polygon in 2D represented as a vertex list.

Intuitively, each vertex corresponds to the tip of an infinite cone. Each cone is equivalent to an unbounded object formed by the intersection of d (assuming d-dimensional data) orthogonal halfspaces that are parallel to the d coordinate axes passing through the vertex. The weights of the vertices have signs which serve to indicate whether the space spanned by their cones is included or excluded from the object being modeled. The space spanned by the union of the cones defines the object being modeled. Formally, the vertex representation is intended as a representation for orthogonal scalar fields. A scalar field is simply a function that maps points of the domain space to scalar values. An orthogonal scalar field is a scalar field where regions of the domain space which are mapped to the same value are delimited by faces orthogonal to the coordinate axes. The overall idea which enables the vertex representation to model orthogonal polygons is to think of these as scalar fields where points inside the polygon are mapped to 1, and points outside the polygon are mapped to 0. A vertex at a point $p = (p_1, p_2, \ldots p_d)$ and weight w has the effect of adding w to the mapping of all points $q = (q_1, q_2, \ldots q_d)$ such that $q_i \geq p_i$ for all $i = 1 \ldots d$ (see Fig. 3a). As a consequence, one can tell if a point q is inside or outside the polygon by adding the weights of all vertices that contribute to its mapping, i.e., a vertex at point p contributes to the mapping of q if and only if $p_i < q_i$ for all $i = 1 \ldots d$ (see Fig. 3b).

3.2 Fitting an Orthogonal Polygon around a Spatial Object

The general procedure for creating an orthogonal polygon $B(S)$ that can serve as a bounding structure for a spatial object S consists of:

1. Partitioning S into smaller objects $s_1, s_2 \ldots s_k$.
2. Finding the minimum bounding rectangle $MBR(s_i)$ for each sub-object s_i.
3. Computing the union of all MBRs: $\bigcup_{i=1}^{k} MBR(s_i)$.

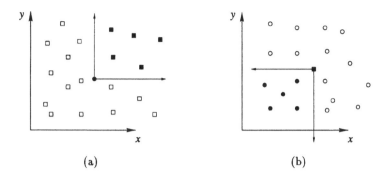

Fig. 3. (a) A vertex (black dot) and some of the points influenced by it (black squares). (b) The point (black square) is only influenced by the vertices shown as black dots.

The problem of partitioning an object into smaller objects is frequently trivial (e.g., a set of line segments can be partitioned by taking each line segment separately) or may require some additional processing. For example, we might use triangulation [19] to split a (simple) polygon into triangles. An MBR for each triangle can be computed easily yielding four vertices which can then be stored in a list. To compute the union of all MBR's, we use the algorithm for performing set operations on vertex list implementations of vertex representations as described in [6]. This is illustrated in Fig. 4.

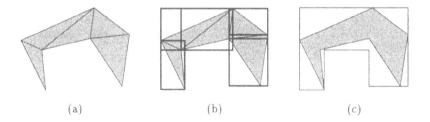

Fig. 4. (a) Polygon is split into triangles, (b) an MBR is fitted around each triangle, (c) taking their union produces an orthogonal bounding polygon.

In some cases, the boundary of a spatial object is amenable to trivial partitioning, while the object itself is not. A general polyhedron, for instance, is delimited by polygonal faces and these can be partitioned with little difficulty by using triangulation as described above. However, the partitioning of its interior, say into tetrahedrons, is rather complicated. In such cases, we may compute an orthogonal bounding polygon for the object by performing a morphological operation known as *closing*[5] on its boundary. The operation of *closing* for rectangular structuring elements can be computed efficiently on orthogonal objects represented by a set of vertices and stored as a vertex list [6]. Loosely speaking, *closing* has the property of filling gaps and holes in polygons. A more precise description of the procedure is as follows:

1. Compute $MBR(S)$, the minimum bounding rectangle of object S.
2. Partition the *boundary* of S into components $s_1, s_2 \ldots s_k$.
3. Find the minimum bounding rectangle $MBR(s_i)$ for each sub-object s_i.
4. Compute the union of the MBRs of all subobjects, i.e., $\bigcup_{i=1}^{k} MBR(s_i)$.
5. Perform a *closing* operation on the orthogonal polygon obtained in (4) using $MBR(S)$ as the structuring element.

This procedure is illustrated in Fig. 5 for a simple polygon in two dimensions. A side effect of this method is that it always produces orthogonally convex polygons,[6] which might be undesirable for non-convex objects.

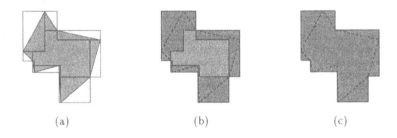

(a)   (b)   (c)

Fig. 5. (a) An MBR for each edge of the polygon is computed. (b) The union of these may produce an orthogonal polygon with holes. (c) A *closing* operation fills eventual holes and gaps.

It is not difficult to see that the orthogonal polygons obtained through the use of the above methods satisfy the bounding structure properties outlined in Section 2. Nevertheless, if $B(S)$ is such a polygon, then property 4 requires us to define a method for obtaining suitable subsets $\partial(B(S), i)$. We notice that both methods for building orthogonal polygons use unions of MBRs as starting points and, although the faces of the MBRs are guaranteed to contain at least one point of the bounded object, the faces of the resulting

---

[5] Due to space limitations, we cannot provide here background information on morphological operations. We encourage readers who might be unfamiliar with the subject to refer to any good text on image processing (e.g., [9]).

[6] A polygon in two dimensions is orthogonally convex if its intersection with any horizontal or vertical line has at most one connected component [19]. This definition can be extended to any number of dimensions by replacing "line" with "hyperplane" and "horizontal/vertical" with "normal to a coordinate axis".

orthogonal polygon do not necessarily do so (see Fig. 6). However, some faces of such polygons do have that property, namely those that constitute local minima or maxima. A face f is said to constitute a local minimum or maximum if all adjacent faces lie to the same side of the plane defined by f. Such faces can easily be extracted from a vertex list implementation of a vertex representation by means of a plane-sweep algorithm [6]. Note that local minimum and maximum faces can then be used as subsets $\partial(B(S), i)$ in the context of property 4 as explained in Sect. 2.

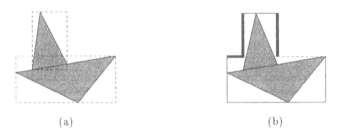

(a)   (b)

Fig. 6. Although faces of minimum bounding rectangles are guaranteed to contain at least one point of the bounded objects (a), the faces of the orthogonal polygon resulting from their union (b), unless they are local maxima or minima (e.g., faces drawn with thin lines), do not necessarily preserve that property (e.g., faces drawn with thick lines).

4   Coarsening Orthogonal Polygons

The primary objective of bounding structures is to provide a good approximation of the bounded object while taking as little space as possible. Thus, when we consider using an orthogonal polygon for this purpose, it is essential to be able to achieve an optimum tradeoff between tightness and space. It is possible to obtain a vertex representation which is as close to a given spatial object as desired. All that is necessary is to partition that object into a suitable number of small fragments and apply the procedure outlined in Section 3.2. However, the vertex representation produced in this way may be too complex (i.e., it may have too many vertices). Therefore, some procedure must be devised which will produce less detailed (coarser) versions of orthogonal bounding polygons while preserving, as much as possible, their usefulness as a bounding structure. In this section, we describe two approaches to solve this problem.

4.1 The "Moving Faces" Approach

This approach derives from the observation that, in general, if a face which is perpendicular to one of the coordinate axes is displaced along that axis, the total number of vertices of the representation will never increase. By choosing an appropriate displacement, it is possible to guarantee that the resulting polygon will not only be coarser (i.e., have less vertices) than the original, but will also preserve the bounding properties outlined in Section 2.

Given an orthogonal polygon in d dimensions represented by a vertex list, it is possible to sweep a plane along the dth coordinate axis and extract the faces perpendicular to that axis by a simple sequential scan of the list. These faces can be divided into two groups, which we term the *positive* and *negative* faces. A positive face corresponds to a region where the sweeping plane *enters* the polygon in its movement from minus- to plus-infinity along the dth coordinate axis, whereas a negative face corresponds to a region where the sweeping plane *leaves* the polygon (see Fig. 7a). Coarsening is achieved

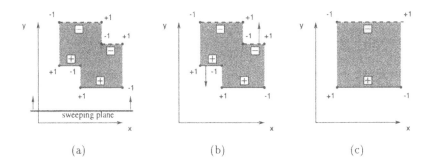

Fig. 7. 2D example of coarsening with the "moving faces" approach. Positive and negative faces are identified (a) and moved away from the polygon until they meet another face of the same sign (b), resulting in a coarser polygon (c).

by moving a positive face until it touches another positive face or, similarly, moving a negative face until it touches another negative face. As we must maintain the bounding properties of the polygon, faces must be moved in directions that will result in polygon expansion. This means that positive faces are moved in the opposite direction of the coordinate axis and negative faces are moved in the same direction of the coordinate axis (see Fig. 7b). Once faces are moved in this fashion, some vertices of the moved face will cancel out vertices of the face to which it is now adjacent, thereby resulting in a coarsened polygon (Fig. 7c).

Notice that since no faces are actually created in the process, faces which were guaranteed to contain a point of the bounded object still do so in the coarsened polygon, thus preserving property 4 as described in Section 2.

A full implementation of the "moving faces" approach requires that the process be repeated for faces perpendicular to all coordinate axes. This can be achieved by reordering the list so that a plane swept along each axis may find the corresponding faces. Also, the choice of which faces to move must take in consideration the size of each candidate face and the amount of displacement required. A rule of thumb is to choose the face displacements which will result in the least enlargement of the original polygon.

4.2 The "Rectangle Pairs" Approach

The moving faces approach will fail to produce a coarsened polygon in some cases. For example, some polygons with holes or having more than one connected component (see Fig. 8) may not contain suitable positive or negative faces. In such cases, a more general approach is needed.

(a)   (b)

Fig. 8. Some polygons with holes (a) or having more than one connected components (b) cannot be coarsened with the "moving faces" approach.

In the "rectangle pairs" approach, the original polygon is first divided into a set of rectangles (see Fig. 9a). ¿From these, a few pairs are chosen and, for each pair, a minimum bounding rectangle (MBR) is constructed (Fig. 9b). The union of the original polygon and MBRs will result in a coarsened polygon (Fig. 9c). The overall idea is that each selected pair of rectangles should contain a number of vertices of the original polygon and, by replacing them with a minimum bounding rectangle, some of these vertices will be eliminated from the representation. A quick calculation shows that, since in d dimensions a rectangle has 2^d vertices, the substitution of two rectangles r_a, r_b by their minimum bounding rectangle $MBR(r_a, r_b)$ may eliminate up to 2^d vertices from the representation.

In an actual implementation of this algorithm, we must address the following points:

1. How to divide an orthogonal polygon into a set of rectangles? This problem can be solved with Shechtman's algorithm for splitting a polygon in maximal horizontal rectangles in two dimensions [24]. This algorithm was

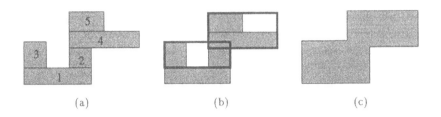

Fig. 9. Coarsening with the "rectangle pair" approach. Polygon is divided into rectangles (a), some pairs are chosen (2-3 and 4-5) and a MBR for each pair is computed (b). Coarsened polygon is union of MBRs and original polygon (c).

extended in [6] to deal with polygons in arbitrary dimensions. Fig. 9a is an example of such a subdivision.

2. Which rectangle pairs to choose? Ideally, rectangles which are close to each other should be considered first. Another consideration is the total amount of "wasted space" introduced when substituting a pair of rectangles by its MBR. The algorithm for dividing polygons into rectangles returns those rectangles in the same order in which the sweeping plane finds them. Although this order does not guarantee that two consecutive rectangles are the closest pair to each other, it does provide a certain amount of clustering. Thus, the selection of pairs of rectangles may consider consecutive rectangles and reject those which are too distant from each other or those whose pair-wise MBR is too large compared with the size of the rectangles.

4.3 The Complete Algorithm

We have implemented a complete coarsening algorithm using both approaches just described. The algorithm takes as input a d-dimensional orthogonal polygon $B(S)$ with n vertices and the desired maximum number of vertices k. Its output is an orthogonal polygon containing no more than k vertices that can be used as a bounding structure for S satisfying the properties outlined in Section 2.

The algorithm consists of a heuristic method for combining the strengths of both approaches in a controlled way. The ultimate goal is to obtain a bounding structure of limited complexity which nevertheless approximates the bounded object as closely as possible. A reasonable strategy involves estimating how much the polygon will grow as each modification prescribed by either approach is applied. In the "moving faces" approach, selecting a face f and displacing it by i units can be seen as computing the union of the original polygon with a "plug" polygon given by sweeping f perpendicularly by i units. Similarly, in the "rectangle pairs" approach, if the MBR of a pair of rectangles $(r1,r2)$ is used in the coarsening process, then we may estimate that the original polygon will grow by an amount represented by the regions of $MBR(r1, r2)$ not already occupied by $r1$ or $r2$. In other words, we can imagine that we applied to the original object a "plug" polygon corresponding to $MBR(r1, r2) \setminus (r1 \cup r2)$[7]. Overall, we can consider each plug shape as a possible modification that will coarsen the original polygon. The amount of growth due to this modification can be estimated as the integral (e.g. area in 2D or volume in 3D) of the plug. This enables us to eliminate plug polygons which are too large.

The complete algorithm is summarized below.

1. If the length of list $B(S)$ is less or equal to k, then return $B(S)$.
2. Initialize array $Plug$ which will hold plug polygons.
3. Repeat the following steps for all coordinate axes:
   (a) Use the "moving faces" approach to compute faces which can be displaced to coarsen $B(S)$. For each candidate face f, compute also the corresponding displacement i. Create a plug polygon by sweeping f by i units and add it to $Plug$.
   (b) Use the "rectangle pairs" approach. Perform a rectangle partition of the polygon. Take every two consecutive rectangles $(r1, r2)$ as returned by the rectangle partitioning algorithm and compute their pair-wise MBR. Compute each plug polygon given by $MBR(r1, r2) \setminus (r1 \cup r2)$ and add it to $Plug$.
   (c) Reorder the vertices of vertex list $B(S)$ so that plane sweeping may occur along another coordinate axis.
4. Compute the integral of each polygon in $Plug$ and sort $Plug$ in increasing order of integrals.
5. If $Plug$ has m polygons at this point, then discard the $\lfloor m/2 \rfloor$ largest polygons. This heuristic aims to consider only "small" plug polygons. Applying "big" plug polygons might lead to a faster coarsening, at the expense of achieving worse approximations. Note that up to $m - 1$ plug polygons could be discarded, but at least one plug polygon must be considered so that the coarsening process terminates.
6. Scan $Plug$ sequentially and apply each plug polygon to $B(S)$ (i.e, compute the union of $B(S)$ and the plug) until the resulting polygon has k vertices or less or $Plug$ is exhausted. In the latter case, return to step 2.

Step 6 deserves a more detailed explanation. Since $B(S)$ and plug polygons are represented by of vertex lists, computing their union takes time proportional to the sum of their (list) lengths [6]. Hence, if $B(S)$ is a fairly long list, then repeatedly computing its union with each plug polygon would be too costly. Fortunately, there is no need to compute the union for each plug polygon, only to count how many vertices the resulting polygon will have. In practice, one may use a data structure, say $Match$, which is suitable for storing and searching points in d dimensions (e.g., a k-d-tree [23]).

---

[7] The symbol "\" is used here to denote set difference.

Initially, *Match* is initialized with all vertices of $B(S)$. Then, as each plug polygon is considered, *Match* is searched for vertices at identical positions. When matching vertices are found and these have opposite signs, the length of the result would decrease by 1; if not, the length of the result increases by 1. Once all plug polygons are processed in this fashion, the resulting coarsened $B(S)$ can be built by collecting all vertices in *Match*.

The time complexity analysis of this algorithm is rather involved and is not given here (see [6]). However, we state without proof that the time complexity of the coarsening algorithm can be estimated at $O(2^d \cdot N \cdot \log^2 N)$ on average, where N is the number of vertices in the original polygon. The $2^d \cdot N$ factor is the average number of vertices in all plug polygons generated in step 3. One $\log N$ factor comes from the search used in *Match* for identical vertices. The remaining $\log N$ factor arises from step 5 which causes the algorithm to be performed $\log N$ times.

It must be observed that the approach described above for obtaining coarse approximations of complex objects is not too attractive if the desired approximation contains much fewer vertices than the original object. For instance, if we are trying to obtain an MBR for a complex object, a straightforward algorithm that computes the limits of the object in each dimension is much more efficient. Thus, in these cases, an alternative approach would consist of starting with an MBR and, by means of successive "erosion" steps, produce a finer approximation until the desired level of approximation is obtained. Such an algorithm is currently being investigated.

5 Experiments

In this section we empirically evaluate both the proposed coarsening algorithm and the usefulness of orthogonal polygons as bounding structures. Whereas bounding structures are used for many purposes in several application areas, we focus on a typical application of bounding structures in the context of a spatial database system.

The experiments were conducted on data extracted from a digital map depicting forest coverage types in Canada. A raster image of this map was provided by Canada's National Atlas Information Service[8] and is shown in Fig. 10. The data consists of a collection of orthogonal polygons delimiting the homogeneous 4-connected regions of the raster image. The algorithm for converting raster images into a set of orthogonal polygons stored as vertex lists is the one described in [6]. A total of 403 polygons ranging from 4 to 1408 vertices (average 50.8) were found.

---

[8] The URL http://elllesmere.ccm.emr.ca/naismap/naismap.html locates this data in the World Wide Web

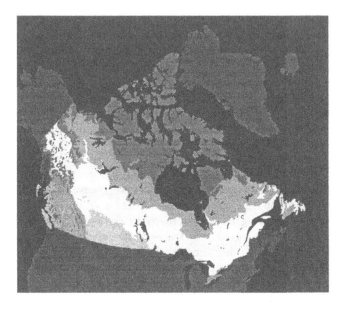

Fig. 10. Forest coverage by type in Canada

5.1 Coarsening experiments

A series of experiments were conducted to determine the general behavior of the proposed algorithm in coarsening the polygons of the sample data set. All polygons were first coarsened to no more than 256 vertices. This limit was repeatedly halved until the coarsened polygons contained exactly 4 vertices, which is a lower bound for polygons in two dimensions. Fig. 11 shows the result of coarsening the polygons to no more than 64, 16 and 4 vertices, respectively.

The graph in Fig. 12 shows the time taken by the coarsening algorithm to produce the 4 vertex MBR of the polygons of different sizes. Although the final 4 vertex MBRs could have been obtained just as easily through the standard procedure of finding the minimum and maximum values of the x and y coordinate values, our goal was to determine how the complexity (i.e., number of vertices) of the original polygon impacts the speed of coarsening. As can be observed from Fig. 12, the graph curve is somewhat irregular due to the small number of polygons of high complexity. Nevertheless, the time complexity of the algorithm seems to be in accordance with the estimate presented in Section 4.3.

We were also interested in evaluating how coarsening affects the area of an orthogonal polygon. The area of a coarsened polygon, when compared with that of the original polygon, provides a measure of how tightly one shape bounds the other. With this in mind, for each coarsening level we computed

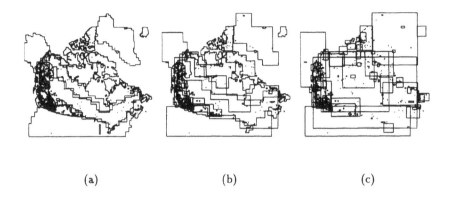

(a)   (b)   (c)

Fig. 11. Map of Canada's forest coverage by type, where each orthogonal polygon was coarsened to no more than 64 (a), 16 (b) and 4 vertices (c).

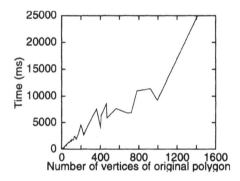

Fig. 12. Plot of times taken by algorithm *coarsen* to reduce to 4 the number of vertices of orthogonal polygons.

the result of taking the sum of the areas of all coarsened orthogonal polygons and dividing that value by the sum of the areas of the original polygons (i.e., their ratio). The plots of these ratios as a function of the coarsening level are shown in Fig. 13a. Note that in this evaluation, total areas for the whole data set were considered, which leaves room for inordinately large polygons to influence the result. This effect can be reduced if we compare relative areas on a per-polygon basis. The relative area consists of the fraction obtained by dividing the area of a coarsened polygon by that of the corresponding original polygon. In Fig. 13b we plot average relative areas.

5.2 Adaptive coarsening

In a spatial database environment, a data set such as the one we have chosen for our experiments could be accessed on the basis of several selection

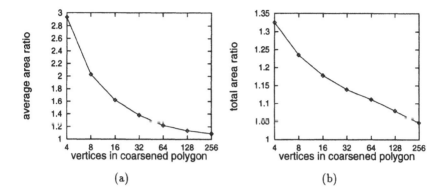

Fig. 13. Plot of ratio between areas covered by orthogonal polygons after and before coarsening. (a) Refers to the ratio between the total area of all coarsened polygons and the total area of all original polygons. (b) Shows the average area ratios between each coarsened and non-coarsened polygon.

criteria. A very common operation in spatial databases is the retrieval of objects that satisfy a condition expressed in terms of spatial proximity. The processing of such queries frequently relies on heuristics employing bounding structures to prune out objects which cannot possibly satisfy the condition. The effectiveness this pruning process depends on two factors:

1. How closely the bounding structure approximates the bounded object.
2. How much extra work must be done to determine if the bounded object satisfies the condition in case the test against the bounding structure is not conclusive.

This suggests that in order to compute a bounding structure of optimal complexity for a given object one has to take into consideration not only the object's complexity but also its extent (i.e., size).

Consider a simplified scenario where a collection of objects is queried in order to select the ones which intersect a single point p (termed the *point intersection* query). We will assume that the position of p is uniformly distributed over a rectangular region A (e.g., the limits of the original map image). Thus, we can estimate the following costs incurred in deciding whether a sample object S intersects p:

1. Determine if $B(S)$ intersects p. This can be done at a cost that is linearly dependent on the complexity (number of vertices) of $B(S)$. If we stipulate that constant c is an estimate of the per-vertex cost of an intersection computation, then the total cost of this first step is:

$$C_1 = c \times length(B(S)).$$

2. Test S against p. In our case, since S is also represented by a vertex list, the cost of such a test is given by $c \times length(S)$. However, this step only takes place if $B(S)$ intersects p. The probability of this event is proportional to area of $B(S)$. In the limit, that is, when $B(S)$ is equal to A, the probability is 1. Therefore, we may estimate that the cost due to this step is

$$C_2 = c \times length(S) \times \frac{area(B(S))}{area(A)}.$$

Cost estimates C_1 and C_2 were used to perform an adaptive coarsening of the polygons in our sample data set. Starting with a very fine approximation of each polygon, this was progressively coarsened until the value of $C_1 + C_2$ ceased to decrease. Since this process only requires that successive cost estimates be compared, the actual value of constant c is of no importance. Notice that if the successive values of $C_1 + C_2$ do not decrease at all, the first (finest) approximation is used. If they decrease monotonically, the coarsest approximation (i.e. the MBR) is used. One could also argue that many inflection points in this sequence of values could exist thereby leading to a false stopping point; however, this did not occur in our experiments (see Section 5).

One last consideration must be made concerning the sampling area represented by A. The assumption that A spans the space covered by all of the map is consistent with a query processing strategy based on sequential search. A more realistic scenario, however, must allow for the use of spatial indices. In other words, if a spatial index is used to filter out values which cannot satisfy the query, then a polygon will be tested only if the index itself was incapable of giving a positive diagnostic. For example, if polygons are indexed by means of an R-tree [11], then any given polygon will be tested only if p falls within its MBR. We can easily modify our adaptive coarsening procedure to take spatial indices into consideration by adjusting the value of A to that of $MBR(S)$.

In our experiments, adaptive coarsening produced the following average number of vertices for $B(S)$:

Average number of vertices	
Sequential search	4.1649
Spatial Index	4.9123

When using a sequential search strategy to test the entire set of polygons against the query point, we find that the average number of vertices at the time the adaptive coarsening process halted was very close to 4 (i.e., 4.1649) which means that the resulting bounding structure was a rectangle (i.e., an orthogonal polygon with 4 vertices). This implies that it is not worth our time to use a bounding structure that has a varying number of vertices. Instead, we can use an MBR. On the other hand when using a spatial index to decide which polygons should be tested against the query point, a tighter bounding structure does lead to better results as can be seen by the fact that the average number of vertices when the adaptive coarsening process halted was close to 5 (i.e., 4.9123). These results are consistent with our intuition in the sense that executing the point intersection query by testing every polygon against the point (i.e., using sequential search) requires testing $B(S)$ for every value of S in the data set, and thus this test should be as fast as possible. In contrast, a spatial index already provides a good way of pruning the search based on spatial proximity, so that fewer objects are tested. Thus, in order to be reduce the set of candidate objects from consideration, it is necessary to use tighter, i.e., more elaborate bounding structures. Loosely speaking, we may think that, in this case, the spatial index provides a rough approximation for the spatial distribution of the objects, while tighter bounding structures are used in a second stage to prune the search even further before testing the actual objects.

5.3   Window query experiments

A series of window query experiments were conducted in order to evaluate the usefulness of adaptive bounding structures in a more realistic spatial database environment. This type of query consists in selecting all objects which intersect a given rectangle, or "window". This query is a two-dimensional generalization of the point intersection query discussed in Section 5.2. Experiments were made for window sizes ranging from 1% to 50% of the total map area using both sequential search and spatial index-based strategies. The same experiments were also conducted using MBRs (as opposed to adaptive orthogonal polygons) and using no bounding structures whatsoever. The experiments involving spatial indices used a PMR-quadtree [23] with a maximum depth of 10, and a splitting threshold value of 16. Fig. 14a shows the total query evaluation times for the experiments using sequential search, and Fig. 14b shows the equivalent results obtained with a spatial index based strategy.

One conclusion that can be drawn from these results is that the more flexible bounding structures derived from our adaptive approach lead to better query performance in all cases. The difference is more evident in the tests using a spatial index, but this was expected since the adaptive scheme only prescribed bounding structures significantly more complex than MBRs when indices were used. Further evidence that adaptive bounding structures are more beneficial when used in conjunction with spatial indices is shown in Fig. 15 which plots the average number of times in which bounding structures failed to determine if the object satisfied the query.

Overall, although the empirical tests do not demonstrate a dramatic improvement when adaptive bounding structures are used, they suggest that bounding structures of variable complexity might lead to significantly enhanced processing of some queries, e.g. window selection queries for large windows (see Fig. 14).

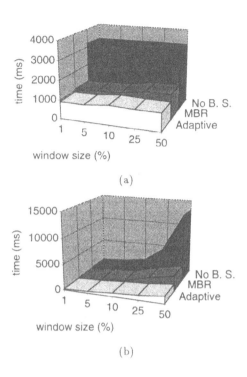

Fig. 14. Comparison of running times for different bounding structure configurations, using the sequential search (a) and the spatial index-based query evaluation plans (b).

6 Concluding remarks

The use of bounding structures in the form of orthogonal polygons with a varying number of vertices in contrast with a minimum bounding rectangle (an orthogonal polygon with just 4 vertices in two dimensions) as an object approximation method is presented. Orthogonal polygons can be used to improve the performance of the refine step in the filter-refine query processing strategy employed in spatial databases. The orthogonal polygons are represented using the vertex representation implemented as vertex lists. The advantage of the vertex representation is that it can be used to represent orthogonal polygons in arbitrary dimension using just their vertices. This is in contrast to conventional methods such as the chain code which only work in two dimensions and cannot be extended to deal with higher dimensional data.

Algorithms were given for varying the number of vertices used to represent the objects. We showed that use of non-trivial orthogonal polygons (i.e., with more than four vertices) is of benefit when a spatial index is used in the filter

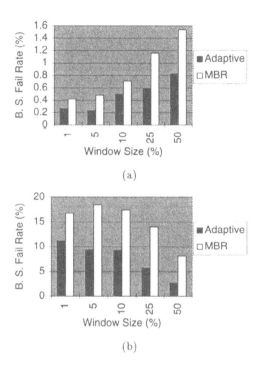

Fig. 15. Relative number of times bounding structure tests fail to give a positive answer; (a) rates for the sequential search plan, and (b) rates for the index-based plan.

step for processing spatial queries such as point-in-object and windowing. If no spatial index is used, then all objects must be examined. In this case, many of the objects are small thereby not benefiting from the variation in the number of vertices that they have as the simple bounding box is adequate.

Future work includes a thorough comparison between the adaptive bounding structures presented in this paper and other fixed-size bounding structures proposed, e.g., in [3]. We also plan to incorporate vertex representations implemented as vertex lists in real spatial database query processor and building query processing strategies around its use. These strategies would be embedded in a query optimizer for queries containing conditions that involve values of both spatial and non-spatial attributes.

References

1. B. G. Baumgart. Geometric modeling for computer vision. Stanford Artificial Intelligence Laboratory Memo AIM-249 STAN-CS-74-463, Stanford University, Stanford, CA, October 1974.

2. N. Beckmann, H. P. Kriegel, R. Schneider, and B. Seeger. The R^*-tree: an efficient and robust access method for points and rectangles. In *Proceedings of the SIGMOD Conference*, pages 322–331, Atlantic City, NJ, June 1990.

3. T. Brinkhoff, H. P. Kriegel, R. Schneider, and B. Seeger. Multi-step processing of spatial joins. In *Proceedings of the 1994 ACM SIGMOD International Conference on Management of Data*, pages 197–208, Minneapolis, MN, June 1994.

4. A. Brodsky, C. Lassez, J. Lassez, and M. J. Maher. Separability of polyhedra for optimal filtering of spatial and constraint data. In *Proceedings of the Fourteenth ACM SIGACT-SIGMOD-SIGART Symposium on Principles of Database Systems*, pages 54–65, San Jose, CA, May 1995.

5. H. Edelsbrunner, J. O'Rourke, and E. Welzl. Stationing guards in rectilinear art galleries. *Computer Vision, Graphics, and Image Processing*, 27(2):167–176, August 1984.

6. C. Esperança. *Orthogonal Objects and their Application in Spatial Databases.* PhD thesis, University of Maryland, December 1995. (also available as Technical Report TR-3566, University of Maryland, College Park, MD).

7. H. Freeman. Computer processing of line-drawing images. *ACM Computing Surveys*, 6(1):57–97, March 1974.

8. V. Gaede. Optimal redundancy in spatial database systems. In M. J. Egenhofer and J. R. Herring, editors, *Advances in Spatial Databases — Fourth International Symposium, SSD'95*, pages 96–116, Portland, ME, August 1995. (also Springer Verlag Lecture Notes in Computer Science 951).

9. R. C. Gonzalez and R. E. Woods. *Digital Image Processing.* Addison-Wesley, Reading, MA, June 1992.

10. R. H. Güting. An introduction to spatial database systems. *Special Issue on Spatial Database Systems of the VLDB Journal*, 3(4), October 1994.

11. A. Guttman. R–trees: a dynamic index structure for spatial searching. In *Proceedings of the SIGMOD Conference*, pages 47–57, Boston, MA, June 1984.

12. G. M. Hunter. *Efficient computation and data structures for graphics.* PhD thesis, Princeton University, Princeton, NJ, 1978.

13. H. V. Jagadish. Spatial search with polyhedra. In *Proceedings of the Sixth IEEE International Conference on Data Engineering*, pages 311–319, Los Angeles, February 1990.

14. A. Klinger. Patterns and search statistics. In J. S. Rustagi, editor, *Optimizing Methods in Statistics*, pages 303–337. Academic Press, New York, 1971.

15. G. D. Knott and E. D. Jou. A program to determine whether two line segments intersect. Department of Computer Science TR–1884, University of Maryland, College Park, July 1987.

16. M. Mäntylä. *An Introduction to solid Modeling.* Computer Science Press, Rockville, MD, 1987

17. J. Nievergelt and P. Schorn. Das ratsel der verzopften geraden. *Informatik-Spektrum*, 11:163–165, 1988. (in German).

18. J. A. Orenstein. Redundancy in spatial databases. In *Proceedings of the SIGMOD Conference*, pages 294–305, Portland, OR, June 1989.

19. J. O'Rourke. *Computational Geometry in C.* Cambridge University Press, 1994.

20. F. P. Preparata and M. I. Shamos. *Computational Geometry: An Introduction.* Springer–Verlag, New York, 1985.

21. N. Roussopoulos, S. Kelley, and F. Vincent. Nearest neighbor queries. In *Proceedings of the ACM SIGMOD Conference*, pages 71–79, San Jose, CA, May 1995.

22. H. Samet. *Applications of Spatial Data Structures: Computer Graphics, Image Processing, and GIS*. Addison-Wesley, Reading, MA, 1990.

23. H. Samet. *The Design and Analysis of Spatial Data Structures*. Addison-Wesley, Reading, MA, 1990.

24. J. Shechtman. Processamento geométrico de máscaras VLSI. Master's thesis, Eng. Elétrica, Universidade Federal do Rio de Janeiro, Rio de Janeiro, April 1991.

25. M. Stonebraker, T. Sellis, and E. Hanson. An analysis of rule indexing implementations in data base systems. In *Proceedings of the First International Conference on Expert Database Systems*, pages 353–364, Charleston, SC, April 1986.

Systems

From GeoStore to GeoToolKit:
The Second Step*

Oleg Balovnev, Martin Breunig and Armin B. Cremers

Institute of Computer Science III
University of Bonn
Roemerstr. 164
D-53117 Bonn, Germany
e-mail: (abc,martin,oleg)@cs.uni-bonn.de

Abstract. Today's geo-information systems are closed systems mainly
supporting geographic tasks in 2D space. In new spatial application areas
such as environmental research, geology, city planning or telecommunications, however, a database and processing support for 3D/4D objects
is required. Thus the development of object-oriented 3D/4D modelling
and data handling components for geo-information systems is a new challenge. We first report on our experience with GeoStore, an information
system already in use for the management of geologically defined geometries. We then follow the way from GeoStore to the design of GeoToolKit,
an object-oriented geo-database kernel system for the development of
3D/4D applications. GeoToolKit supports object-oriented modelling of
geo-applications, spatial data maintenance within an ODBMS including
spatial indexing and the 3D-visualization of spatial objects and query
results. Conversely, we also present the reconstruction of a geological
application on top of GeoToolKit. Finally we give an outlook on our
GeoToolKit research within an open component based environment of
geo-information services.

1 Introduction

Todays geo-information systems are closed systems mainly supporting geographic
tasks in 2D space. In new spatial application areas such as environmental research, geology, city planning or telecommunications, however, a database and
processing support for 3D/4D-objects is required [Rap89], [MJ94], [Wor94], [Bre96].
Thus the development of object oriented 3D/4D modelling and data handling
components for geo information systems is a new challenge for the spatial database
community. To our knowledge, there are no comprehensive experiences with the
management of 3D/4D-data in ODBMS.

Since 1993 our group is a member of the collaborate research centre "Interactions between and Modelling of Continental Geosystems". In this interdisci-
plinary project about 20 groups of the University of Bonn, mainly geoscientists,

---

* This work is supported by the German Research Foundation (DFG) within the
collaborate research centre (CRC) "SFB 350".

are examining the interactions of "fast" processes (time units of about $10^{-2} - 10^3$ years) of the atmosphere and the hydrosphere and "slow" processes (time units of about $10^3 - 10^7$ years) of the pedosphere and the lithosphere, i.e. the outer strata of the earth, which influenced the geology of the Lower Rhine Basin. The long time goal in the GIS subproject is the development of an object-oriented 3D/4D geo-information system managing the thematic, spatial and temporal data of the Lower Rhine Basin.

The paper is organized as follows: starting from the requirements of the interactive geological modelling in the next section, our experiences with GeoStore, an information system for the management of geologically defined geometries, are presented. Section three shows the concepts of GeoToolKit focusing on the architecture and spatial index support. In section four we follow the way back from GeoToolKit to GeoStore, constructing a geological application on top of GeoToolKit. In the last section we give an outlook on our future work for open geo-information systems in the collaborative research centre.

2   GeoStore, an Information System for the Management of Geologically Defined Geometries

Modelling geological structures and history always starts with a careful geometric analysis of the present assembly of geological surfaces, bodies and property distributions, this being the key to reveal the nature and interaction of earlier processes.

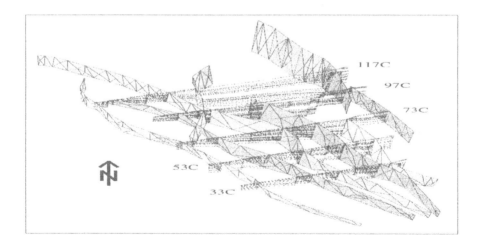

Fig. 1. 3D-modelling of geological sections and faults in the Lower Rhine Basin.

A process of computer-aided development of a consistent geometric model is known in geology as *interactive 3D/4D modelling*([Sie93]). The model is further tested by backward restoration and balancing. The final model is used for the specification of boundary conditions for 3D transport models, or for the production and consistent updates of geological maps.

As we outlined in [BBC94a] the starting point for the interactive geological 3D/4D-modelling within the CRC was the digitalization of geological *sections* gained from open-cast workings in the Lower Rhine Basin (Figure 1). A section is a geological abstraction obtained as a result of an intersection of a vertical plane with geological strata and faults. Geological strata are sheet-like structures in earth with different mineral composition, texture and/or grain size. A fracture in earth materials along which the opposite sides have been displaced is known as a fault. A stratum is modelled as a list of volumes which are usually specified by their bounding surfaces. A fault is modelled as a simple surface. Within a section strata and faults are represented as point sets grouped in startigraphic and fault lines. Every point in the section contains 3D-coordinates complemented with geologically-specific data like stratigraphy.

The second step in the interactive 3D-modelling is the generation of *triangulated surfaces* from the point sets spread between two sections. The final step, which is a part of our current work, is the transition from stratigraphical bounding surfaces to *volumes*.

2.1 Object Oriented Modelling of Geometry and Topology

The intention of GeoStore was to supply geo-scientists with a tool which would provide a consistent storage and efficient access for the data involved in all stages of the interactive 3D-modelling using modern database technology. We began the development of GeoStore[1] with the elaboration of an object data model. The object-modelling technique [RBP+91] provided an appropriate environment for the communication between geo- and computer scientists. Geologists not-experienced in software design quickly adopted this technique and successfully applied it. Figure 2 shows a part of the object model realized in GeoStore. It supports the geometries gained by the first and second step of the interactive 3D-modelling. There are three dominant classes: *Section*, *Stratum* and *Fault* with extensions used as entry point to the database.

A stratum in a section (*StratInSection*) contains an ordered list of stratigraphical lines. A stratigraphical line (*StratLine*) is a specialization of a geometric line, extended for the efficient computations with the additional topological information, namely a list of stratigraphical lines above the given one (*Hanging*) and the bounding faults (*Left Fault* and *Right Fault*).

The geometry and topology of strata and faults consist of a *Surface*, which are represented by simplicial 2-complexes [EFJ89] extended by their 3D-geometry. Each triangle is represented by its three points (the geometry), and its three opposite neighbour triangles (the topology of each triangle). Earlier tests outlined the simplicial complex approach to be a good working basis for the realization of efficient geometric and topological operations on 3D-objects [BBC94b].

---

[1] GeoStore resulted from a cooperation between the Institute of Computer Science III and the Geological Institute at the University of Bonn.

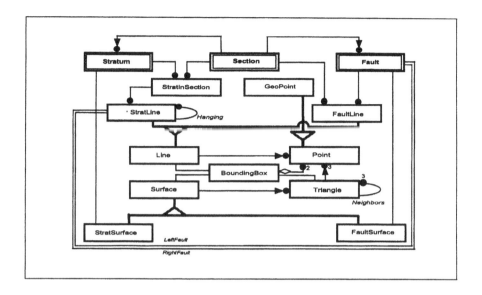

Fig. 2. Object model of GeoStore

We first implemented GeoStore on top of the RDBMS Oracle.[2] However, taking into account the complex structure of data with multiple interlinks between objects the object-oriented data-model turned out to be more suitable for the development of GeoStore because it avoided the "impedance mismatch" between data as stored in the database and the representation required by the geometric algorithms for the efficient computing in main memory. As a result GeoStore was completely re-implemented with the ODBMS ObjectStore[3].

2.2 Queries on Geologically Defined Geometries and Topologies

Currently GeoStore supports two types of spatial queries:

- *Queries on strata and faults within a section.*
  The user, for example, may select all strata with a certain hanging stratum in a given depth interval located between specified faults. This type of queries uses primarily the topological relationships explicitly stored in the database.
- *Queries on triangulated stratigraphic and fault surfaces.*
  This type of queries deals primarily with geometric 3D-operations like horizontal or vertical slices through stratigraphic and fault surfaces, an intersection between surfaces or clipping of a part of a surface within a bounding box.

Naturally, GeoStore supports non-spatial queries as well: the user can retrieve an arbitrary thematic attribute related to a geological object.

---

[2] Oracle is a registered trademark of Oracle Inc.
[3] ObjectStore is a registered trademark of Object Design Inc.

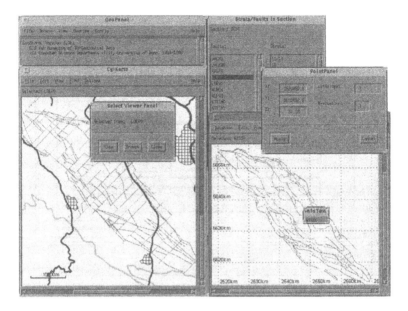

Fig. 3. Lower Rhine Basin map with section and fault data and a horizontal slice.

2.3 Visualization and Spatial Integrity Checking

GeoStore is characterized by its tight coupling between database and vizualization. There were two alternatives how to integrate the visualization within GeoStore. The first one precludes the use of external viewers.Though external viewers may provide sufficient manipulation facilities (move, rotate, zoom, select, etc.) truly interactive interaction with a database is hardly possible. Having loaded objects from a file generated by GeoStore a viewer operates further absolute independently from the database. Apart from this most of the viewers are closed for the extensions and do not permit to integrate application-specific control facilities. The second approach is oriented on the use of program interfaces (graphical libraries). The developer in this case is responsible for the implementation of all control facilities. Being more expensive this approach allows the implementation of a fully interactive communication with a database. Due to this the GeoStore user is able to switch at any moment between the object browser and one of the internal 2D/3D graphical viewers.

For example, GeoStore provides 2D-surface maps of the examination area (see figure 3). Geological sections and faults are drawn there as lines. A user can select with a mouse an object he wants to be further visualized or browsed. He can get a series of *horizontal slices* (specified by z-coordinates) within a specified region. The query results are visualized within a new surface map. The user can also specify on the map a *vertical slice* with strata and faults surfaces stored in GeoStore's database. The results of are presented in graphical form and then by demand they can be stored in the database.

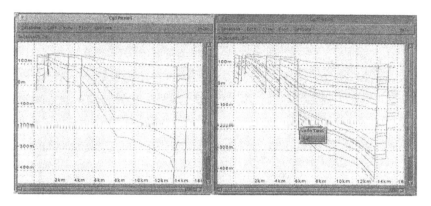

Fig. 4. Comparison of a computed (left) and a digitized (right) section.

The geologists used to compare computed slices with the digitized sections as stored in GeoStore (see figure 4). Such relative simple "visual" spatial consistency checks turned out to be a very effective method in the search for digitalization errors within geologically defined geometries. Several lost faults and discrepancies in the geometry of stratigraphic lines were successfully found using this technique. The investigation of methods which would permit us to check the geometric and topological integrity constraints without or with minimal participation of geoscientists is part of our ongoing work.

The embedded 3D-viewer allows for the user to view from different perspectives selected stratigraphic and fault surfaces as geometries in 3D space (see figure 5).

2.4 Motivation for a Geo-Toolkit

GeoStore originated from the tight cooperation with one of the CRC groups. Soon it became obvious that other CRC researchers such as geographers, hydrologists, paleontologists, etc. were interested to get their own GeoStore-like systems. We already launched the development of a new application for the management of well data. A subsequent analysis of the related domains convinced us that to a large degree they use common data sources (well data, geological sections, digital elevation models) viewing them from different sides. They also share a lot of functionality primarily concerning spatial data managing and geometric operations. The idea to reuse already implemented components (sources) in the development of new applications was obvious. This was the starting point for designing a "geo" toolkit rather than implementing n-1 GeoStore-like systems from scratch.

3 GeoToolKit - an Object-Oriented Database Kernel System for 3D/4D Geo-Applications

GeoToolKit is a collection of software tools and methods evolved from the Geo-Store experiences. It is intended to facilitate the development of 3D/4D geo-

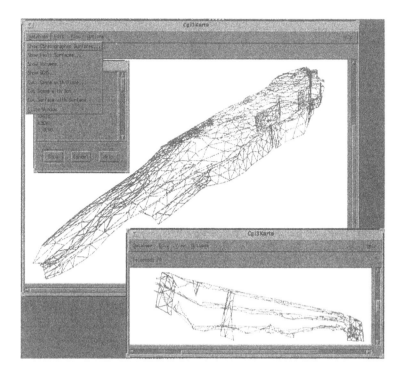

Fig. 5. 3D-visualization and object browser for selected geological strata and faults.

applications, especially those having to deal with the managing of large amounts of spatial data. GeoToolKit is not a GIS-in-a-box package - it is rather a library of C++ classes that allows the incorporation of spatial functionality within an application. Thus it is primarily oriented on software engineers with the C++ experience involved in the development of special-purpose geo-applications which can be hardly modelled within standard GISs. Being a component toolkit, it encourages the development and deployment of reusable and open software. Following the object-oriented modelling technique, not only abstract geometric primitives, such as points, curves, and surfaces but real world entities such as drilling wells, geological sections and strata, can be modelled and maintained. Applications developed with GeoToolKit simply inherit as much geometric functionality from GeoToolKit as needed, extending it with the application-specific semantics.

Our goal was, on one side, to make the toolkit general enough to be used in various applications. As a result a special attention by the development of GeoToolKit classes was paid to the generality and extensibility of the class library so that it could be easily adopted for diverse applications. At the same time it was obvious that a geo-toolkit should be more than just an empty interface specification, i.e. it should do something. We focused primarily on the efficient maintenance of spatial objects within a database. Since often objects need to be retrieved according to their relative position in space (e.g., an object which intersects a given one), we have to offer a set of basic spatial operations as well. We started with the operations needed at first hand for the interactive geological modelling, keeping our set open for the extensions.

GeoToolKit is intrinsically three-dimensional. We tried to avoid dealing with the two-dimensional issues as far it was not necessary for the three-dimensional data structures and algorithms.

Another principal design feature of GeoToolKit is its tight coupling with the object-oriented DBMS ObjectStore. The main problem with any "universal" database-independent toolkit it that its integration with a still non-standardized ODBMS may demand a cardinal re-design of both interfaces and sources. Instead of this we tried from the very beginning to extract maximum benefits from ObjectStore specific features.

We designed GeoTolKit classes with special respect to the persistent nature of objects. Many problems concerning primarily object migration in time are extremely critical in the database context. At the same time they are practically not known in the object-oriented programming community at large since most of the problems are eliminated by simple re-compilation of sources. A database may already contain gigabytes of data. An introduction of a new class, any change in layout or an introduction of a new method within classes already stored in the database automatically results in the exhaustive database re-organization. We tried to design our classes as tolerant to the persistence as possible. Our design technique allows the introduction of new classes and methods to the spatial objects already existing in the database.

3.1   GeoToolKit Architecture

GeoToolKit deals primarily with two basic notions: a *SpatialObject* and a collection of spatial objects referred to as a *Space*. Figure 6 presents their relationships within the data model of GeoToolKit shown in the OMT-like notation[RBP+91].

On the abstract level a spatial object is defined as a point set in the three-dimensional Euclidean space. Diverse geometric operations can be applied to a spatial object. However, they cannot be implemented until a spatial object has a concrete representation. There is a direct analogy with the object-oriented modelling capabilities. An abstract spatial object class specifies exclusively the interface, inherited by all concrete spatial objects. A concrete object is modelled as a specialization of the abstract spatial object class. It provides an appropriate representation for the object as well as the implementation for the functions.

The geometric functionality involves geometric predicates returning *true* or *false* (e.g. contains), geometric functions (e.g. distance) and geometric operations (e.g. intersection). Geometric operations are algebraically closed. The result of a geometric operation is a spatial object which can be stored in the database or used as an argument in other geometric operations. Naturally every spatial object class precludes a set of service facilities required for the correct maintenance of objects (clone, dynamic down cast).

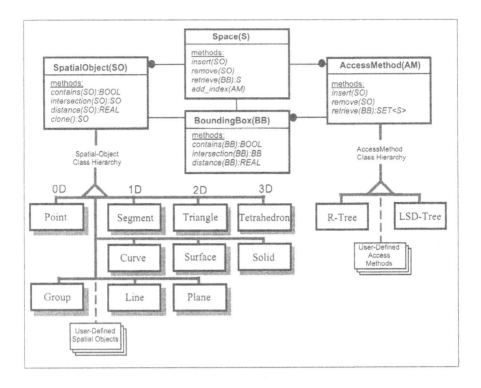

Fig. 6. Data model of GeoToolKit

Currently the GeoToolKit class hierarchy includes following classes:

- 0D-3D spatial simplexes: *Point, Segment, Triangle, Tetrahedron;*
- 1D-3D spatial complexes: *Curve, Surface, Solid;*
- Compound objects: *Group;*
- Analytical objects: *Line, Plane.*

Usually complexes are approximated (digitized) and represented as homogeneous collections of simplexes. A *Curve* (1D complex) is approximated through a polyline, a *Surface* and a *Solid* - as a triangle and tetrahedron network, respectively. However, we do not intend to restrict users only with the representations supplied with GeoToolKit. Complex spatial objects are designed in such a way that they do not predefine a physical layout of objects. They contain a reference to a dependent data structure referred to as a representation. The two layer architecture allows for the object to have multiple representations (e.g. one representation for the compact storage, another more redundant one - for efficient computations). An object can change its representation without changing the object identity. This feature is of extreme importance in the database context since an object can be referred to from multiple sources. Following certain design patterns a user is able to integrate his own special-purpose representation within GeoToolKit's standard classes.

Spatial objects of different types can be gathered into a heterogeneous collection - a *Group*, which is further treated as a single object. A group is a necessary construction for the representation of the results of geometric operations.

A space is a special container class capable of efficient retrieval of its elements according to their location in space specified either exactly by a point or by a spatial interval. A spatial interval, often referred to as a bounding box, is defined in 3D-space as a cuboid with the sides parallel to the axes. Since all operations in the Cartesian coordinate system are considerably faster for cuboids than for other objects, the approximation of spatial objects by their bounding boxes is intensively used as effective pre-check by geometric operations and spatial access methods in GeoToolKit.

A space serves both as a container for spatial objects and as a program interface to the spatial query manager which is realized as an internal GeoToolKit library function linked to a geo-application. The spatial query manager is invoked by member functions of the Space class. Practically a call of any space method means a call of the spatial query manager. A user can add/remove a spatial object to/from a space syntactically in the same way as in the case of usual object collections. However, all changes will go through the spatial query manager. A spatial retrieval is performed through the family of *retrieve* methods. The task of various retrieve methods is to provide a convenient interface for the spatial query manager. In the simplest case a retrieve member function takes a bounding box as a parameter and returns a set of spatial objects contained in or intersected by this bounding box. A user can also formulate his query in the ObjectStore style as a character string. Such interface may be useful for the implementation of interactive retrieval. However, it needs multiple conversions of data and therefore it is not convenient for the internal use within a program. A spatial retrieval involving "indirect" spatial predicates (e.g. intersects) is usually decomposed into two sequential steps. On the first step (pre-selection) the query manager retrieves all objects which intersect the bounding box of the given object. On the second step it checks whether the pre-selected objects really intersect the given object.

To provide an efficient retrieval a space must have a spatial index (see below). In order to make an arbitrary user-defined index known to GeoToolKit, it must fit the interface defined within the abstract *AccessMethod* class. GeoToolKit supports two indexing methods: the R-Tree [Gut84] for a pure spatial retrieval and the LSD-Tree [HSW89] for a mixed spatial/thematic retrieval.

In addition to the database support and geometric functionality GeoToolKit offers some other services, involving the 2D-/3D-visualization of spatial objects, class interfaces to 2D/3D graphical viewers and a remote access to the database for heterogeneous applications.

3.2 Spatial Index Support

A spatial index organizes a set of spatial objects for the efficient search according to a multidimensional key - the object's position in space. The indexing methods used within DBMSs are based on the relationships of absolute (lexicographic) order which can be always established between alphanumeric values. Since the lexicographic order in common sense does not exist for the multidimensional keys the standard indexing methods are no more relevant.

Most of spatial indexes were designed and implemented for the autonomous operating. While spatial data are often already located in the database and managed by the DBMS, the spatial indexes have still to be maintained by standard file management systems. In emergency situations (crashes, multiple updates) the consistency between data and indexes cannot be provided.

Some commercial database providers already offer spatial indexes. Unfortunately none of spatial indexes did succeed to prove its superiority over others relating to various distributions of objects in space and different query types. There is no guarantee that embedded within a DBMS spatial index will fit well enough your particular distribution of objects.

It is more promising to make a DBMS open for new indexing methods. We developed a general technique which allowed the integration of arbitrary spatial indexing methods within object-oriented DBMSs. We define a new container type for the maintenance of spatial objects (space) and implement for this class a spatial query manager. The spatial query manager overloads the ODBMS native query manager. It parses a query, extracts spatial predicates and checks whether a spatial index is available. If no spatial index is associated with the space it simply forwards the query to the native query manager. If a spatial index is found the spatial query performs index-based retrieval. The results (if not empty) together with the rest of the query are forwarded to the native query manager. Since the spatial query manager performs the role of a preprocessor the syntax of the native query language can be preserved or even extended.

To enable the cooperation between the spatial query manager and spatial indexing methods they must have a common interface. This requirement is not as restrictive as it seems to be at the first glance since the majority of the spatial indexes deals exclusively with the bounding box approximation of spatial objects. Index developers do not even need to modify the sources to fit the required interface. A usual technique is to develop an adapter class which carries out all necessary conversions and then it simply calls corresponding functions according to the pre-defined interface.

The situation is more complicated in the case of so called combined indexing. Combined indexes may be beneficial when separately neither a spatial nor a thematic component of the search criterion is selective enough. Distinct to the bounding box the number and types of non-spatial attributes cannot be pre-defined. To combine spatial and heterogeneous non-spatial subkeys GeoToolKit offers a special construction. The multikey class provides a uniform access to an arbitrary subkey according to its number (dimension). The retrieval with the multikey is similar to the spatial retrieval with the only difference that a generic multikey substitutes the bounding box.

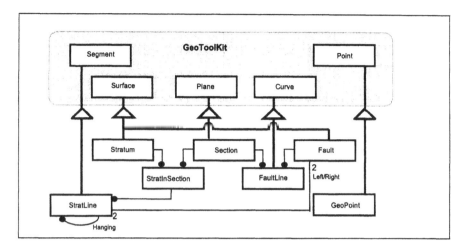

Fig. 7. Geological object model of GeoStore constructed on GeoToolKit.

A query can be formulated and called either in the same way as a usual ObjectStore query or by means of the additional retrieve function. The retrieve function demands from the user to separate explicitly the spatial and non-spatial parts of the query. The spatial part is represented as a bounding box or as a multidimensional key. If a space has more than one spatial index and a user wants a particular index to be used for the retrieval he can explicitly specify it in the retrieve function.

To maintain the spatial indexes consistently within GeoToolKit every member function (dealing with spatial updates) incorporates additional checks before and after performing updates. The pre-check function tests whether an object is contained in at least one indexed space. If yes, the bounding box is stored. The post-check function activates re-indexing only when the bounding box has been changed. To eliminate re-indexing by serial updates GeoToolKit provides an update block.

Before starting the development of new applications we decided to re-design GeoStore using GeoToolKit. The main design pattern we followed was to inherit from the geometric kernel as much functionality as possible. For every geological entity we tried to find a geometric "ancestor" (see figure 7). For example, a geological stratum is represented now as a specialization of the geometric entity *Surface*. It was extended with geology-specific features (e.g. stratigraphy) and relationships (e.g. the *Hanging* stratum).

By using the inheritance we achieved that every geological entity could be treated as a geometric object. GeoToolKit's geometric kernel was able to operate directly on the collections of geological objects without intermediate transformations. In the case of the incorporation we would have to create a temporary space and to copy there the geometric ingredients of geological objects - the task which may be too expensive in the database context.

Topological relationships (e.g. neighborhood) used to provide very useful "hints" not available earlier in the "pure" geometric world. In order to benefit from this additional information we simply re-implemented virtual functions. An overridden function either substitutes the geometric function completely or performs a kind of pre-selection for the geometric function calling it with the restricted subset of spatial objects.

Another application developed from very beginning on top of GeoToolKit deals with the maintenance of spatio-temporal objects used for the backward animation of geological processes.

The partial re-design of GeoStore and development of new applications on top of GeoToolKit proved the advantages of this approach. GeoStore classes contain now only geology-specific members. Geometric relationships between geometric objects are hidden within GeoToolKit. This resulted in the considerable reduction of the written code and made the sources more clear. The developers could focus on the application semantics instead of such "creative" tasks as optimal assembling of spatial objects from multiple tables or the implementation of routine geometric algorithms.

GeoToolKit classes could be lightly extended to satisfy special requirements. The GeoToolKit functionality (spatial retrieval, indexing, etc.) is available in a full volume for the extended objects as well.

Apart from this we expect to provide with standard GeoToolKit classes a basis for the non-redundant and consistent maintenance of shared data in CRC thus enabling the future integration of different applications.

4   Conclusions and Outlook

We have followed the way from a specific geological application to the design of a general geo-database kernel system for 3D/4D-applications. After this we took the way back and presented our experiences during the construction of the geological application on top of GeoToolKit. First geo-database kernel systems like PROBE [JO88] or the DASDBS-Geokernel [SW86], [WB92] have already been realized 10 years ago. However, these systems were not widely used in the geoscience community. Our experience shows that the object-oriented modelling approach helped to overcome the gap between computer- and geo- worlds.

We intend to make GeoToolKit less programmer-oriented. We are going to develop a set of interactive tools which would allow for geo-scientists to produce and to maintain their own objects on top of GeoToolKit without a mediator (software engineer) and without (or with minimal) C++ knowledge. These tools will include at least an object class generator, a database browser with graphical visualization capabilities and a spatial indexes repository manager.

The web infrastructure as one of the preconditions for open software systems [AKD94], [BRS96], [BM96], [Bre96] has developed very quickly. As a consequence we are about to integrate GeoToolKit within an environment of spatially distributed geo-information services. We are planning to extend GeoToolKit with geological and geophysical 3D modelling tools by the use of object-oriented software architectures (CORBA). GeoToolKit could then be regarded as a 3D/4D geo-service in a multilayer architecture of an open geo-information system as it is described in [BM96], [VS94].

5 Acknowledgements

The authors would like to acknowledge the excellent cooperation with the group of Agemar Siehl at the Geological Institute, University of Bonn. We are also indebted to Norbert Klein, Wolfgang Müller, Dirk Möbius and the students of the database practical courses under the guidance of Thomas Bode for the valuable implementation support during the development of GeoStore and GeoToolKit.

References

[AKD94] D.J. Abel, P.J. Kilby, and J.R. Davis. The Systems Integration Problem. *Int. J. Geographical Information Systems*, 8:1–12, 1994.

[BBC94a] Th. Bode, M. Breunig, and A.B. Cremers. First Experinces with GeoStore, an Information System for Geologically Defined Geometries. In *J. Nievergelt, Th. Roos, H.-J. Schek, P. Widmayer (Eds.): IGIS'94': Geographic Information Systems, LNCS No. 884*, pages 35–44. Springer-Verlag, 1994.

[BBC94b] M. Breunig, Th. Bode, and A.B. Cremers. Implementation of Elementary Geometric Database Operations for a 3D-GIS. In *Proceedings of the 6th International Symposium on Spatial Data Handling*, pages 604–617. Edinburgh, 1994.

[BM96] K. Buehler and L. McKee. The OpenGIS Guide - Introduction to Interoperable Geoprocessing. Technical report, Open Geodata Interoperability Specification (OGIS), Open GIS Consortium Inc., 1996.

[Bre96] M. Breunig. Integration of Spatial Information for Geo-Iinformation Systems. *Lecture Notes in Earth Sciences, Vol. 61, 171 pp.*, 1996.

[BRS96] S. Blott, L. Relly, and H.-J. Schek. An Open Abstract-Object Storage System. In *Proceedings ACM Sigmod'96, Montreal, Quebec, Canada*, pages 330–340, 1996.

[EFJ89] M.J. Egenhofer, A.U. Frank, and J.P. Jackson. A Topological Data Model for Spatial Databases. In *Proceedings of the Annual Meeting ACM SIGMOD*, pages 271–285. LNCS 409 Springer Verlag, 1989.

[Gut84] A. Guttman. R-Trees: A Dynamic Index Structure for Spatial Searching. In *Proceedings of the Annual Meeting ACM SIGMO*, pages 47–57, 1984.

[HSW89] A. Henrich, H.-W. Six, and P. Widmayer. The LSD Tree: Spatial Access to Multidimensional Point and Non Point Objects. In *Proceedings of the 15th VLDB Conference*, 1989.

[JO88] F. Manola J. Orenstein. Probe Spatial Data Modeling and Query Processing in an Image Database Application. *IEEE Transactions on Software Engeneering*, 14, 1988.

[MJ94] C. Bauzer Medeiros and Genevieve Jomier. Using Versions in GIS. In *DEXA*, pages 465–474. Springer-Verlag, LNCS, 1994.

[Rap89] J. Raper, editor. *Three Dimensional Applications in Geographical Information System*. Taylor& Francis, London, 1989.

[RBP+91] J. Rumbaugh, M. Blaha, W. Premerlani, F. Eddy, and W. Lorensen. *Object Oriented Modelling and Design*. Prentice Hall, 1991.

[Sie93]   A. Siehl. Interaktive geometrische Modellierung geologischer Flächen und Körper. *Die Geowissenschaften*, 11:343–346, 1993.

[SW86]   H.J. Schek and W. Waterfeld. A Database Kernel System for Geoscientific Applications. In *Proceedings of the 2nd Symposium on Spatial Data Handling*. Seattle, 1986.

[VS94]   A. Voisard and H. Schweppe. A Multilayer Approach to the Open GIS Design Problem. In *Proceedings of the Second ACM Workshop on Advances in Geographic Information Systems*, pages 23–29. Gaithersburg, Maryland, 1994.

[WB92]   W. Waterfeld and M. Breunig. Experiences with the DASDBS Geokernel: Extensibility and Applications. In *From Geoscientific Map Series to Geo-InformationSystems*, pages 77–90. Geolog. Jahrbuch, A(122), Hannover, 1992.

[Wor94]   M. F. Worboys. A Unified Model of Spatial and Temporal Information. *Computer Journal*, 37:26–34, 1994.

Geo-Opera: Workflow Concepts for Spatial Processes

Gustavo Alonso   Claus Hagen

Database Research Group
Institute of Information Systems
ETH Zentrum
CH-8092 Zürich, Switzerland
E-mail:{alonso,hagen}@inf.ethz.ch

Abstract. A Process Support System provides the tools and mechanisms necessary to define, implement and control processes, i.e., complex sequences of program invocations and data exchanges. Due to the generality of the notion of process and the high demand for the functionality they provide, process support systems are starting to be used in a variety of application areas, from business re-engineering to experiment management. In particular, recent results have shown the advantages of using such systems in scientific applications and the work reported in this paper is to be interpreted as one more step in that direction. The paper describes Geo-Opera, a process support system tailored to spatial modeling and GIS engineering. Geo-Opera facilitates the task of coordinating and managing the development and execution of large, computer-based geographic models. It provides a flexible environment for experiment management, incorporating many characteristics of workflow management systems as well as a simple but expressive process modeling language, exception handling, and data and metadata indexing and querying capabilities.

1 Introduction

The notion of *process* is currently being used in a variety of application areas to refer to complex sequences of computer programs and data exchanges controlled by a meta-program (the process). This idea has proven to be very helpful in solving a number of problems associated with environments involving heterogeneous platforms and applications. Some of these problems are: (1) interoperability, which is solved on a process basis instead of tackling the general case, (2) distribution, addressed by providing adequate support for programming distributed applications out of pre-existing tools, (3) forward-recovery, guaranteed by making every step of the process persistent using a database, (4) monitoring, based on the current and past states of the process stored in a database, and (5) history tracking, solved by providing a querying and data mining tool over the database storing the states of the processes. Surprisingly, these problems are pervasive and appear in many different application areas, from virtual enterprises and business environments [Fry94, LA94], to software engineering [CKO92, TKP94, BDMQ95] and scientific data management [ILGP96, BSR96].

Such pervasiveness explains the success of *workflow management* concepts and products, which are the most recent incarnation of process support systems.

Existing workflow tools, however, are often tailored to a particular domain. In particular, existing systems target in most cases either business processes [Hsu95] (this is the case of IBM FlowMark [MAGK95], for instance) or imaging systems (like OmniDesk [AAEM97]), with a few research prototypes addressing other areas [BK94, ILGP96]. Such narrow purpose design is combined with severe limitations related to performance and functionality [AS96, AAEM97], limitations that further restrict the applicability of these systems (see for example [MVW96, BSR96] reporting on attempts to use commercial workflow products to support scientific applications).

The project *OPERA* (Open Process Engine for Reliable Activities), currently ongoing at the Database Research Group of ETH Zürich, has as its main objective to paliate such glaring limitations of the state-of-the-art. OPERA is a workflow tool in the sense that it provides functionality that today only exists in workflow systems. OPERA is unlike existing workflow products in that it supports, from the same application platform, processes of very different nature. As an example of the potential offered by OPERA, this paper discusses in detail *Geo-Opera*, a tool, built as an extension of OPERA, supporting the development, execution and management of complex geo-models. Geo-Opera does not address requirements of spatial data handling such as indexing, storage, or efficient retrieval. For such purposes, Geo-Opera relies on existing systems. Geo-Opera is intended instead as a tool for the management of geo-processes, i.e., complex sequences of programs and transformations over spatial data. Geo-Opera supports all aspects of geo-process management, from the definition of geo-processes to their execution over heterogeneous hardware and software platforms, including as well querying and indexing capabilities over meta-data related to the geo-processes. An important contribution of Geo-Opera is bringing together under a single system functionality that was previously available only as part of isolated tools. It also introduces new concepts in geo-process management. For instance, the language used has a sophisticated exception and event handling mechanism. Since Geo-Opera also controls the execution of such geo-processes over distributed, heterogeneous environments, exception handling and events allow a greater flexibility when dealing with failures. Moreover, since a geo-process is a persistent entity in Geo-Opera, forward recovery and persistent execution is always ensured, which adds considerable guarantees when executing large and complex models involving many steps and, possibly, many computers. Finally, Geo-Opera provides a sophisticated mechanism for dependency tracking that can be used to derive the lineage of data sets, modifications to models, and perform automatic change propagation and notification. This mechanism also allows geo-processes to be modified creating new versions or to be executed repeatedly overwriting previous results or creating new versions of the resulting data.

With such functionality, Geo-Opera is a very powerful and flexible environment for the development, management, and execution of large geo-processes. This functionality is described in more detail in the sections that follow. Sec-

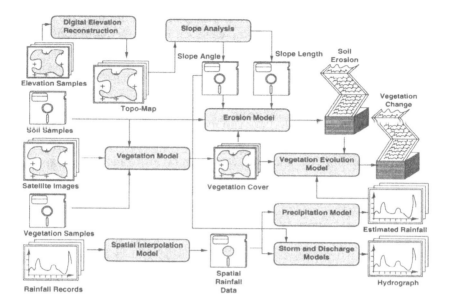

Fig. 1. Example of geo-process representing a geo-model of the hydrographic characteristics of a region

tion 2 presents an example of the applications to be supported by Geo-Opera as motivation for the rest of the paper. Section 3 discusses the architecture and functionality of OPERA. Section 4 describes how geographic process management can be performed using the facilities provided by Geo-Opera. Section 5 discusses related work and Section 6 concludes the paper.

2 Motivation and example

Most of the research activities in GIS and spatial data are centered around issues such as data representation, indexing, storage, and retrieval. To certain extent, this is the result of an attempt to extend database technology beyond traditional data sets (banking, insurance, financial) [vOV91, ZG91, Kea93]. These, however, are not the only problems related to spatial models [AE94a, AE94b, SSE+95], a point that is best illustrated with an example[1].

Figure 1 shows a combination of typical spatial models as sequences of transformations applied to different data sets (in what follows, the terms geo-model and geo-process will be used interchangeably. From an abstract point of view, a model is a representation of a geographic phenomenon while a geo-process is the

---

[1] This example is taken from [AE94b] where a solution to a related problem is proposed. Although the solution presented in this paper derives in part from that earlier work, the functionality and capabilities provided by Geo-Opera are more comprehensive.

model described in terms of programs and data sets). For instance, the elevation samples are used as input to a digital elevation reconstruction algorithm that produces a topographic map with interpolated elevations. This topographic map is then used as input to a slope analysis program that extracts the length, orientation and angle of the different slopes. These results, along with samples of the soil in the area and information about the vegetation cover, are used as the input for a model that tries to predict the future erosion patterns in the area. The vegetation cover is obtained by combining the information of the soil samples, vegetation samples and satellite images of the area. The vegetation cover is also used as one of the inputs of a vegetation evolution model that, given the predicted erosion, estimates the possible changes in vegetation cover. Similar step-wise procedures are followed to derive the required information regarding precipitation levels. The figure shows all these spatial models combined together into a complex set of operations representing different geographic phenomena.

This example illustrates several requirements that are very similar in nature to those related to process management. In what follows and for reasons of space, the paper will be focussed on a necessarily simplified list of requirements and how Geo-Opera provides functionality to meet those requirements. This list can be be summarized in three large areas: *modeling language*, *distribution and parallelism*, and *querying capabilities*.

Modeling Language . In any process support system, one of the key components is the language in which to express processes (geo-processes in this case). Given the intrinsic modularity of geo-processes (the example contains several geo-processes within a larger one), the language used must be structured, allow nesting, and facilitate reusability. Any geo-process should be suitable to be used as a building block for a larger geo-process. In addition, and on account of the complex execution environment, the language must provide support for event and exception handling. Events can be used, for instance, to notify the system of changes in the input data sets and trigger the execution of the model (or parts thereof) to produce more up-to-date results. Similarly, regarding exceptions, a reliable mechanism must be in place to cope with them to avoid that any deviation from the prescribed behavior results in having to abort the execution of the entire model. As in workflow systems, the language must allow defining and registering with the system external objects and applications since, especially in the case of geo-processes, both the algorithms and the spatial data will be external to the system. This registration of external entities is the basis for addressing interoperability issues (Point (1) above).

Distribution and Parallelism . To alleviate the cost of executing such complex geo-processes, their different steps should be executed in parallel whenever possible (Point (2) above). To avoid losing generality, we will assume the underlying platform is a cluster of PCs or workstations, not necessarily using the same operating system. Each step of the geo-process can be assigned to a different node in the cluster, thereby exploiting the parallelism inherent to the model. The same ideas can be applied in a multiprocessor machine (in which homogeneity greatly simplifies the problem). As a direct consequence, and due to the cost of the models and the time it takes to complete their execution, a mechanism must be in place to avoid losing all computations when a failure occurs. This is the notion of forward recovery (Point (3) above), in which execution can be resumed from the point where it was interrupted by a failure. Moreover, it does not help to provide a system in which complex geo-process are executed from beginning to end without being able to stop execution at intermediate points, or to dynamically modify the geo-process to correct errors. Often, a geo-process is built not so much to test its results but to test the validity of the corresponding geo-model. For these cases, the system should support step-by-step execution and provide the ability to, at any point, stop, examine the geo-process, make changes, and resume execution. Such functionality can only be provided through close monitoring of the geo-process execution (Point (4) above).

Querying Capabilities . Geo-processes result in derived data that cannot be interpreted without knowledge about the model used for its creation and the initial data used in the model [Arm88, Lan88, LV90]. This leads to the well known problems of lineage-tracking, change propagation and versioning [Tob79, GG89, Rad91, SSAE93] which are basically no different from those of tracking the history of more general processes (Point (5) above). Questions such as "which models use algorithm X", "which results may change if dataset Y is updated", and "which data sets are used to derive result Z" are typical of such environments. Moreover, operations such as automatic change propagation (re-execution of a model when the input data is modified producing new versions of the output data), change notification, and process control (maintaining a copy of the model used to create a data set for as long as the data set exists, and maintaining copies of subprocesses for as long as there are other processes using them) must be supported by the system to make it truly useful. This functionality can be provided only if there are adequate mechanisms to track dependencies and efficient ways of retrieving information about these dependencies.

3 OPERA: a Process Support System

The OPERA project is an example of the ongoing efforts towards generalizing workflow concepts. OPERA is an open and flexible process support system intended to work as an specialized distributed operating system for large, heterogeneous, distributed environments, in which OPERA takes over the tasks of scheduling, synchronization and load balancing complex applications.

3.1 System Architecture

The architecture of OPERA is intrinsically modular. Parts of it have been implemented using TP-monitors and other middleware solutions. The current prototype relies on Oracle and ObjectStore as the underlying databases, and on Encina as the transaction engine. Future versions may incorporate functionality from CORBA related products, and queuing systems. For reasons of space, only the aspects of OPERA relevant to this paper are discussed.

The system architecture of OPERA consists of three service layers, shown in Figure 2: *database services, process services* and *interface services.* The idea is that each service layer should be independent and easily replaceable, an important point since OPERA is a generic process support system that needs to be tailored to specific applications. The components within each layer are also built as separate modules so they can be discarded or included in the final system depending on the requirements.

The database service layer provides persistent storage. It is divided into five *spaces*: template, instance, object, history, and configuration spaces, each one of them storing a different type of process data. This approach enhances scalability as it allows to install each space on a different physical database and on a different machine to prevent, for instance, the traffic generated during process execution from interfering with the traffic generated by history analysis. The logical organization of these spaces corresponds to the main entities handled in OPERA.

Templates are similar to the process code image. The template space is used to store the process defined and to create new processes, effectively separating development of processes from process execution. Instances are running processes. For each running instance of the same process, the instance space contains a copy of the process image that is updated as the process executes. Storing this information persistently guarantees forward recoverability in case of failures, since the system can resume operations as soon as failures are repaired. If more availability is needed, there are several techniques that can be used [KAGM96].

Objects are used to store information about externally defined data. This space forms the basis to allow OPERA to interact with other applications and it also stores part of the information related to the dependencies among objects. The history space is used to store information about already executed instances, which allows to query the system for any result of past executions. Finally, the configuration space is used to record system related information such as configuration, internet addresses, program locations, and so forth.

An important additional aspect of OPERA is that it is database independent thanks to the *DB abstraction layer* which translates a canonical representation [KAGM96] to the private representations of the underlying repositories. OPERA handles the contents of the five spaces using its own internal representation (see below), which has been optimized for performance and expressibility. The *DB abstraction layer* translates this internal representation to the appropriate language (SQL, C++, system calls) as required by the physical implementation of the underlying database.

The process service layer provides the tools necessary for coordinating and monitoring the execution of processes, as well as the tools for defining and creating new processes. The most relevant components are the *dispatcher*, the *navigator*, the *object manager*, and the *query manager*.

The dispatcher plays the role of resource allocator, determining which system

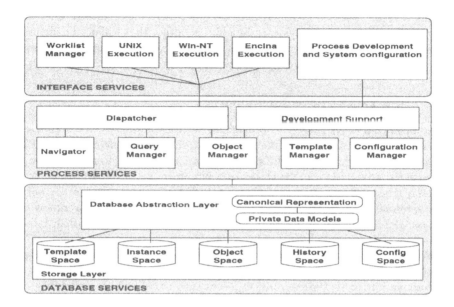

Fig. 2. System architecture of OPERA

will execute the next step, locating a machine with the appropriate characteristics, checking whether it is currently available, and managing the communications with remote system components. The navigator acts as a scheduler: it
"navigates" through the process description establishing which steps should be
executed next, which ones are to be delayed, and which ones have finished their
execution. During this navigation procedure, the related variables in the process
instance are updated so as to have a faithful record of every step taken in the
execution of the process.

The object manager is used for maintaining up to date the information related to external objects. As it will be explained in more detail below, this
includes keeping track of which processes use a particular object, the lineage of
the object if it is a derived one, and avoiding inconsistencies in the version being
used. The query manager has as its main role to provide a suitable interface
for querying information about past and current processes. The query manager
reroutes queries to the space involved and merges results to present a coherent
view to the requesting entity. The query manager is actively used not only by the
final user but also by the system itself to find out existing dependencies between
processes and objects.

The interface service layer provides the mechanisms necessary to interact
with users (through the *worklist managers*), and applications (through the *execution interfaces* supporting several operating systems and specific tools). The
execution interfaces are divided into two parts, one internal to OPERA and used
for establishing the correct interaction protocol with external applications, and
one external to OPERA and residing in the same machine where the application

being invoked resides. In this, OPERA follows closely the architecture of existing workflow management systems. In addition, although it will not be discussed further in this paper, a number of tools are provided for defining new processes as well as for monitoring the system's configuration.

3.2 OPERA Execution Environment

The main goal of OPERA is to provide an environment in which complex processes can be executed. Thus, the main hardware platform targeted by OPERA is a cluster of workstations or PCs, linked by a LAN or WAN. Such a platform is used as a shared nothing multi-processor environment in which OPERA plays the roles of scheduler and resource allocator. OPERA relies on a database for storing the information it needs to perform its tasks but, unlike existing workflow systems, OPERA does not rely on a unique, centralized database. OPERA takes advantage of the underlying hardware platform to distribute its functionality. Thus, all OPERA components can be moved from node to node in the system, including the database with the system information. This makes OPERA very robust and helps to address difficult issues such as reliability, availability, and scalability. OPERA is capable of relying on a variety of databases both relational and object oriented, and it can invoke applications under both UNIX and NT. It can also invoke applications developed in specific systems such as TP-Monitors (Encina, for example) and provides the necessary hooks to create additional execution interfaces for other environments, thereby maintaining an open architecture. In this, OPERA is no different from existing workflow management system. OPERA, however, has generalized many of the existing solutions and paid special attention to avoiding the pitfalls and design limitations found in current workflow products [AS96, AAEM97].

4 Geo-Opera: a Geo-Process Support System

Geo-Opera is a customization of OPERA for geo-processes, or using recent terminology, for experiment management in geographic applications [ILGP96]. The particular characteristics of geo-processes make it necessary to extend OPERA with functionality not strictly related to generic processes. This section discusses such functionality and also illustrates how Geo-Opera can be used to address the issues discussed in Section 2.

4.1 From User Representations to Data Models

Geo-Opera takes advantage of the open architecture of OPERA to provide a modeling language suitable for geographic applications. OPERA contains a hierarchy of process representations rather than a single model. Figure 3 describes this hierarchy of languages and representations, along with its current status in Geo-Opera. At the top, and used at the interface service layer, is the application specific language. This is the language that has been customized for Geo-Opera.

As in most process support systems, this language has a strong graphic component and, from the user's point of view, a geo-process is represented through icons not very different from those shown in Figure 1. In addition to this mostly graphical language, there are a number of interfaces to allow the user to specify any additional information necessary to execute a geo-process. This ranges from registering external objects to indicating in which machines a particular algorithm can be invoked.

User representations, however, are not suitable for handling processes efficiently. As an intermediate representation, used at the process service layer to facilitate interoperability across heterogeneous platforms, there is the *Opera Canonical Representation* (OCR). OCR is not so much a language but a data model or internal representation that provides a unique way to identify all entities used in the process support system. It provides features such as rules, event handling, exception handling, sophisticated programming constructs, and complex representations,, not all of which are necessarily used in a given application. For instance, many of the programming constructs built in OCR are not used in Geo-Opera (for instance, business processes often need to describe complex and open-ended negotiations between tasks that do not occur in geographic models). Finally, since the different storage spaces correspond to actual databases with their own private data models, OCR must be translated into these data models (ObjectStore and Oracle [KAGM96]).

4.2 Expressing Geo-processes in Geo-Opera

In its first version, the Geo-Opera language is a simplification of the languages traditionally found in commercial workflow management systems (see [KAGM96] for an example). The simplification is possible due to the intrinsic nature of geographic modeling in which most of the modeling logic is in the individual algorithms. Thus, constructs such as start and end conditions, as well as control flow conditions are not strictly necessary although, since they are supported by OCR, they could be easily incorporated if needed. The main components of the Geo-Opera language are:

Models (also called *projects* in GOOSE [AE94b], or *processes* in most workflow systems). A model corresponds to a geo-process and consists of a collection of *tasks* linked by *connectors* along with the so-called *blackboard*. The tasks are either activities (basic units of execution), blocks (groupings of tasks with special properties), or submodels (references to other model descriptions, late binding is used to instantiate submodels of a running model). The connectors specify order precedences between the different tasks. The blackboard is used to store intermediate results and exchange data between tasks. It fulfills the same role as the control data in workflow systems, the only difference being that all control variables are available in a single place, the blackboard, while in workflow systems this data is often found in the private data containers of each task.

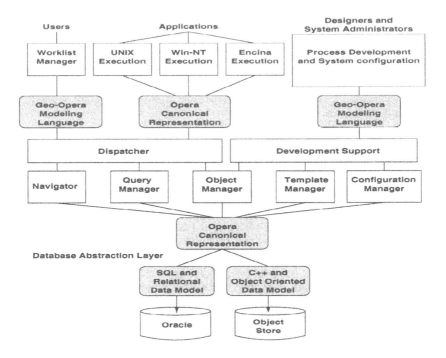

Fig. 3. Language and Representation Hierarchy in Geo-Opera

Tasks (also called *activities* or *steps* in some workflow systems). Tasks are internally represented in Geo-Opera as objects. This provides a well-defined interface to allow starting and stopping execution, to provide call parameters, to query execution state and to return values. Each task has a well defined set of attributes and methods in order to facilitate basic navigation. This set is defined through the *task interface* and inheritance is used to allow hierarchical structuring of task interfaces. This allows to extend the system by adding new task interfaces. At the root of the tasks inheritance tree there is a special interface called BasicTask, which provides a minimal set of functionality. The structure of the basic task interface is given in Table 1. The different task types (submodels, blocks, and activities for our purposes here) allow nesting and modular design of models. Submodels correspond to other models which have been independently defined (similar to a library call). Blocks correspond to logical constructs within a model, they have no name and can only be used once (similar to bracketing a set of instructions between *BEGIN* and *END*). Activities are the actual programs to be invoked. These programs are external to the system and must be registered before they can be used (as it is done in any workflow system).

Control Flow . In the Geo-Opera language, control flow is specified through simple connectors between tasks. Thus, the user only needs to link two tasks with a connector to establish the order of execution. Internally, this control connectors

Name	Type	Description
Guard	Guard	Control flow description
InBox	Parameterlist	Call parameters
OutBox	Parameterlist	Allows to access return values
State	ExecutionState	Allows to access the current state of task execution
ExceptionDecl	List of Exception	Declares the exceptions that can possibly be thrown by the task
EventQueue	List of Event	Allows to query the list of events raised by the task
Type	TaskType	Distinguishes between submodels, blocks, and activities
Reference	Object	A reference determining how the task is to be executed

Table 1. Basic Task Interface

do not exist, control flow inside a model is specified only through the guards of the corresponding tasks. A *guard* describes when the task has to be executed. The paradigm used is similar to that known as *ECA rules* in active databases [WC96]. The guard consists of an *activator* that specifies when the task has to be considered for execution, and a *start condition* (the latter is not used in the case of Geo-Opera, but is part of OCR). The activator is a predicate that can reference the execution state of other tasks, events raised by other tasks, and imported events of the process. Note that only tasks within the same process are visible to the guard. This principle of locality guarantees efficient evaluation of guards and is an important difference to active database systems, where all events are visible to all rules. Figure 4 shows how the example of Figure 1 looks like using tasks and control connectors.

External Objects . Data items not residing within Geo-Opera must be registered before they can be used (in the same way that in a workflow management system, applications and users must be registered with the system before they can be involved in process execution). The registration process involves specifying all the necessary parameters so as to allow the system to access and extract the corresponding data. This includes the user provided filter (see below for more details). Additional attributes of the external object allow the system to keep track of the dependencies established with other objects and models. These attributes are maintained by the system and are not directly accessible to the user. The list of attributes of a basic external object is shown in Table 2.

Exception Handling . Exception handling is necessary to be able to cope with all the possible errors that may occur during the execution of a model. It is not reasonable to expect that the programmer of a model will foresee all possible

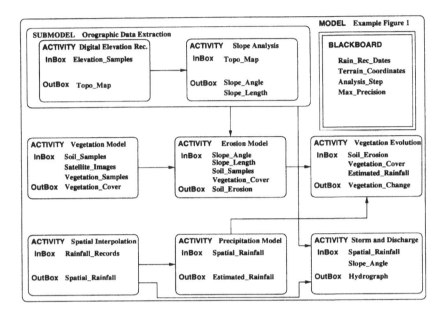

Fig. 4. The Example of Figure 1 in terms of activities, submodels and control connectors.

exceptions and hence, a mechanism must be in place to cope with such cases. OPERA incorporates an extension to workflow languages to support the flexible, transparent modeling of fault-tolerant workflows and has a strong focus on reusability. This extension is based on exception handling concepts from programming languages and uses a default hierarchy of exception handlers to allow an exception to always be trapped, if by nobody else, at least by the system. This approach has the advantage of always providing an opportunity to correct the exception and resume execution before having to abort the execution.

Events . Events are used both to interact with external entities and to allow interactions between models executed concurrently. As mentioned above, events can be associated with the guards of a task. In this way, a modification to an external object triggers an event that can be captured by the system which will proceed accordingly (see below the discussion on active mechanisms). Events also allow external applications to control the process support system: by generating different events, an application can control the execution of a model. For reasons of space, these two issues (exception handling and events) will not be discussed further.

4.3 Distribution and Parallelism

The execution of geo-processes requires expensive computations. Often, the data does not fit in main memory (as it is the case with many raster images) which

Name	Type	Description
ObjectName	String	User given name
OID	Long Integer	System generated object identifier
System	String	System under which the data item is located (UNIX, Win-NT, etc.)
Location	String	System address to use when locating the object
Access	String	Access mask to use to retrieve the data (the user provided filter)
Sources	List of OID	Other objects used to create this object, only the immediate ancestors are stored (search is recursive)
Versions	List of OID	Other versions of this object
Algorithm	Task	The task used to create this object
Usage	List of Task	Tasks that use this object as input (OID is in Task.InBox)

Table 2. External Object Interface

implies expensive I/O, and the operations tend to be complex transformations. Geo-Opera allows to alleviate this problem by supporting the execution in parallel of as many activities as the data flow dependencies in the model allow. In this, Geo-Opera follows the trend towards using clusters of workstations as the basic hardware infrastructure. The intrinsic distributed nature of processes, exploited in business applications to increase the decentralization of the organization, is used in Geo-Opera to separate the execution of every single task. As long as the hardware resources and necessary communications are available, each activity of a model can be executed in a different machine. This idea is shown in Figure 5 in which the model of Figure 1 is executed in a cluster of 5 workstations using a shared file system. The execution of the model requires only three steps, as activities that can be executed concurrently are assigned to different workstations and run in parallel. The use of a common file system (such as NFS, found in most UNIX environments) simplifies a great deal the problem of data handling. It is not necessary, however, to have such shared file system. Geo-Opera can operate and execute activities in parallel also over wide area networks (as most workflow management systems do) but this implies that data will have to be accessed remotely which can have a considerable impact on performance (see [ARM97] on how to cope with data handling problems in these environments).

4.4 External Objects

Handling external data is crucial in geo-processes. Most geographic and spatial data resides in files and not in databases. This creates an additional problem for there is no obvious way to interact with data outside the system. To address

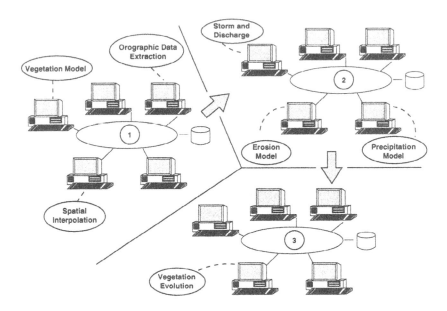

Fig. 5. The model of Figure 1 executed in a distributed fashion in a cluster of workstations; in each phase, several tasks may be executed in parallel.

this interesting issue, Geo-Opera makes use of the ideas described in the GOOSE system [AE94b] and recent contributions from the area of externalized database functionality, in particular from the Concert prototype [BRS96]. The basic idea is to treat external objects as black boxes over which database functionality can be applied. From the Concert prototype, Geo-Opera uses the ability to export storage management services such as indexing, replication, and basic query processing over externally defined data. From the GOOSE system, Geo-Opera has borrowed the object representation mechanism as well as most of the object handling capabilities (*object base* in GOOSE). Accordingly, the *external objects* supported by Geo-Opera are data objects located in external repositories such as file systems, databases, or the internal storage of a particular applications. An external object is a *view* over external data, i.e., it is possible to have several Geo-Opera external objects that refer to the same actual data item. A standard example is that of one external object defined over the chemical properties of a set of soil samples, and a second object defined over the physical properties of that same set of soil samples. Such views are computed through a user-provided *filter* that is activated when the object is accessed.

External objects must be registered in the instance space. A user wishing to define an external object has to provide a unique name (under which it is referenced in process descriptions) and the filter that describes how the object is constructed from the external data source. A system address is also necessary to indicate where Geo-Opera can find the data set (this can be either a network address or a symbolic name such as those used in environments like CORBA or

DCE). The user provided filter can be a database query or a sequence of operating system commands extracting data from a series of files. This is irrelevant from the point of view of Geo-Opera since the only role of the system is to invoke the filter to get the data. Note that this filter is no different from an algorithm or a task as defined in geo-processes. In fact, Geo-Opera uses the same mechanisms to invoke the filter of an external object and the program corresponding to a geo-process' activity. Also note that, through this filters, Geo-Opera can incorporate external repositories without having to address the general interoperability problem. If there is a uniform way to extract information from an external repository (like an SQL interface, for instance), this can be used for all accesses to the repository. For legacy systems, the filter option provides a quick and *ad-hoc* solution for those cases in which there is no need to implement a uniform access mechanism.

To allow the system to keep track of modifications produced by the execution of geo-processes, external objects can be versioned. This is accomplished by assigning a new version number to the objects resulting from executing a task and maintaining the corresponding attributes of all related objects up-to-date (see the "Versions" attribute of an object in Table 2). In the same way, Geo-Opera keeps track automatically of an object's *lineage*, recording by which processes it has been produced and which inputs have been used for its computation (see the "Algorithm", "Usage", and "Sources" attributes in Table 2). This information is heavily used by the active mechanisms and by the query manager.

4.5   Active Mechanisms

One of the most complex tasks of an experiment management system is to automate the tedious operations of versioning and updating. In most cases this is done manually, which adds considerable overhead and is an error prone process. Geo-Opera provides the notion of *active objects* and *active models* to address this issue. Active objects are automatically recomputed if some object they depend on has been modified. To avoid expensive checks, the update takes place in a lazy manner. Instead of doing a search of all related objects every time an object is modified, the update takes place when an object is accessed. The underlying mechanism is based on the attributes added to both objects and models. When an object is modified, a flag is set in all related objects. This takes only a few operations and does not interfere with the actual computation. When an active object is accessed, the flag is checked. If the flag is set, it means there has been a modification of the sources used to create the object, i.e., the current version is not the most up-to-date version. In such cases, the lineage chain is retraced and executed again to produce a new version. Non-active objects use this flag to notify the user of possible inconsistencies. This mechanism is similar to *backward chaining* in Marvel [TKP94]). Figure 6 shows the lineage of an object (from the model shown in Figure 1) and how this lineage can be used for the active object mechanism. Note the double linked list effect of the attributes, a characteristic exploited for more effective querying and mining.

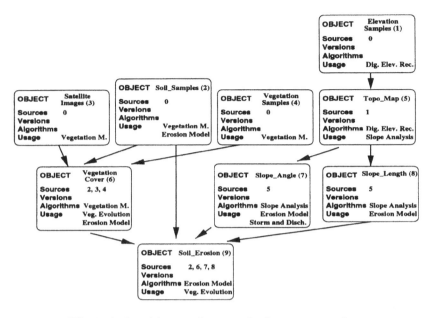

Fig. 6. Active objects and automatic change propagation.

Similarly, active models are those that must be recomputed when some of their inputs are modified. This allows to get derived data automatically every time new input is available. As with active objects, the mechanism is recursive but, unlike active objects, is based on eager propagation, i.e., the model is re-computed every time one of the inputs changes. If an active model produces new versions of objects that are input to other active models, these projects are triggered as well (similar to *forward chaining* in Marvel [TKP94].) The active object and active project facilities are complementary and give the process designer fine-grain control over the system's behavior. The active object mechanism provides a form of lazy evaluation and is thus well-suited for objects that are seldom read but whose inputs change very often. The active project mechanism, on the other hand, supports fast updates to derived objects. Combined with the event handling mechanisms, this is a very powerful tool to perform complex computations over a stream of data provided by external sensors.

Geo-Opera also adds a versioning mechanism *for programs*. Versioning has been used in a variety of systems for manipulating data. In Geo-Opera, each individual task within a process corresponds to a program. In business applications these programs do not change often and, if they do, the older versions can be immediately discarded. In scientific applications, modeling is done not only to obtain data but also to test the model itself, i.e., whether the perception of natural processes reflected in the model corresponds to reality. This implies running many versions of the same geo-process adjusting individual tasks until the results are satisfactory. For the same reasons that the dependencies between a process and its related data must be kept, the different versions of a task and a geo-process must be preserved so as to allow the user to repeat experiments, readjust parameters, and iterate over several design phases. In this regard, Geo-Opera is no different from a CASE environment in which versioning of code is a common feature. The way in which this versioning takes place is very similar to the procedures followed for external objects (in fact, some information about external programs is stored in the same form as the external objects).

4.6 Querying Capabilities

An important feature of Geo-Opera is its query management facilities. Aside of the standard process query capabilities common to all process support systems, Geo-Opera also tracks dependencies and versions of both data (internal and external) and processes. Most of these mechanisms have been outlined above and are based on the information stored about the external objects and the geo-process instances. Unlike in business applications where the process itself is usually of no importance once the final result is achieved, in many scientific applications, including the ones targeted by Geo-Opera, data cannot always be understood without the process that was used to create it. Examples of this problem in genetic experiments are well known [BSR96]. There are very strong ties between a process and the data resulting from the process, ties that must be maintained for as long as the data or the process remains in the system. In Geo-Opera, most operations result in a number of checks being performed to avoid deleting data or processes related to other data or other processes. This capability is not present in neither workflow management systems or GISs. The querying capabilities are extensively used to find information about object lineage (what objects are ancestors of another object), process and object dependencies (which processes use which objects, which objects are used in a process), and, in general, by any of the active mechanisms. Needless to say that these facilities are also made available to the end user, but it is important to point out that they are also used by the system itself to avoid inconsistencies.

As an example of the role played by the querying facilities, consider the problem of calculating the error associated to a particular data set. This error is derived by some complex manipulations and combinations of the errors associated with all the ancestors of the object. An algorithm trying to establish this error, needs to be able to access the lineage of the object. Thus, the user first retrieves the lineage of the object (which is calculated by the system by following the chain of "Source" attributes), then it retrieves the algorithms used in each step (following the algorithm attribute of each object). With this information, it can then try to establish the resulting error from the overall computation. A second example, this time on how Geo-Opera uses this query facilities, is the active mechanism. If a geo-process modifies an object, this is recorded in the system by creating a new version. When an object is notified that a new version exists, the "Usage" attribute is used to determine which tasks have this object as an input. Accessing the information about these tasks, Geo-Opera can determine whether it is necessary to execute them again. It can also set a flag in all objects derived from the original object indicating that a new version exists. The lists of derived objects is obtained by recursively following the "Usage" attribute of the object and the "OutBox" attribute of the tasks. In both examples, the interaction with the corresponding spaces where this information is stored (instance, history, and configuration spaces) is handled by the query manager. To increase performance, the query manager provides a number of entry points (for operations such as lineage, process recalculation, and inter-process dependencies) that are used to trigger all the queries necessary to solve the request in question.

5 Related Work

Geo-Opera is motivated by the similarity between geo-processes and generic processes. Some of the common issues were already hinted at by early work in the area [SSE+95, AE94a, AE94b, SW93, HGM93, HQGW93] but it has been only recently that workflow systems have been suggested as an adequate tool for scientific data management [MVW96, BSR96]. In addition, some work has been done in the experiment management area [ILGP96] with close ties to workflow notions. All of this research has provided insights on how process support systems may be used in scientific applications. In particular, ZOO [ILGP96] is a *desktop experiment management environment* for different scientific disciplines. It consists of a generic tool that can be extended by customized enhancements to make it suitable for different application domains. ZOO is based on a scientific database for experiment data and provides services for modeling, automation of experiment execution, and analysis of results. Conceptually and from an application point of view, Geo-Opera would be similar to an extension of ZOO for geographic modeling. In practice, however, there are significant differences between both approaches. OPERA is a generic process support system while ZOO is a dedicated system supporting scientific experiments. ZOO does not provide a generic notion of process and the tailoring refers to adapting the system to a particular scientific discipline, not to a given type of processes. Since much of the functionality provided in Geo-Opera applies to other scientific disciplines, it is conceivable to develop modifications of Geo-Opera for other type of scientific modeling. Moreover, ZOO models processes implicitly through dependencies between objects in an object-oriented data model. There is no explicit modeling of control flow and a centralized database is used for storage. Thus, ZOO lacks the extendibility and modularity of OPERA, which can employ arbitrary databases (even simultaneously) for storage and to manage external objects through the view mechanism.

The WASA project [MVW96, MVW95] is an attempt to use a commercial workflow management system for supporting scientific work. WASA allows the modeling, execution, and analysis of experiments in different scientific disciplines. Like GEO-Opera, WASA supports access to data stored in repositories external to the system (through so-called object brokers). There is, however, no support for the modeling of dependencies between objects and the automatic recomputation of models. The Geo-Opera architecture differs from the WASA approach further in its focus on distribution, heterogeneity, and scalability - issues which we believe are of crucial importance in these applications.

Geo-Opera has borrowed a number of solutions from GOOSE [AE94b] and CMS [SSE+95]. GOOSE is a tool for scientific experiment management, less developed than ZOO from an architectural point of view, but with a more powerful modeling mechanism in terms of dependency tracking. The notion of external objects and the way they are handled in Geo-Opera follows closely the ideas described in GOOSE. GOOSE, however, was never intended as a generic tool and it would be very difficult to adapt it to applications beyond those for which it was originally intended

Finally, Process Centered Environments, PCE, support process modeling and execution in software engineering applications [BK94, DG90, TKP94]. While these systems provide some of the data handling facilities present in Geo-Opera (namely the modeling of object dependencies and the automatic recomputation), the generality of the view mechanism is missing. In addition and due to the very different nature of the application domain, Geo-Opera provides functionality that cannot be expected from a software engineering tool (although it might be possible to extend these tools to provide functionality resembling that of Geo-Opera).

6 Conclusions

Geo-Opera is an extension of a generic process support system, OPERA, tailored to geographic and spatial modeling. Geo-Opera takes advantage of workflow technology to provide a fully distributed, heterogeneous computing platform in which to develop, execute, and manage complex geographic models. Geo-Opera, however, goes well beyond workflow management systems by providing ad-hoc functionality for tracking data dependencies and adapting them to the particular needs of scientific modeling. We see Geo-Opera as a first step towards bridging the gap between the potential offered by clusters of workstations and real applications. In many ways, Geo-Opera can be seen as an specialized distributed operating system for scientific applications in general and for geographic modeling in particular. Geo-Opera is currently in the prototyping stage, although much of the functionality described in the paper has already being tested and implemented. Future work includes enhancing the functionality of Geo-Opera by building explicit links to commercial GIS. We also plan to explore extensions of both OPERA and Geo-Opera to other application areas.

References

[AAEM97] G. Alonso, D. Agrawal, A. El Abbadi, and C. Mohan. Functionality and Limitations of Current Workflow Management Systems. *IEEE Expert*, 12(5), September-October 1997.

[AE94a] G. Alonso and A. El Abbadi. Cooperative Modeling in Applied Geographic Research. In *Proceedings of the International Conference on Cooperative Information Systems, CoopIS'94, Toronto, Canada*, May 1994.

[AE94b] G. Alonso and A. El Abbadi. Cooperative Modeling in Applied Geographic Research. *International Journal of Intelligent and Cooperative Information Systems*, 3(1), May 1994.

[Arm88] M.P. Armstrong. Temporality in spatial databases. In *Proceedings GIS/LIS*, pages 880–889, November 1988.

[ARM97] G. Alonso, B. Reinwald, and C. Mohan. Distributed Data Management in Workflow Environments. In *Seventh International Workshop on Research Issues in Data Engineering (RIDE'97)*, Birmingham, England, April 1997.

[AS96] G. Alonso and H.-J. Schek. Database Technology in Workflow Environments. *INFORMATIK/INFORMATIQUE (Journal of the Swiss Computer Science Society)*, April 1996.

[BDMQ95] A. Bernstein, C. Dellarocas, T.W. Malone, and J. Quimby. Software Tools for a Process Handbook. IEEE Computer Society, March 1995.

[BK94] I.Z. Ben-Shaul and G.E. Kaiser. A paradigm for decentralized process modeling and its realization in the oz environment. In *Proceedings of the 16th International Conference on Software Engineering*, Sorrento, Italy, 1994.

[BRS96] Stephen Blott, Lukas Relly, and Hans-Jörg Schek. An open abstract-object storage system. In *Proceedings of the ACM SIGMOD International Conference on Management of Data*, Montreal, Canada, June 1996.

[BSR96] A. Bonner, A. Shrufi, and S. Rozen. LabFlow-1: A Database Benchmark for High Throughput Workflow Management. In *Proceedings of the Fifth International Conference on Extending Database Technology (EDBT96)*, Avignon, France, March 1996.

[CKO92] B. Curtis, M.I. Kellner, and J. Over. Process Modelling. *Communications of the ACM*, 35(9):75–90, September 1992.

[DG90] W. Deiters and V. Gruhn. Managing software processes in the environment MELMAC. In *4th ACM SIGSOFT Symposium on Software Development Environments*, 1990.

[Fry94] C. Frye. Move to Workflow Provokes Business Process Scrutiny. *Software Magazine*, pages 77–89, April 1994.

[GG89] M. Goodchild and S. Gopal. *Accuracy of Spatial Databases*. Taylor and Francis Ltd, 1989.

[HGM93] N.I. Hachem, M.A. Gennert, and Ward M.O. The Gaea System: A Spatio-Temporal Database System for Global Change Studies. In *AAAS Workshop on Advances in Data Management for the Scientist and Engineer, Boston, Massachussetts*, pages 84–89, February 1993.

[HQGW93] N.I. Hachem, K. Qiu, M. Gennert, and M. Ward. Managing Derived Data in the Gaea Scientific DBMS. In *Proceedings of the 19th International Conference on Very Large Databases, Dublin, Ireland*, 1993.

[Hsu95] M. Hsu. Special Issue on Workflow Systems. *Bulletin of the Technical Committee on Data Engineering, IEEE*, 18(1), March 1995.

[ILGP96] Y.E. Ioannidis, M. Livny, S. Gupta, and N. Ponnekanti. ZOO: A desktop Experiment Management Environment. In *Proceedings of the 22nd VLDB Conference, Mumbai (Bombay), India*, September 1996.

[KAGM96] M. Kamath, G. Alonso, R. Günthör, and C. Mohan. Providing High Availability in Very Large Workflow Management Systems. In *In Proceedings of the Fifth International Conference on Extending Database Technology (EDBT'96)*, Avignon, France, March 1996. Also available as IBM Research Report RJ9967, IBM Almaden Research Center, July 1995.

[Kea93]   R.H. Katz and et al. Design of a Large Object Server Supporting Earth System Science Researchers. In *AAAS Workshop on Adavances in data Management for the Scientist and Engineer, Boston, Massachussetts, USA*, pages 77–83, February 1993.

[LA94]   F. Leymann and W. Altenhuber. Managing Business Processes as an Information Resource. *IBM Systems Journal*, 33(2):326–348, 1994.

[Lan88]   G. Langram. Temporal GIS design tradeoffs. In *Proceedings GIS/LIS*, pages 890–899, November 1988.

[LV90]   D.P. Lanter and H. Veregin. A Lineage Meta-Database program for propagating error in Geographic Information Systems. In *Proceedings GIS/LIS*, pages 144–153, November 1990.

[MAGK95]   C. Mohan, G. Alonso, R. Günthör, and M. Kamath. Exotica: A research perspective on workflow management systems. *Bulletin of the Technical Committee on Data Engineering, IEEE*, 19(1), March 1995.

[MVW95]   C.B. Medeiros, G. Vossen, and M. Weske. WASA: A workflow-based architecture to support scientific database applications. In *Proc. 6th DEXA Conference*, London, 1995.

[MVW96]   J. Meidanis, G. Vossen, and M. Weske. Using Workflow Management in DNA Sequencing. In *Proceedings of the 1st International Conference on Cooperative Information Systems (CoopIS96), Brussles, Belgium*, June 1996.

[Rad91]   F.J. Radermacher. The Importance of Metaknowledge for Environmental Information Systems. In *Proceedings of the 2nd Symposium on the Design and Implementation of Large Spatial Databases, Springer Verlag*, volume 1, pages 35–44, August 1991.

[SSAE93]   T.R. Smith, J. Su, D. Agrawal, and A. El Abbadi. Database and Modeling Systems for the Earth Sciences. *IEEE Bulletin of the Technical Committee on Data Engineering*, 16(1):33–37, March 1993.

[SSE⁺95]   T. Smith, J. Su, A. El Abbadi, D. Agrawal, G. Alonso, and A. Saran. Computational Modeling Systems. *Information Systems*, 20(2), 1995.

[SW93]   H.J. Schek and A. Wolf. From Extensible Databases to Interoperability between Multiple Databases and GIS Applications. In *Proceedins of tch 3rd Int. Symposium on Large Spatial Databases*, Singapore, June 1993.

[TKP94]   A.Z. Tong, G.E. Kaiser, and S.S. Popovich. A flexible rule-chaining engine for process-based software engineering. In *Proceedings of the Ninth Knowledge-Based Software Engineering Conference*, Monterey, CA, USA, 1994.

[Tob79]   W.R. Tobler. A transformational view of cartography. *The American Cartographer*, 6(2):101–106, 1979.

[vOV91]   P. van Oosterom and T. Vijlbrief. Building a GIS on top of the open DBMS Postgres. In *Proceedings of EGIS'91, Brussels, Belgium*, pages 775–787, April 1991.

[WC96]   J. Widom and S. Ceri. *Active Database Systems*. Morgan Kaufmann Publishers, 1996.

[ZG91]   Q. Zhou and B.J. Garner. On the integration of GIS and Remotely sensed data: towards an integrated system to handle the large volume of spatial data. In *Proceedings of the Second Symposium on Advances in Spatial Databases*, pages 63–72, August 1991.

Physical Database Design for Raster Images in CONCERT

Lukas Relly[1], Hans-J. Schek[1], Olof Henricsson[2], and Stephan Nebiker[2]

[1] Institute of Information Systems, ETH Zentrum, 8092 Zürich, Switzerland,
{relly,schek}@inf.ethz.ch
[2] Institute of Geodesy and Photogrammetry, ETH Hönggerberg, 8093 Zürich,
Switzerland, {ohenri,sn}@geod.ethz.ch

Abstract. In order to extend database technology beyond traditional applications a new paradigm called "externalization of database functionality" has been proposed in research and development. Externalization of database functionality is a radical departure from traditional thinking. Traditionally, all data is loaded into and owned by the database, whereas in the new paradigm data may reside outside the database in external repositories or archives. Nevertheless, database functionality such as query processing, optimization, physical database design, indexing and more general replication support is provided. In this paper we explore the applicability of our CONCERT approach when raster image data from photogrammetry and cartography are used. We report on investigations of a joint project where engineering applications such as 3D building reconstruction and raster image mosaics are supported by database functionality. We specifically mention advantages with respect to tertiary storage integration and database memory mapping thus directly supporting application specific algorithms.

1 Introduction

Todays Database Management Systems (DBMS) make the implicit assumption that their services are provided only to data stored inside the database. All data has to be imported into and being "owned" by the DBMS in a format determined by the DBMS. *Traditional database applications* such as banking usually meet this assumption. These applications are well supported by the DBMS data model and its query and data manipulation language. *Engineering applications* such as GIS, CAD, PPS or document management systems however differ in many respects from traditional database applications. Individual operations in engineering applications are much more complex and not easily expressible in existing query languages. Powerful specialized systems, tools and algorithms exist for a large variety of tasks in every field of engineering applications requiring data to be available in proprietary or data exchange formats.

We conjecture that one of the important reasons for engineering applications not to use general purpose database systems is that database systems restrict their services to data *stored in the database* and in the *database internal format*. All the complex specialized application systems and tools would have to be rewritten using the data structures enforced by the DBMS, or at least complex transformations would have to take place to map the DBMS representation into the application representation. These observations led to a radical departure from traditional thinking as it is expressed in [15,14].

In the database research group at ETH, we focus on *exporting database functionality* by making it available to engineering applications instead of requiring the application to be brought to the DBMS. Exporting DBMS functionality has different facets according to the different services provided by the DBMS. We have explored some of them over the past years with particular focus on the management of spatial data [18,12,13,4,2]. In this paper, we report on a joint project of the Institute of Information Systems (IS) with the Institute of Geodesy and Photogrammetry (IGP) at ETH targeting raster image management. We focus on the aspects of physical database design and query processing for raster image management, making novel contributions in the following areas:

- External tertiary storage typical for large raster image collections can be included in an elegant way.
- Replication by selectively caching pieces of large images is enabled.
- Efficient and direct access can be offered to complex image analysis algorithms such as feature extraction through database controlled memory mapping.

The paper is organized as follows: In Sect. 2, we describe two application projects and their requirements towards the underlying storage system. Section 3 presents different possibilities for managing raster images in the context of a database system. In Sect. 4, we introduce the architecture of our database system prototype CONCERT, which is capable in providing database functionality to non-database data. Section 5 shows, how CONCERT can be used in the raster image management context.

2 The Raster Data Management Example

In the areas of digital photogrammetry, remote sensing and cartography raster images play a prominent role. Here, we sketch two engineering applications, i.e. automatic building reconstruction and raster image mosaics, both requiring a flexible DBMS. Raster images are primarily used as data source often derived from digitized aerial photographs, satellite images, digitized maps, or laser scanning data. Raster images are large, sometimes more than a gigabyte, thus imposing new requirements to the underlying storage system.

2.1 Automatic 3-D Building Reconstruction

The automatic reconstruction of buildings from aerial images is a complex hierarchical process requiring raster image handling, image processing, matching procedures, texture and color modeling, to mention a few. The employed imagery is assumed to be multiple overlapping, digitized color photographs, however also other raster data is frequently used such as Digital Terrain Models (DTM).

The most promising approach to this general image understanding problem consists in separating detection from reconstruction. *Building detection* can either be automatic or manual, depending on the complexity of the scene. The main objective of building detection is to obtain a set of windows enclosing the same building in all overlapping images. A zoomed image (thumbnail preview) of the scene is presented to the operator for manually identifying the buildings. Their locations are stored in the database for further processing. This process is shown in the upper part of Fig. 1.

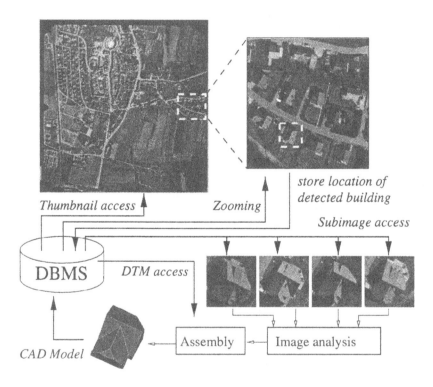

Fig. 1. The entire process from detection to reconstruction of buildings from aerial images.

Our approach to automatic *building reconstruction*, as it is shown in the lower part of Fig. 1, processes three (intensity) raster subimage (approximately [300 × 300]) and one color (RBG) subimage. In order to create the complete building, the vertical walls are inferred from the DTM. The final result is a complete CAD model which is subsequently stored in the DBMS to be used for example in 3-D city model applications. For more details on building detection and reconstruction we refer to [6,5].

In terms of DBMS access, this requires retrieving the subimages of the same geographical area from aerial color images together with the corresponding DTM. The building detection is performed in two steps: first locating the regions of interest, and subsequently zooming in to locate and identify single buildings. For identifying areas of interest, a thumbnail preview of the scene is needed, that allows the operator to stepwise zoom in (see Fig. 1). In the building reconstruction phase, the data of an individual house is required at full resolution from all overlapping images. Due to their size, whole aerial color images can not be held in main memory during the detection and reconstruction. Therefore, efficient partial image access at different resolutions is essential.

2.2 Management of Raster Image Mosaics

The management of raster image mosaics addresses the increasing demand in the fields of cartography and remote sensing for merging individual raster images into potentially very large raster image mosaics. Raster image mosaics or 'seamless' map databases provide arbitrary selections of spatial or thematic data subsets, thus freeing client systems from the complex tasks of searching and joining matching data sets. However, up until now raster image mosaics were generated and managed by specialized systems with no or only limited DBMS support.

In a cartographic application we built a multi-Gigabyte database with a seamless representation of all Swiss National Topographic Maps. We have investigated the required metadata and algorithms for managing raster image mosaics necessary for building a seamless map database[9]. Even though there is comprehensive metainformation associated with each image and mosaic, the metadata volume is small compared with the raster data. Metadata can typically be represented with standard data types and is therefore a quite typical "traditional" application supported by most DBMS's.

In terms of DBMS access, beside the efficient subimage access already described in Sect. 2.1, we conclude from this example that building 'seamless' map databases involving gigabytes or even terabytes of image data requires large parts of the data residing on tertiary storage. Therefore, good integration of tertiary storage in the DBMS is essential. The fact that complex metadata has to be managed together with the raster images suggests that specialized systems only focusing on the raster data management aspects are not very useful, because all the 'standard' database functionality would be missing.

2.3 Storage Management System Requirements

Summarizing the above requirements and also those of other engineering applications, we find four fundamental storage management system requirements for handling large raster data:

1. Due to the large data volumes involved, it is not possible to build a substantial image database using only secondary storage. Therefore, *good integration of secondary and tertiary storage* is essential. Raster image data is often shipped on tapes or CDROM's. Instead of copying the data onto the tertiary storage managed by the DBMS, it is desirable to directly integrate the original tape or CDROM into the DBMS's tertiary storage management. Such an approach avoids excessive copying and also allows (non-database) applications to access the data in its original form.

2. In many situations, a *quick display of a thumbnail preview* is essential for human operators to browse sets of images. In 3-D building reconstruction, the operator quickly identifies urban areas of interest. Likewise, accessing a seamless map database requires a thumbnail preview of the whole mosaic. Although it is possible to compute a thumbnail preview on demand from the original picture, in most cases such an image will be precomputed replicating some of the image information. This replication should be done under the control of the DBMS as part of the physical design process, not separately in each individual application.

3. Starting from thumbnail images, *stepwise zooming* is usually the next step an operator performs. For detecting buildings in a rural area, an overview image might be sufficient, while in urban areas a higher resolution is required. As a rule of thumb, independently of the scale of the image, "a screen full of data" is required. Depending on the size and the information contents of an image, this "multi-resolution" support might involve storing preview images at different scales. This again is a data replication issue, which should be dealt with in the physical design component of the DBMS.

4. After specifying the region of interest, the operator applies to these partial images algorithms such as the ones mentioned above. Often, the full resolution image data has to be accessed only in this last step requiring tertiary storage access. Since access to tertiary storage is slow, *efficient partial image access* on tertiary storage and *effective caching strategies* (in main memory as well as in secondary memory) integrated with the database and operating system storage access and caching strategies is important.

These requirements are fundamental to such engineering applications and they directly influence the design of our DBMS prototype system CONCERT.

3 Raster Data Management in Database Systems

In this section, we present different possibilities for using a DBMS as a storage system for managing raster images. First, we discuss the "traditional" approach using standard database technology. Then, we present some issues of more advanced approaches that are also relevant for our own system architecture.

3.1 The Traditional Approach

Although not designed for storing large, complex objects such as raster images, todays commercial DBMS can be used as storage management subsystems to store such data. Most of them provide a binary large object (BLOB) data type for storing arbitrary byte sequences. Some of them are restricted in size, making it necessary to store a raster image using several BLOB objects. Basically, the only operations these data types provide is setting and retrieving this arbitrary byte sequence. However, some systems additionally can set or retrieve a substring of a BLOB. This technique is used in one of our projects using standard relational DBMS technology[1] (Oracle Universal Server and its data type "long raw") implementing a raster image mosaic management system. Also commercial products such as SICAD use a similar approach building a specialized GIS system on top of a relational DBMS.

Instead of storing a raster image in a single "long raw" attribute, we store partial images in an image table providing faster access to image parts. Comparing different strategies [9], the best results are achieved using the tiling method described in [8]. The required image index associating the image parts with their relative position within the whole image is implemented as an additional database table. The query processing, mapping a user request for retrieving a subimage, is done in the application environment in the following way: Based on the user query, the application retrieves the relevant information using the index table to determine all required tiles. These are fetched in a second step from the disk through the DBMS buffer into the application, where they are composed and clipped to the requested size. This second step is illustrated in Fig. 2.

With this approach, physical design and query processing has to take place in the application, which leads to inefficient data access. The application initially storing the image in the database has to decide on the appropriate tile size for the subimages. A good choice for this size is essential. If the tiles are too big, far too much data will be read from disk and transferred from the database into the application. If on the other hand, they are too small, each query will retrieve many individual subimages. A substantial overhead is involved with every subimage retrieval due to the client–server architecture

---

[1] In most of our systems however, we use the file system directly for storing raster images.

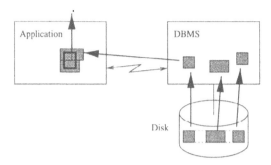

Fig. 2. System architecture using ORACLE as a raster image repository

being used . Since different query sizes are used depending on the application and applications may vary over time, it is almost impossible to find a good tile size for all applications for all times. Therefore, a repartitioning might be necessary requiring all applications to be able to cope with the new tile sizes. One of the benefits using database systems is described by the term 'data independence', which says the application can be implemented independently of the current physical representation of the data. With such an approach, data independence is lost. In addition, main memory is wasted by the fact that the same data requires buffer space in the DBMS buffer as well as in the application.

3.2 Advanced Approaches

In the last years, techniques have been developed to better support engineering applications. In contrast to the traditional approach, where engineering application support is built on top of the DBMS, they integrate it inside the DBMS.

Manufacturer Enhanced DBMS: Because of the increasing demand for better engineering application support, DBMS vendors start integrating more and more new data types into their systems. They are accessible through the query language and often have index support. With this method, support for a "Raster Image" base data type and operations such as the rectangular subimage retrieval can be implemented. However, many raster image formats exist and, depending on the application, different of them are used. It is not feasible to support all of them as database base types requiring expensive format conversions.

Extensible Database Systems: The above problems have fostered the development of extensible database systems, and commercial products (Illustra[7], Informix[17]) start becoming available. The idea is to allow the user to extend the system by incorporating user-specific code into the DBMS kernel. These user extensions are often referred to as "data blades" allowing the subimage to be composed in the server environment. This avoids copying unnecessary data to the application. As Fig. 3 illustrates, the same byte sequences as in the ORACLE approach are fetched into the DBMS buffer. But instead of transferring these byte sequences to the application context, the user-defined operation composing and clipping the image parts is executed in the DBMS environment. For each raster data format, a user extension is created implementing this operation for the particular format.

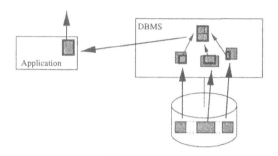

Fig. 3. Subimage access using extensible DBMS

However, an important problem, shared by both advanced and traditional approaches, remains: images shipped on tertiary storage media have to be read into the database before being transformed into the database representation and finally stored in the databases own tertiary storage format. This procedure involves unnecessarily moving around gigabytes of data. None of the approaches offers a satisfactory solution to the tertiary storage integration requirement presented in Sect. 2.3.

3.3 Exporting Physical Database Design Functionality

As we have shown, all approaches managing raster image data have advantages and disadvantages. Our method combines their advantages while avoiding the disadvantages. It provides solutions to the four identified raster data management requirements.

The problem originating in the client–server paradigm involving expensive copying and interface crossing operations, as illustrated in Fig. 2, is solved using the techniques of extensible database systems. We define a general framework handling different (user defined) data types, as long as these data types comply to certain rules. These rules will be discussed in the presentation of the overall CONCERT architecture in the next section. In contrast to the extensible systems approach, we do not allow arbitrary extensions to be built into the CONCERT kernel. Instead we identify a limited set of fundamental physical design concepts. The query optimizer is well aware of these concepts and their semantics. Because every extension is a concrete instantiation of a combination of these fundamental concepts, the query optimizer is able to deal with these extensions, even if the actual implementation is user-defined as in an extensible system.

Furthermore, we show how, using its framework integrating user defined data types and operations, the CONCERT approach allows the DBMS to manage external objects using DBMS functionality. In particular in the raster image context, CONCERT can integrate data shipped on tertiary storage directly without copying or reformatting the data on tape or CDROM.

4 The General CONCERT Architecture

This section describes some of the aspects of the CONCERT architecture, focusing on two issues particularly relevant to raster data management. A more general discussion of other CONCERT aspects can be found in [1,11,2]. The first issue concerns the CONCERT buffer management, the second discusses the Abstract Object Manager, the component defining the limited set of fundamental physical design concepts and its integration into the kernel.

4.1 The CONCERT Memory Mapped Buffer

As in traditional database systems, the CONCERT kernel manages a main memory area used as a cache for data on persistent storage. The persistent storage is abstracted by a set of segments consisting of a set of contiguous byte sequences each. In contrast to other systems, these byte sequences are not fixed size pages, but arbitrary length sequences of fixed size pages. We refer to them as page ranges.

For large raster images, it is important to allow arbitrary large byte sequences. In the ORACLE BLOB approach, each BLOB has to be mapped to a set of database pages. Each database page contains a few bytes of header information. The remaining part of the page can be used for the image data. If a BLOB is larger than a page, which is always the case for large raster images, the image data has to be composed in the database kernel copying the data part from each page into contiguous address space. In CONCERT, raster images or raster image tiles are stored in page ranges of appropriate size to hold the whole image or tile. Because there is no header information required on each page, no copying is required. The page range provides direct access in one single linear address space. Figure 4 illustrates these two possibilities.

A problem for buffering page ranges is obvious: If the page range is larger than the available main memory, it is not possible to read the whole page range at once. Even if the page ranges are a lot smaller than the available main memory, buffer management becomes very difficult due to buffer fragmentation resulting from different sized page ranges. CONCERT solves this problem

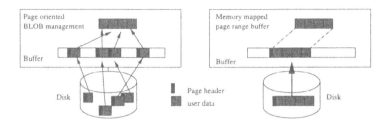

Fig. 4. Traditional BLOB management vs. memory mapped page ranges

by exploiting the operating systems paging mechanism through memory mapping. Instead of issuing a read or write request bringing pages from disk to the database buffer and back, CONCERT memory maps the corresponding disk pages. Using it's normal paging mechanism, the operating system takes care of the database buffer solving many problems associated with traditional buffer management in database systems.

It is beyond the scope of this paper to discuss all the problems and their solutions with respect to CONCERTS memory mapped buffer. A detailed discussion can be found in [1].

4.2 The Abstract Object Manager

To explain the general CONCERT idea of exporting physical design to non-database data, let us first look at a standard btree index [3], as it is implemented in most database systems. Relational systems provide a user level statement "CREATE INDEX", that stores a certain attribute of each tuple of a table in a btree together with a reference (physical address) of the tuple. The btree stores the entries in ascending order. Depending on the data type of the attribute, the ordering operation is chosen: for data type "NUMBER", the standard ordering over numbers is used, for "STRING" data types, lexicographic order is used, for "DATE" data type, the date ordering is used. Some systems provide other data types with data type specific ordering operations. Other data types do not provide ordering operations and consequently can not be indexed with a btree index. The important observation is that a btree requires an ordering operation for being able to be created over an attribute. But the btree does not depend on any other property of the data type other than the ordering operation.

This observation was discussed in [16] and led to the central idea of generalized access structures, that is also the basis of the *abstract object storage type* concept in CONCERT: Instead of defining the data types suitable for a btree index, CONCERT requires so called *concept typical operations*[2] for all

---

[2] By concept typical operations, we mean the methods the user must provide to allow CONCERT to interpret a given abstract object storage type

objects to be indexed by a CONCERT btree. In CONCERT, we denote this concept as SCALAR. Every user defined data type claiming to behave *like* a SCALAR and therefore implementing its concept typical operation, the ordering operation, can be indexed with a btree.

In the same way, as we argue for the btree, it is possible to identify all concept typical operations needed for other methods of physical design. An other well known example is the inverted file index often used in full text retrieval. Information retrieval systems support a specialized data type "TEXT", that is understood by the inverted file index. When a text attribute is indexed, the text is broken up into individual tokens, and for each token, an entry is made in the inverted file index data structure. Again, it is irrelevant to the inverted file index, that it is built on a "TEXT" data type. It is only important, that the argument attribute can be broken up into subcomponents, over which the index can iterate. In CONCERT, this concept of iteration over a set of components is denoted as LIST requiring the iteration operation as its concept typical operation. See Fig. 5 for an example using the LIST concept in order to build an inverted file index. Such an index could equally well be used to index, for example, feature vectors describing images, if the image type would be defined to behave *like* a LIST and therefore provide the feature extraction as its concept typical iteration operation.

```
Token    = SCALAR
Document = LIST (Token)
d: Document
o: RECORD (key: SCALAR, data: ANY_TYPE)
...
foreach t in d.ITERATE
    o.COMPOSE (key=t, data=OID(d))
    tree_insert (inv_file.root, o)
```

Fig. 5. Generic inverted-file index

The statement made for the above two examples holds for other methods of physical design: horizontal partition requires a discriminating operation on each object determining the partition the object belongs to, vertical partitioning relies on the possibility dividing each object in smaller parts (the corresponding CONCERT concept is RECORD). Figure 6 illustrates insertion into and retrieval from a simple vertically partitioned collection.

The vertical partitioning can also be used to replicate partial objects, if in one partition, the whole object is stored and in the other partition, one part is replicated. This idea can be used for storing a thumbnail preview picture replicating the preview information in secondary storage while the primary data resides on tertiary storage.

```
Data = RECORD (d1: ANY_TYPE, d2: ANY_TYPE)
...
insert_partition (c1, c2: collection; d: Data)
    c1.insert (d.PROJECT (d1))
    c2.insert (d.PROJECT (d2))
...
get_partition (c1, c2: collection; d: Data; id: OID)
    d.COMPOSE (d1=c1.get(id), d2=c2.get(id))
```

Fig. 6. Generic vertical partition

Within the context of our example, expressing spatial relationships for physical design is particularly interesting for storing raster images as a set of partial images and recombining them at query time. CONCERTs built-in physical design concept SPATIAL is used to express the fact, that an instance of this concept can be accessed with a subwindow query defined by a rectangle bbox[3]. Figure 7 shows, how such a subwindow query WINDOW_PROJ (bbox) can be transformed into a query over a set of partial images allowing the DBMS to access only the required partial images instead of accessing the whole image. In order to perform this optimization, the objects implementing the SPATIAL concept must provide, beside the WINDOW_PROJ operation, the concept typical operations PARTS returning a set of partial spatial objects, the predicate OVERLAPS over the partial objects determining, whether a given partial object overlaps the query window and the operation COMPOSE combining partial objects to a single spatial object.

```
img: SPATIAL
...
img.WINDOW_PROJ(bbox) ≡
img.COMPOSE ( { i | i ∈ img.PARTS ∧
                    i.OVERLAPS(bbox) } ).WINDOW_PROJ(bbox)
```

Fig. 7. Expressing an image as a combination of partial images

Generally speaking, the fundamental idea is to distinguish on the one hand between knowledge of a concrete object representation, that is not necessary for physical design and is completely up to the application, and on the other hand, the concept typical properties such as ability to being ordered or partitioned, that are relevant for physical design and query processing purposes. While [16] presents this idea in the context of a btree, we generalize

---

[3] Note that the concept is not limited to two-dimensional objects. In n-dimensional space, the application object bbox represents an n-dimensional polyhedron.

and claim that all low-level physical database design decisions are governed by the four concepts SCALAR, RECORD, LIST and SPATIAL. Notice, that we allow arbitrary implementations for each of these concepts. But we separate the application specific implementations from the four generic concepts. This allows the query optimizer to base its decisions on the four built-in concepts. Although some uncertainty remains for the optimizer calling external operations, the four concepts give important hints about the semantics of the operations.

5 Application in the Raster Image Management Context

Exploiting the CONCERT functionality, we present here solutions to the storage management requirements stated in Sect. 2.3. We start with general observations concerning efficient subimage retrieval before going into specific details about tertiary storage management and exploitation of the memory mapped buffer.

5.1 Efficient Subimage Retrieval

Many different image formats exist for raster images mapping the two-dimensional pixel space onto a linear address space. Common to most of them is the property that contiguous spatial subareas of the whole image are stored on contiguous substrings in linear address space. Figure 8 shows examples of different possible linearizations: scanline linearization, space filling curves such as a Hilbert curve and space tiling linearization. Usually, byte sequences are clustered together for example onto the same page of the underlying storage medium. Areas having the same shade illustrate the pixels associated with such a cluster.

As we have stated in Sect. 2.3, one of the predominant requirements for the storage management component in the raster image context is efficient rectangular subimage retrieval. In order to satisfy such queries, n sequences of bytes out of the whole image byte stream stemming from different blocks have to be retrieved. If the image is stored in a clever way, n is small, however it is not possible to find a structure, that can satisfy arbitrary queries with only byte sequence containing exactly the data required. To achieve good performance, n has to be small for *typical* queries.

It is the task of good physical design to choose from all possibilities the one that satisfies this condition. For different image formats, this optimal spatial division of the the whole image space might be completely different. The subimage retrieval however always results in a set of continuous byte sequences representing the subimage in a particular application format. Therefore, physical design and query processing for efficient subimage retrieval is based on two important steps.

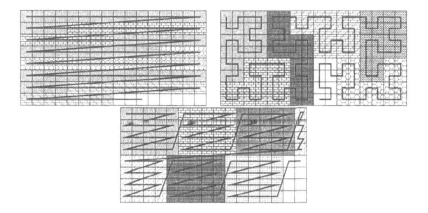

Fig. 8. Example possibilities for linearizing raster images

1. The two-dimensional raster image has to be linearized storing partial images in contiguous byte sequences.

2. Based on these partial images, at query time the relevant byte sequences needed for the query have to be identified copying them and restructuring them to form the subimage required.

As a consequence, the concept typical operations for physical design of raster images are the operations to divide images into partial images and to reconstruct images from partial images. Although the "natural" scanline approach is not very efficient for large raster images, it can be viewed as a partition of the whole image into partial images of one pixel height consisting of one pixel row. The subimage to be composed for a given query consists of a combination of subsequences of the pixel rows. From a physical design point of view, better storage structures such as tiled images can be used for representing the image. Obviously, the algorithm for constructing the subimage required on retrieval looks different, but the general mechanism remains the same: substrings of the partial images are combined to form the subimage. These algorithms are instances of the generic operations PARTS and COMPOSE shown in Fig. 7 and can be used in the CONCERT internal handling of SPATIAL objects.

What we have said so far looks similar to approaches taken in extensible DBMS as we have shortly discussed them in Sect. 3.2. However there is an important difference greatly improving the physical design and query optimization, that results from the limitation to only a few fundamental concepts. By giving concrete examples in the following, we want to motivate these differences.

5.2 Integration of Tertiary Storage Using Concept Typical Operations

In raster data management systems based on standard database technology, image data shipped on tertiary storage media such as tapes or CDROM's has to be brought into the system by loading it off the tape with an application program inserting the data into the DBMS. For large image archives, sooner or later the data has to be stored onto the DBMS's own tertiary storage media, because the volume of image data is too large for keeping all data in secondary storage all the time. As a result, gigabytes of data are loaded off tape, being piped trough the DBMS and ending up on a different tape in a slightly different format. This process is illustrated in Fig. 9.

In addition, unless the DBMS offers a special loader program integrated in the DBMS kernel, the loader program is a standard DBMS application being subject to the DBMS transaction management. As a consequence, redo information will be generated for the loader process storing the whole image on the redo log as well.

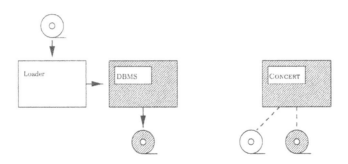

Fig. 9. Tertiary storage management

With the CONCERT approach, tertiary storage media can be used directly without loading all the images into the database. The only requirement is, that the concept typical operations retrieving the image parts are able to deal with the tertiary storage format. Instead of initially executing a loader program for the image and, at query execution time, using the built-in tertiary storage management operations of the DBMS, CONCERT uses a slightly modified loader program at query execution time avoiding the expensive initial loading process. Only the metadata describing which loader program and which tape to use, has to be initially brought into CONCERT.

We have mentioned the subimage retrieval being one of the major storage management requirements. Assume such a query in an SQL-like syntax, as it could be generated by an application program such as one of the image analysis programs in Fig. 1.

```
SELECT subimage(Rasterdata, Pixelcoordinates(x0,y0,x1,y1))
FROM    ImageRelation
WHERE   ...
```

given the image relation having a complex attribute 'Rasterdata' defined in any syntax of for example an object-relational system. We are not focusing on the user interface aspect but rather, we would like to show, how such a query is executed at the storage management layer.

Independent of the system being considered, first the image data has to be loaded from tertiary storage into the DBMS. For traditional DBMS approaches maintaining their own tertiary storage, the image data will be read using the DBMS's built-in tertiary storage management operations. In the CONCERT approach, the DBMS only knows about the existence of the external image, but the image data itself still resides on the shipping media. In database terminology, this can be viewed as a vertical partition storing some metadata of the image in an internal table and the image data on a 'virtual' table on tertiary storage. Using the CONCERT concept for a vertical partition (a RECORD), the following pseudo-code expresses the CONCERT-internal storage concept:

```
External_Image = RECORD (header: Description, image_data: Image, ..)
```

The concept typical operations for a RECORD are the projection operations for its components. The projection operation ext_img.PROJECT(image_data) is generating the "computed attribute" by executing the image load program reading the image from tape.

The name Image in the RECORD definition denotes a concrete instantiation of the internal concept SPATIAL in the following way

```
Image = SPATIAL
```

expressing the concept of a spatially extended object supporting window queries. Therefore, the SQL-like query above will be executed inside CONCERT as

```
ext_img.PROJECT(image_data).WINDOW_PROJ(bbox)
```

invoking the user-defined *subwindow extraction* operation on the result of the *loader program*. In this way, the operation of extracting part of an external image is made explicit to the query engine using the two known concepts RECORD - PROJECT and SPATIAL - WINDOW_PROJ.

Often not only a single query will be issued for one image but rather several regions of the same image will be retrieved. 3D house extraction is usually done for all elements of a given area. Because it is known to the query engine that the loader program and the subwindow extraction are not arbitrary operations possibly having side effects but a RECORD projection and a SPATIAL subspace projection operation, the result of the loader program can be cached in secondary storage. This can be expressed by

```
cached_img = ext_img.PROJECT(image_data)
```

performing a standard vertical partition / replication operation on a RECORD component. Note the importance of the query engine understanding this operation as a RECORD projection also with respect to transaction management: no redo information is needed for writing the image cache. For repeated partial image retrieval of the same area, the query engine can now access the cached image

```
cached_img.WINDOW_PROJ(bbox)
```

in the same way, a standard query optimizer is able to rewrite queries against a normal vertical partition with replication. In this situation, three physical design possibilities can be considered: (1) The image data can be loaded "on demand" every time calling the image_data operation, (2) the image can be permanently replicated in secondary storage (eg. as a result of a system administrator issuing CREATE INDEX REPLICATE img ..) or (3) the image can be loaded when it is accessed for the first time and then cached in secondary storage for a certain time. Which of these three variants is chosen is dealt with in the physical design component of the DBMS, and not in the application: *Standard DBMS heuristics can be used for this physical design issue on data stored outside the DBMS.*

Instead of caching the whole image, individual partial images could be cached as

```
cached_imgs = ext_img.PROJECT(image_data).PARTS
```

enabling the DBMS to use the optimization shown in Fig. 7. Therefore, the internal query accessing subimages used for 3D house extraction would be executed as follows:

```
subimg.COMPOSE ( { i | i ∈ cached_imgs ∧
                       i.OVERLAPS(bbox) } ).WINDOW_PROJ(bbox)
```

In the following, we show how these physical design variants can be used for efficient subimage access in a multiuser environment exploiting the CONCERT memory mapped buffer.

5.3 Exploiting the Memory-Mapped Buffer in CONCERT

As we have discussed in the raster image mosaic example in Sect. 2.2, this type of data is especially suited for corporate or public geospatial image and map databases. Therefore, many users will retrieve subimages possibly stemming from the same source image at the same time. As we have shown, the rectangular subimage retrieval is mapped to a retrieval of n byte sequences out of the whole raster image. In a standard extensible database system approach, the operation generating a rectangular subimage would therefore retrieve each of these byte sequences into a buffer space, from where the resulting subimage is created. If several different subimage requests happen at the same time, overlapping byte sequences have to be buffered more than once unless exactly the same byte sequence is needed for both requests. This is illustrated in Fig. 10.

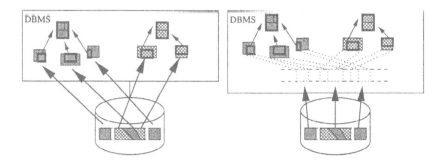

Fig. 10. Data access by copying and by memory mapping

Alternatively, longer byte sequences could be retrieved hoping that all the necessary requests can be satisfied identifying for each request an appropriate substring of the larger byte sequence. Due to fragmentation problems and due to the fact that future requests are not known in advance, this approach is not possible. This idea however leads directly to the memory-mapped buffer approach. If the larger byte sequence is identical to the whole image, it is always possible to identify appropriate substrings for all requests. Obviously, it is a bad idea to copy the whole image into main memory. But using CONCERTS memory mapping, a virtual copy can be brought into the address space. The operating system then takes care using its normal demand page swapping to bring in the pages actually needed. Only the pages accessed are retrieved, and no page is copied twice.

5.4 Exploiting Application Semantics

Obviously, a memory-mapped buffer could be implemented in a standard extensible system the same way we did in CONCERT experiencing the same benefits. But in extensible systems allowing to run arbitrary operations in the kernel, the optimizer has no idea about the semantics of these operations. If it knew about the operation being a subimage retrieval operation, appropriate prefetching strategies for the mapped area could be used, for example. The subimage retrieval operation is known by the optimizer to be a read-only operation making it possible allow concurrent access to the shared mapped area without risk of a deadlock in case of a lock upgrade due to the arbitrary operations starting to modify their mapped areas.

Another example of exploiting the operation semantics based on the limited set of available operations is the following. Consider two concurrent subimage retrieval requests extracting different areas of the same image. The extensible DBMS would see twice the same function call with different parameters. Because this could be arbitrary operations, the DBMS has little possibilities exploiting the fact, that these operations are reading the same tape building a secondary memory cache for the image and finally extracting different subimages. In CONCERT however, the subimage retrieval is defined in terms of the computed attribute disk_image. This allows a sort of "common subexpression elimination" when transferring data from tertiary to secondary memory.

6 Conclusion

In this paper we have shown how it is possible to meet the fundamental storage management requirements for efficient physical design and query processing identified together with our project partners using *concept typical properties* instead of concrete (built-in) data types for physical design in CONCERT.

We have used traditional database technology for raster image mosaics and have built a prototype. Starting from this experience we have shown how the new paradigm of externalizing database functionality will overcome the deficiencies of traditional databases. Specifically we stress the need to concentrate on tertiary storage management of external data, by selectively and gradually replicating images into the database and — by doing so — applying database memory mapping in order to support complex computations.

We have reported on an ongoing project. Many questions still deserve more investigations including the evaluation of the concept by continuing the CONCERT prototype implementation. We also did not describe coordination mechanisms between external data and CONCERT (meta-) data as well as replica and indexes. Here, we want to apply similar techniques as we have applied in a previous project [10] in the area of computer integrated management.

Acknowledgment

We thank Gustavo Alonso and Helmut Kaufmann as well as the anonymous referees for valuable comments to improve this paper.

References

1. Stephen Blott, Helmut Kaufmann, Lukas Relly, and Hans-Jörg Schek. Buffering Long Externally-Defined Objects. In M.P. Atkinson, V. Benzaken, and D. Maier, editors, *Proceedings of the Sixth International Workshop on Persistent Object Systems (POS6)*, pages 40–53, Tarascon, France, September 1994. British Computer Society, Springer-Verlag.

2. Stephen Blott, Lukas Relly, and Hans Jörg Schek. An open abstract-object storage system. In *Proceedings of the 1996 ACM SIGMOD Conference on Management of Data*, June 1996.

3. D. Comer. The ubiquitous B-tree. *ACM Computing Surveys*, 11(2):121–137, 1979.

4. Gisbert Dröge. *Eine anfrage-adaptive Partitionierungsstrategie für Raumzugriffsmethoden in Geo-Datenbanken*. PhD thesis, Eidgenössisch Technische Hochschule (ETH), Zürich, CH-8092 Zürich, Switzerland, 1995.

5. O. Henricsson. *Analysis of Image Structures using Color Attributes and Similarity Relations*. PhD thesis, Swiss Federal Institute of Technology (ETH) Zurich, 1996. No. 11663, also published in *Mitteilungen Nr. 59* at the Institute of Geodesy and Photogrammetry.

6. O. Henricsson, F. Bignone, W. Willuhn, F. Ade, O. Kübler, E. Baltsavias, S. Mason, and A. Grün. Project Amobe, Strategies, Current Status, and Future Work. In *International Archives of Photogrammetry and Remote Sensing*, volume XXXI, Part B3, pages 321–330, 1996.

7. Illustra Information Technlogies, Oakland, CA. *Illustra User's Guide*, October 1995.

8. Peter Lamb. Tiling very large rasters. In Thomas C. Waugh and Richard G. Healey, editors, *Advances in GIS Researches, Proceedings of the Sixth International Symposium on Spatial Data Handling*, volume 1, pages 449–461, London, 1994. Taylor and Francis.

9. Stephan Nebiker. Datenbankverwaltungssystem für kartographische Rasterbilder und Rasterbildmosaiken. In *Kartographie im Umbruch – neue Herausforderungen neue Technologien, Kartographiekongress Interlaken*, number 14 in Kartographische Publikationsreihe, pages 169–179. Schweizerische Gesellschaft für Kartographie, 1996.

10. Moira Norrie, Werner Schaad, Hans-Jörg Schek, and Martin Wunderli. CIM through database coordination. In *Proc. of the Int. Conf. on Data and Knowledge Systems for Manufacturing and Engineering*, Hongkong, May 1994.

11. Lukas Relly and Stephen Blott. Ein Speichersystem für abstrakte Objekte. In *Proceedings of the 6th German Conference on Database Systems for Office, Engineering and Scientific Applications (Büro, Technik, Wissenschaft, BTW)*, Lecture Notes in Computer Science, pages 338–347. Springer Verlag Berlin Heidelberg New York, March 1995.

12. Hans-Jörg Schek and Andreas Wolf. Cooperation between autonomous operation services and object database systems in a heterogeneous environment. In David K. Hsiao, Erich J. Neuhold, and Ron Sacks-Davis, editors, *IFIP, DS-5, Semantics of Interoperable Database Systems*, pages 245–271, Lorne, Victoria, Australia, November 16–20 1992.

13. Hans-Jörg Schek and Andreas Wolf. From extensible databases to interoperability between multipel databases and GIS applications. In *Advances in*

Spatial Databases: Proceedings of the 3rd International Symposium on Large Spatial Databases, Lecture Notes in Computer Science. Springer Verlag Berlin Heidelberg New York, June 1993.

14. Hans-Jörg Schek. Improving the role of future database systems. In *http://www.cs.brown.edu/ people/ sbz/cra/ positions/ schek1.html*, 1996.

15. Avi Silberschatz and Stan Zdonik. Database systems — breaking out of the box. In *http://www.cs.brown.edu/ people/ sbz/ cra/ paper.ps*, September 1996.

16. Michael Stonebraker. Inclusion of new types in relational database systems. In *Proceedings of the International Conference on Data Engineering*, pages 262–269, Los Angeles, CA, February 1986. IEEE Computer Society Press.

17. Michael Stonebraker. Architectural options for object-relational dbmss. Technical report, Informix White Paper, November 1996.

18. Walter Waterfeld. *Eine erweiterbare Speicher- und Zugriffskomponente für geowissenschaftliche Datenbanksysteme*. PhD thesis, Technische Hochschule Darmstadt, 1991.

Spatial Data Models

Topological Error Correcting in GIS

Thierry Ubeda[1] and Max J. Egenhofer[2]

[1] Laboratoire d'Ingénierie des Systèmes d'Information (LISI).
Institut National des Sciences Appliquées de Lyon (INSA).
Bât 404, 20 avenue A. Einstein
69621 Villeurbanne Cedex - France
e-mail : ubeda@if.insa-lyon.fr

[2] National Center for Geographic Information and Analysis (NCGIA) and
Department of Surveying Engineering, Department of Computer Science,
University of Maine, Orono ; ME 04469-5711, U.S.A.,
e-mail : max@spatial.maine.edu

Abstract. It is agreed upon that topological relations are of great importance regarding to GIS data sets consistency. A lot of errors that can be found in GIS data sets are coming from a lack of knowledge about topological relations between the geographical objects stored in the database. Consequently, topology can help to find errors in GIS data sets, and can help to correct them. The topic of this paper is to present how topological relation can be used to define, to detect and to correct errors in GIS data sets. Such an approach required three parts : the definition of errors using topological integrity constraints, how to check the database and how to correct errors. This paper focuses on the first and the third part. Errors will be described using topological integrity constraints. This method allows one to define the constraints that fit its own data set, that allows to take the semantics of data into account. Correction will be made by applying transformations to the data. For each error detected, a set of possible corrections will be compute and the end-user will have to choose the appropriate one.

1 Introduction

Spatial Analysis in GIS is often hindered by erroneous information among data sets. Answers to spatial queries and spatial reasoning are then not reliable. Different kinds of spatial errors in GIS data sets can be defined. Geometric errors, like non polygon closure and self-intersecting lines, are coming from data models and structures. Two main origins can be drawn: the data model is weak and is not able to capture all the characteristics of geographical objects, or the data do not respect the data model requirements. Semantics errors, like a road within a lake and a building described as a line, are coming from the real world description. Such errors can not be found without using the semantics of real world entities.

Geometric errors depend on data models and data structures of GIS (Geographical Information Systems). Several attempts to define properties of geographical objects have been made in [9]; in [10] and in [11]. The goal of those properties was to detect and to avoid structuring errors. The list of properties defined in [10] and in [11] is not complete for all kinds of data models, but can be used as a starting point for most of data structures. The list of properties given in [9] was complete for the data model he studied, namely planar graph. It appears that a classification of data model in GIS is required to design a set of properties suitable for all spatial databases. Unfortunately, such a classification doesn't exist yet. In [6], Franck proposed a classification for model using only points and arcs. Area objects have to be added to such a classification if we want to cover all kind of data models.

Semantics errors are defined using the meaning of geographical objects, that is to say the real world entities described by the object. Topological errors are a kind of semantic errors. Topological relations are based only on the shape of objects, but semantics of objects have to be taken into account to decide whether a topological scene is consistent or not. Topological relations are of great importance in GIS [1]. A lot of errors contained in GIS came from erroneous topological relations among geographical [8]. Most of GIS do not deal with topological relations, or consider only few relations such as adjacency and inclusion.

The goal of this paper is to define a correcting process for topological errors in GIS. A complete topological error correcting process should contain the three following parts:

- A topological error definition process. Because topological errors are semantics errors, such a process must provide the end-user with an easy way to defined topological integrity constraints for its own database.
- A topological error checking process.
- A topological error correcting process.

In this paper, topological errors will be defined as topological integrity constraints. Each time a constraint is not respected, an error is found. A scene in which a topological constraint is not respected is an error.

In a first part, a topological integrity constraint definition method will be introduced. It is based on topological relations defined by the 9-intersection model [5]. It allows to define a constraint using all possible topological relations (and not only inclusion and adjacency).

Secondly, a topological error correcting method will be presented. The goal of such a process is to compute a set of possible corrections among which the end-user has to choose.

The way to present correcting scenarios will be discussed in the last part, then we will conclude.

2 Topological Integrity Constraints Definition

The goal of topological integrity constraints is to define rules on the data stored in the database and to provide a mean to detect topological errors.

2.1 The 9-intersection model

This topological model has been designed by Max J. Egenhofer in [3] and in [5]. In this model, binary topological relations between two objects A and B are defined in terms of the nine intersections of A's boundary (∂A), A's interior (A°) and A's exterior (A⁻) with the boundary (∂B), interior (B°) and exterior (B⁻) of B (see figure 1).
Each object A and B can be a point, a line or a polygon.
Definition of each part of each kind of geometric object is the following:

P is a point : $P = \partial P = P°$.
L is a line : ∂L = the two ending points of L.
                      $L° = L - \partial L$.
Po is a polygon : ∂Po = the intersection of the closure of Po and the closure of the exterior of Po.
                      $Po°$ = the union of all open sets in Po.

For each intersection, the value empty (ϕ) or non-empty ($\neg\phi$) is compute and store into a 9x9 matrix:

$$\begin{pmatrix} \partial A \cap \partial B & \partial A \cap B° & \partial A \cap B^- \\ A° \cap \partial B & A° \cap B° & A° \cap B^- \\ A^- \cap \partial B & A^- \cap B° & A^- \cap B^- \end{pmatrix}$$ ϕ if the intersection is empty
$\neg\phi$ if the intersection is non-empty

Fig.1. The 9-intersection Matrix

2.2 Group of relations

The 9-intersection model can be applied to all kinds of geometric objects. Considering points, lines and polygons, it leads to six groups of relations: point/point, point/line, point/polygon, line/line, line/polygon, polygon/polygon.

In [4], the authors gave the list of relations that can be realized in each group, if objects are embedded in 2-D (see Table 1).

The results given in table 1 consider two converse relations as only one since it is possible to change A in B and B in A. Converse relations can only happen

Group of relations	Number of relations
point/point	2
point/line	3
point/polygon	3
line/line	23
line/polygon	19
polygon/polygon	6

Table 1. Number of relations per group

between two objects of the same kind, namely in point/point, line/line and polygon/polygon groups.

2.3 Design of topological integrity constraints

Topological integrity constraints are defined using topological relations described by the 9-intersection model. The topological relation between two objects is the main part of the constraint. Considering the shape of objects, it is possible to compute all possible topological relations between two objects (according to the 9-intersection model). Considering the semantics of objects (their meaning), it is possible to define which topological relation is consistent and which one is inconsistent.

A topological constraint is defined as the association of two geographical objects, a topological relation between them and a specification (see Figure 2) which can be one of the following:

1. Forbidden
2. At least n times
3. At most n times
4. Exactly n times

CONSTRAINT = (Entity class1, Relation, Entity class2, Specification).

Fig.2. The definition of a topological constraint

The specification *forbidden* is the most interesting and usable one. Topological integrity constraints defined using this specification are a mean for end-users to describe topological situation they do not want to occur in their database.

The number of possible topological relations between two objects is an impediment to the design of topological integrity constraints. To provide an easy way to use interface, end-users must be given a set of topological relations described by names and not by mathematical definitions.

Unfortunately, it is almost impossible to give an expressive name to each relation. In addition, some topological relations are so closed that to avoid a situation to happen one will have to create several constraints (one for each topological relation matching some characteristics). For example, to forbid a river (a line) to cross a building (polygon) six constraints have to be defined.

To provide a more usable interface, topological relations sharing common attributes have been grouped in subsets. Such subsets have been built in each group of relation (points/points, points/lines, etc.). For example, in the line/polygon group, a subset called *cross* have been created. It contains all topological relations where the line's interior intersect the polygon's interior and exterior : $(L°∩P°=¬\phi) ∧ (L°∩P^-=¬\phi)$. For more details see [11].

In [7] Hadzilocas and Tryfona have proposed a model for expressing topological integrity constraints. The model presented here is less expressive than the model proposed in [7] but is more easy to manipulate. We want to point out that the topological integrity constraints designed with our model can be easily translated into the model given in [7]. Therefore, it is possible to combined both models to take advantage of the two approaches.

2.4 A visual interface to define topological constraints

In this part, we present a visual interface to define topological integrity constraints. Specifically, a dialogbox in which the user can choose a pair of entities, a topological relation or a set of topological relations, and a specification (see Figure 3).

Topological constraints are defined following the list of operations given here:

1. Choose a first class of entities.
2. Choose a second class of entities.
3. Choose a relation or a set of relations among the list proposed.
4. Define the specification.

In the case shown on Figure 3, the constraint defined is:

    (Road, Inside, Building, Forbidden)

The dialogbox shows a schema that illustrates the topological relation chosen in the constraint definition.

This interface has been designed using VisualC++.

Examples of topological constraints

C1(Road, Cross, Building, Forbidden)

C2(Sluice, Joint, Waterpipe, Exactly 2 times)

This visual interface allows one to define constraints based on topological relations, as a first step. The next step is to translate them into a language capable of checking them.

Such a language is out of the scope of this paper, but is not out of the scope of the whole study on topological consistency of spatial data.

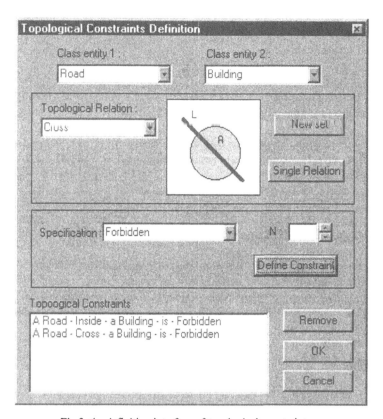

Fig.3. the definition interface of topological constraints

3 Correcting Scenarios Computation.

The goal is to define a model to compute corrections to topological integrity constraint violations (topological errors). **The model described in this part deals only with constraints defined using the *forbidden* specification**. This specification is the most useful, and the most common one. In addition, such constraints are very easy to describe. It is easier to define a case that should not happen than to define a case that must exists. A topological constraint is then defined as an inconsistent topological relation between two geographical objects.

Since an error is defined as a forbidden topological relation between two objects, the way to correct an error will be to change the topological relation between those objects. A set of correcting scenarios will be computed by applying several kinds of changes to both objects involved in the forbidden topological

relation (together or one after each other). The changes proposed are the following :
- Objects modification :
  - Moving the objects.
  - Reshaping the objects.
- Deleting one object.
- Object splitting (creating an new object).

To compute and to propose correcting scenarios have two main advantages. The first one is to facilitate and to accelerate the end-user work. The second one is to control the correcting process so that it can be ensured that the correction doesn't create a new error regarding to the constraints.

3.1 Objects modification

We present in this part two different kinds of modifications. Changes, as *moving an object*, ensure that the surface area of object remains unchanged. Other changes, as reshaping, have been designed to leave the topological relation between the reshaped object and the other objects of the databases unchanged (objects not involved in the forbidden relation).

Moving

One of the two objects involved in the forbidden relation will be moved according to *main directions*. We use the letter A for the moving object and the letter B for the other object.

The main directions are :
- X axis
- Y axis
- perpendicular to B (when possible)
- parallel to B (when possible)
- along A (when possible)

For the moves *perpendicular to*, *parallel to* and *along*, if the object is not a straight line, the direction has to be defined regarding to the boundary of the object :
- for a line : the two end points,
- for a region : the boundary segment that is the closest to the other object.

Such a correction does not change the length of a line, or the area of a region. The relative position of the two object is the only change. Consequently, determining all the possible moves of only one of the two objects leads to all the possible new scenes.

For each direction, A can be moved according two ways. An ending condition to the move is when the topological relation between the two objects changes. The new scene is stored into a list of correcting scenarios and the object A is moved again until the relation become `disjoint`. Each time a new topological relation is reached, the scene is added to the list.

As `disjoint` is as well a topological relation, the last scene is stored. Nevertheless, since a lot of positions are available for A (in this particular case), a minimum distance between the two objects is defined. This distance depends on the precision of point coordinates in the database. The stored scene based on the `disjoint` relation is the one in which the distance between the two objects is the minimal distance set for the database.

The following correction algorithm computes the correcting scenarios based on main directions :

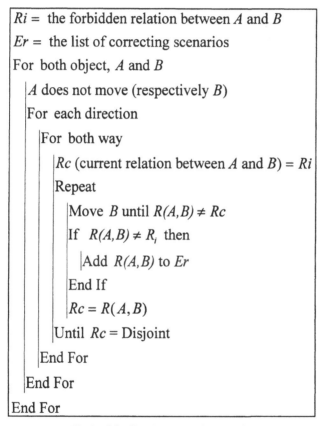

Fig.4. Main direction correcting algorithm

Reshaping

Reshaping means moving a part of an object, leaving the other part unchanged. The goal of such a correction is to change the topological relation between the two objects without changing the relations with the other objects of the database. This kind of correction will affect the length of a line or the area of a region. Consequently, it will have to be applied to both objects (one after each other, always keeping one object unchanged) in order to determine all the possible new scenes. Such corrections will be used to adjust the borders of two closed regions, a line and the border of a region, or two lines. The adjustment will be made by a force-fitting algorithm that will snap characteristic points of A onto characteristic points of B. A characteristic point is a point used in the shape definition of the object (for example to describe the boundary of a region or a line). Homologue points are defined as a pair of very closed characteristic points, one belonging to A and the second to B, that will be snapped onto each other by the force-fitting algorithm.

There are two steps in the reshaping process :
- homologue points finding
- force-fitting algorithm

Homologue points finding

The goal is to find which point of A and B will be matched. This process is defined by the two following steps :

1. Compute the distance between all points of A and all points of B. Store the results in a *distance matrix*.
2. The homologue points are each couple of points for which the distance is minimum and under a value d, set by the end-user.

Force-fitting algorithm

The goal of this algorithm is to defined how to adjust very closed objects. It works on homologue points that have been computed by the previous process. The algorithm is described on Figure 5.

Applying these algorithm to all the *homologue points* provides all the possible scenarios of correction according to the reshaping process.

For a line, only one of the ending points is allowed to move. Some other characteristic points can also be moved at the same time.

3.2 Deleting one object.

This correction is useful when an object have been digitized twice. Two objects very closed to each other can then be found. Two corrections are possible :
- keeping A and removing B
- keeping B and removing A.

This leads to two correcting scenarios.

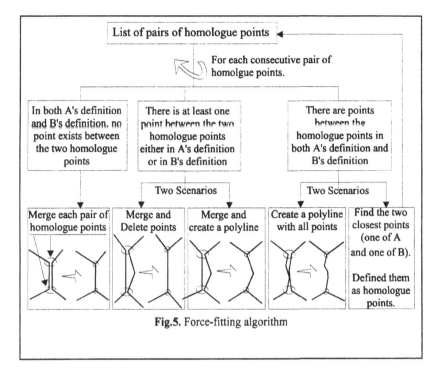

Fig.5. Force-fitting algorithm

3.3 Object splitting (Creating an object).

The last way to change the topological relation between two objects is to split one of them into two different sub-objects. The only condition to check is that the two new topological relations are different from the previous one. This kind of correction is useful to keep the planarity of a map. For example, when two lines crossing each other is a forbidden topological relation, a way to correct such an error is to create a point at the intersection of the lines and to split one of the line into two parts.

Such a correction can be proposed when the forbidden topological relation is such as one of the two objects shares a part of its interior with the interior or the boundary of the other object :

$$(O_1^\circ \cap \partial O_2 = \neg\phi) \vee (O_1^\circ \cap O_2^\circ = \neg\phi)$$

The corrections are :
- to split one of the two objects into several parts (2 or more)
- to create a new object based on the shared part and removing this part from each other object.

The tables 2 and 3 give the possible corrections for lines and regions.

The forbidden Topological relation	Correction schema	New relations
Line - Line $L_1^\circ \cap L_2^\circ = \neg\phi$	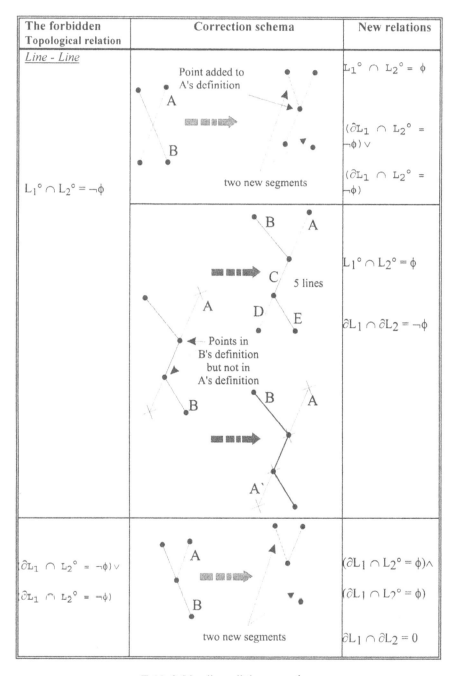	$L_1^\circ \cap L_2^\circ = \phi$ $(\partial L_1 \cap L_2^\circ = \neg\phi) \vee$ $(\partial L_1 \cap L_2^\circ = \neg\phi)$
		$L_1^\circ \cap L_2^\circ = \phi$ $\partial L_1 \cap \partial L_2 = \neg\phi$
$(\partial L_1 \cap L_2^\circ = \neg\phi) \vee$ $(\partial L_1 \cap L_2^\circ = \neg\psi)$		$(\partial L_1 \cap L_2^\circ = \phi) \wedge$ $(\partial L_1 \cap L_2^\circ = \phi)$ $\partial L_1 \cap \partial L_2 = 0$

Table 2. Line-line splitting corrections

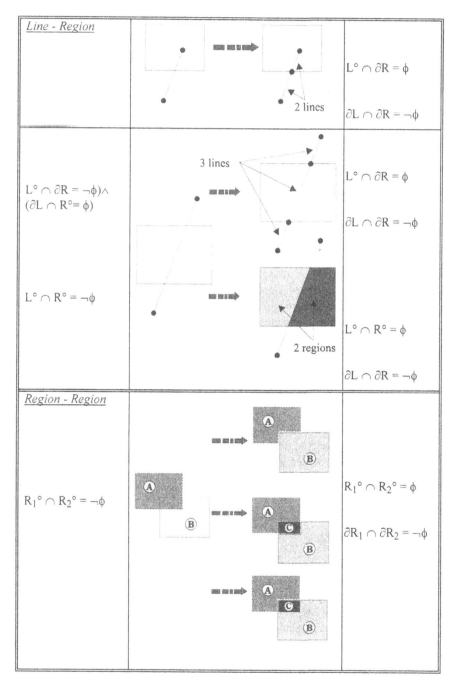

Line - Region		$L^{\circ} \cap \partial R = \phi$
		$\partial L \cap \partial R = \neg\phi$
$L^{\circ} \cap \partial R = \neg\phi) \wedge$ $(\partial L \cap R^{\circ} = \phi)$	3 lines	$L^{\circ} \cap \partial R = \phi$
		$\partial L \cap \partial R = \neg\phi$
$L^{\circ} \cap R^{\circ} = \neg\phi$	2 regions	$L^{\circ} \cap R^{\circ} = \phi$
		$\partial L \cap \partial R = \neg\phi$
Region - Region		
$R_1^{\circ} \cap R_2^{\circ} = \neg\phi$		$R_1^{\circ} \cap R_2^{\circ} = \phi$
		$\partial R_1 \cap \partial R_2 = \neg\phi$

Table 3. Line-polygon and polygon-polygon splitting corrections

4 Correction Scenarios Presentation and Application

For each topological error, a list of correcting scenarios will be computed. The last step of the correcting process is then to choose which one to apply. To help the end-user to select the appropriate correction, the list of correcting scenarios will be presented using filtering and sorting process.

Filtering process
1. All the correcting scenarios in which the topological relation is used in a topological constraint are removed from the list of corrections. This will be applied when there is more than one constraint defined for a given pair of geographical objects.
2. For correcting scenarios obtained by moving one object, a maximum range is defined. All corrections for which the moving distance is over the threshold are removed from the list of corrections.

Sorting process
The end-user can specify one parameter that will help him to find the appropriate correction. Correcting scenarios where this parameter is verified are proposed first. The possible values for this parameter are :
1. keeping the area of the object unchanged
2. minimum distance move
3. border adjustment (result of force-fitting first)
4. keeping two objects
5. specifying the new topological relation

Those two process will facilitate the choice of the correcting scenario to apply. If the end-used cannot find an appropriate correction among the list proposed, a set of tools will be provided in order to let the end-user modify the geographical object.

5 Conclusion

The constraint definition process and the correcting scenarios computation process are both part of a general framework designed to improve data quality in GIS. In this paper, data quality is defined as geometric and topological correctness. The goal is then to detect and correct geometric and topological errors and also to compute new topological relations to be added to the database.

Several other processes must be added to obtain the general framework. First of all a geometric correction process must ensure that each geometric object of the database (points, lines, polygons) is consistent [11]. Between the two processes presented in this paper, another one that searches for all constraint violations in the database has to be applied. Such a process is currently studied. It is a topological constraint language that can express all the constraints defined by the end-user.

Some situation cannot be corrected by the algorithms proposed in this paper. Figure 6 presents one of those situation. In this case, Andorra can disappeared or be placed partially in France and partially in Spain.

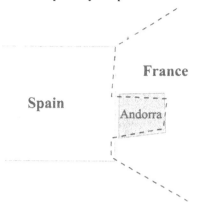

Fig.6. The case of Andorra.

To allow the end-user to correct such a case, a manual correcting tool has to provided.

One of the main impediment to consistency checking and error correcting in GIS is the vast amount of data stored in such databases. Spatial indexing methods must be added to the processes presented in this paper in order to accelerate data access.

6 Bibliography

1. Cui Z., Cohn A.G., et Randell D.A., 1993 Qualitative and Topological Relationships in Spatial Databases, *in Proceedings of the Third Symposium on Large Spatial Databases*, Singapore, June 23-25, 1993 (SSD'93). Lecture Notes in Computer Science n°692. pages 296-315.
2. Egenhofer M. J. & Franzosa R. D., 1991 Point-set topological spatial relations, *in International Journal of Geographical information Systems*, volume 5, number 2, April-June 1991. pages 161-174.
3. Egenhofer M. J. et Herring J. R., 1990 A Mathematical Framework for the Definition of Topological Relationships, *in Proceedings of the 4th International Symposium on Spatial Data Handling*, Zurich, 1990 (SDH 90). pages 803-813.
4. Egenhofer M. J. et Herring J. R., 1991, Categorizing Binary Topological Relations Between Regions, Lines, and Points in Geographic Databases.

Technical Report, Department of Surveying Engineering, University of Maine, Orono.

5. Egenhofer M. J., 1991 Reasoning about Binary Topological Relations, *in Proceedings of the Second Symposium on Large Spatial Databases*, Zurich, 1991 (SSD'91). Lecture Notes in Computer Science n°525. pages 143-159.

6. Frank, A. U, 1983. Data Structures for Land Information Systems: Semantic, Topological and Spatial Relations in Data of Geo-Sciences - phD thesis, in German, Swiss Federal Institute of Technology.

7. Hadzilacos T. and Tryfona N., 1992, A Model for Expressing Topological Integrity Constraints in Georaphic Databases, *in Proceedings of Theories and Methods of Spatio-Temporal Reasoning*, Pisa, september 21-23, 1992 (GIS 92), pages 253-368.

8. Laurini R. et Milleret-Raffort F., 1994, Topological Reorganization of Inconsistent Geographic Databases : a Step Toward their certification, *in Computer and Graphics*, volume 18, number 6, nov-dec 94, p. 803-813.

9. Plümer L., 1996, Achieving Integrity and Topology in Geographical Information Systems, In: *First International Conference on Geographic Information Systems in Urban Regional and Environmental Planing*, Samos, Greece - April 1996, pages 45-60.

10. Ubeda T. & Servigne S., 1996a, Capturing Spatial Object Characteristics for Correcting and Reasoning, In: *Proceeding of Joint European Conference and Exhibition on Geographical Information*, Barcelona, Spain, March 1996 (JEC-GI 96), pages 24-33.

11. Ubeda T. & Servigne S., 1996b, Geometric and Topological Consistency of Spatial Data, in the Proceedings of the First International Conference on Geocomputation, Leeds, UK, 17-19 September 1996, pages 830-842.

Vague Regions

Martin Erwig & Markus Schneider

FernUniversität Hagen
Praktische Informatik IV
D-58084 Hagen, Germany
[martin.erwig | markus schneider]@fernuni-hagen.de

Abstract: In many geographical applications there is a need to model spatial phenomena not simply by sharp objects but rather through indeterminate or vague concepts. To support such applications we present a model of vague regions which covers and extends previous approaches. The formal framework is based on a general exact model of spatial data types. On the one hand, this simplifies the definition of the vague model since we can build upon already existing theory of spatial data types. On the other hand, this approach facilitates the migration from exact to vague models. Moreover, exact spatial data types are subsumed as a special case of the presented vague concepts. We present examples and show how they are represented within our framework. We give a formal definition of basic operations and predicates which particularly allow a more fine-grained investigation of spatial situations than in the pure exact case. We also demonstrate the integration of the presented concepts into an SQL-like query language.

1 Introduction

In the literature about spatial database systems and geographical information systems (GIS) that advocates an entity-oriented view of spatial phenomena, the general opinion prevails that special data types are necessary to model geometry and to efficiently represent geometric data in database systems, for example [Eg89, GNT91, GS93, GS95, Gü88, LN87, OM88, Sc95, SV89]. These data types are commonly denoted as *spatial data types* such as *point*, *line*, or *region*. We speak of *spatial objects* as occurrences of spatial data types.

So far, spatial data modeling implicitly assumes that the extent and hence the boundary of spatial objects is precisely determined and universally recognized. This leads exclusively to *exact object models*. Spatial objects are represented by sharply described points, lines, and regions in a defined reference frame. Lines link a series of exactly known coordinates (points), and regions are bounded by exactly defined lines which are called *boundaries*. The properties of the space at the points, along the lines, or within the regions are given by attributes whose values are assumed to be constant over the total extent of the objects. Examples are especially man-made spatial objects representing engineered artifacts (like highways, roads, houses, and bridges) and some predominantly immaterial spatial objects exerting social control (like countries and districts with their political and administrative boundaries or land parcels with their cadastral boundaries). We will denote this kind of entities as *determinate spatial objects*.

Increasingly, researchers are beginning to realize that there are many spatial objects in reality which do not have sharp boundaries or whose boundaries cannot be precisely determined. Examples are natural, social, or cultural phenomena like land features with continuously changing properties (such as population density, soil quality, vegetation), oceans, biotopes, deserts, an English speaking area, or mountains and valleys. The transition between a valley and a mountain usually cannot be exactly determined so that the two spatial objects "valley" and "mountain" cannot be precisely separated and defined. Frequently, the indeterminacy of spatial objects is associated with temporal changes; for example, clouds and sandbanks dynamically change their shapes in the course of time. We will denote this kind of entities as *vague* or *indeterminate spatial objects*.

This paper presents an object model for defining vague regions[1] which rests on "traditional" (that is, exact) modeling techniques. This modeling strategy simultaneously expresses the authors' opinion that it is unnecessary to begin from scratch when modeling vague spatial objects. On the contrary, it is possible to extend, rather than to replace, the current theory of spatial database systems and GIS. Furthermore, moving from an exact to a vague domain does not necessarily invalidate conventional geometry; it is merely an extension. Consequently, the current exact object models that are restricted to determinate spatial objects can be considered as simplified special cases of a richer class of models for general spatial objects. It turns out that this is exactly the case for the model to be presented.

Section 2 gives a characterization of the various meanings of indeterminacy, discusses the notion of "boundary", and presents a classification of the approaches proposed so far. Section 3 informally introduces the concept of vague regions and motivates the necessity of vague topological predicates and vague spatial operations. Section 4 formalizes these concepts and discusses the problem of adequately defining numerical operations on vague regions. Section 5 demonstrates an embedding into an SQL-like query language, and Section 6 draws some conclusions and gives a prospect of future research activities.

2 Classifying Models for Vague Spatial Objects

A first attempt of a taxonomy of vague spatial objects has been given by Couclelis [Co96]. She proposes to examine the essence of vague spatial objects from three different perspectives: the empirical nature of the object, the mode of observation, and the user's purpose. All three perspectives are based on the intuitive meaning of the notion "boundary". The nature of the object (for example, whether it is homogeneous or heterogeneous, continuous or discontinuous, solid or fluid, fixed or moving) influences how we become aware of the boundaries and their degree of sharpness. The mode of observation (given, for example, by scale, resolution, time, error) affects the knowledge about the position of the boundary. The user's purpose for which a model is designed leads to a preference for one model over the other, since different user categories have

---

1. Concepts for vague points and vague lines are currently not taken into account.

different requirements and conceptual views. Administrators, for instance, demand precisely defined objects; scientists, however, strive to integrate the vagueness of boundaries into their models.

The entity-oriented view of spatial phenomena, which we will take in this paper, considers spatial objects as conceptual and mathematical abstractions of real-world entities which can be identified and distinguished from the rest of space. For example, a region divides space into three parts: one part inside the object, another part on the border of the object, and the remaining part outside the object. The three parts form a partition of space, that is, they are mutually exclusive and covering the whole space. Hence, the notion of a region is intrinsically related to the notion of a boundary, be it sharp or indeterminate.

So far, in spatial data modeling boundaries are considered as sharp lines that represent abrupt changes of spatial phenomena and that describe and thereby distinguish regions with different characteristic features. The assumption of crisp boundaries harmonizes very well with the internal representation and processing of spatial objects in a computer which requires precise and unique internal structures. Hence, in the past, there has been a tendency to force reality into determinate objects. In practice, however, there is no apparent reason for the whole boundary of a region to be sharp or to have a constant degree of vagueness. There are a lot of geographical application examples illustrating that the boundaries of spatial objects can be indeterminate. For instance, boundaries of geological, soil, and vegetation units (see for example [Al94, Bu96, KV91, LAB96, WH96]) are often sharp in some places and vague in others; many human concepts like "the Indian Ocean" are implicitly vague.

The treatment of spatial objects with indeterminate boundaries is especially problematic for the computer scientist who is confronted with the difficulties how to model such objects in his database system so that they correspond to the user's intuition, how to finitely represent them in a computer format, how to develop spatial index structures for them, and how to draw them. He is accustomed to the abstraction process of simplifying spatial phenomena of the real world through the concepts of conventional binary logic, reduction of dimension, and cartographic generalization to precisely defined, simply structured, and sharply bounded objects of Euclidean geometry like points, lines, and regions.[2]

In reality, there are essentially two categories of indeterminate boundaries: sharp boundaries whose position and shape are unknown or cannot be measured precisely, and boundaries which are not well-defined or which are useless (for example, between a mountain and a valley) and where essentially the topological relationship between spatial objects is of interest.

---

2. Ironically, this abstraction process itself mapping reality onto a mathematical model implicitly introduces a certain kind of vagueness and imprecision.

Spatial objects with indeterminate boundaries are difficult to model and are so far not supported in spatial database systems. According to the two categories of boundaries, two kinds of vagueness or indeterminacy concerning spatial objects have to be distinguished: *Uncertainty* relates either to a lack of knowledge about the position and shape of an object with an existing, real border (*positional* uncertainty) or to the inability of measuring such an object precisely (*measurement* uncertainty). *Fuzziness* is an intrinsic feature of an object itself and describes the vagueness of an object which certainly has an extent but which inherently cannot or does not have a precisely definable border.

The subject of modeling spatial vagueness has so far been exclusively treated by geographers but rather neglected by computer scientists. At least three alternatives are proposed as general design methods:

- *fuzzy models* [Al94, Ba93, Bu96, Ed94, KV91, LAB96, Us96, Wa94, WHS90] which are all based on fuzzy set theory and predominantly model fuzziness,
- *probabilistic models* [Bl84, Bu96, Fi93, Sh93] which are based on probability theory and predominantly model positional and measurement uncertainty, and
- *exact models* [CF96, CG96, Sc96] which transfer data models, type systems, and concepts for spatial objects with sharp boundaries to spatial objects without clear boundaries and which predominantly model uncertainty but also aspects of fuzziness.

Fuzzy sets were first introduced by Zadeh [Za65] to treat imprecise concepts in a definable way. Fuzzy set theory is an extension or generalization (and not a replacement) of classical boolean set theory and deals only with fuzziness, not with uncertainty. Fuzziness is not a probabilistic attribute, in which the grade of membership of an individual in a set is connected to a given statistically defined probability function. Rather, it is an admission of the possibility that an individual is a member of a set or that a given statement is true. Examples of fuzzy spatial objects include mountains, valleys, biotopes, oceans, and many other geographic features which cannot be rigorously bounded by a sharp line.

Probability theory can be used to represent uncertainty. It defines the grade of membership of an entity in a set by a statistically defined probability function. Examples are the uncertainty about the spatial extent of particular entities like regions defined by some property such as temperature, or the water level of a lake.

The main difficulty of fuzzy and probabilistic models is that their use with spatial data is still a non-trivial application. On the one hand, our current computational technology does not allow efficient processing of uncertain and fuzzy spatial data. On the other hand, it is an open problem how to integrate and transform these models into the concept of spatial data types.

A benefit of the exact object model approach is that existing definitions, techniques, data structures, algorithms, etc., need not be redeveloped but only modified and extended, or simply used. The currently proposed exact methods model vague regions by using some kind of *zone* concept, either without holes [CF96, CG96] or with holes [Sc96]. The central idea is to consider determined zones surrounding the indeterminate boundaries of a region and expressing its minimal and maximal extension. The zones serve as a description and separation of the space that certainly belongs to the region and the space that is certainly outside.

While [CF96] and [CG96] are mainly interested in classifications of topological relationships between vague regions for which a simple model is assumed, [Sc96] proposes a model of complex vague regions with vague holes and focusses on their formal definition. Unfortunately, the three approaches are limited to "concentric" object models and have problems with geometric closure properties. The model described in this paper also pursues the exact model approach but is much more general and much simpler than the approaches suggested so far.

3 What are Vague Regions?

Our goal to base a concept of vague regions on traditional modeling techniques first necessitates a general exact object model for determinate regions. We will introduce this model only informally here. A formal definition of this model based on the point set paradigm and on point set topology is given in the Appendix. Each alternative model should fulfill the properties described there. Possible candidates are the models described in [ECF94, WB93], and the discrete model of the ROSE algebra [GS93, GS95, Sc95].

A (*determinate*) *region* is a set of disjoint, connected areal components possibly with disjoint holes (see the picture below). This model is very general and closed under (appropriately defined) geometric union, intersection, difference, and complement operations. It allows regions to contain holes and islands within holes to any (finite) level. The requirement of disjointedness is not meant in a strict sense; components of regions as well as holes of a component may be neighbored in a common boundary line or in common single boundary points.[3] We only require that the employed model satisfies the requirements defined in the Appendix.

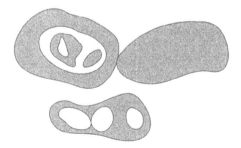

---

3. Usually, common boundary lines make no sense, since then adjacent components and adjacent holes, respectively, could be merged together by eliminating the common boundary parts. For our purposes, this aspect is not relevant.

Our concept of vague regions mainly deals with the aspect of uncertainty but also includes some aspects of fuzziness. Frequently, there is uncertainty about the spatial extent of phenomena in space, that is, objects can shrink and extend. An example is a lake whose water level depends on the amount of precipitation or on the degree of evaporation and which has thus a minimal and maximal extent. Another example is a map of natural resources like iron ore. For some areas experts definitely know the existence of iron ore because of soil samples. For other areas experts are not sure and only assume the incidence of this mineral. These are the kinds of vague regions we are especially interested in. On the other hand, our concept is also able to model the aspect of fuzziness that areal objects have an extent but cannot be bounded by a precise border, for example, the transition between a mountain and a valley. Continuous changes of features (like air pollution continuously decreasing from city centers to rural areas) cannot currently be modeled by this concept (but see Section 5).

A *vague region* is a pair of disjoint regions. The first region, called the *kernel,* describes the determinate part of the vague region, that is, the area which definitely and always belongs to the vague region. The second region, called the *boundary,* describes the vague part of the vague region, that is, the area for which we cannot say with any certainty whether it or parts of it belong to the vague region or not. *Maybe* the boundary or parts of it belong to the vague region, *maybe* this is not the case. Or we could say that this is *unknown.* It is important to notice that boundaries need not necessarily be one-dimensional structures but can be regions, and that the semantics of the boundary of a vague region is not fixed by our model but depends on the meaning the application associates with it.

The figure below gives an abstract example of a vague region *v.* The blank areas annotated with *v* depict kernels, the shaded areas annotated with *v* denote the boundaries of the vague region *v,* and the blank areas that are not annotated describe holes. The example demonstrates the complexity of the model. Kernels and boundaries may be adjacent; they may have holes which themselves can contain a hierarchy of kernels and boundaries with holes.

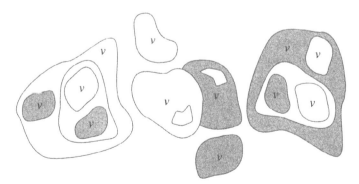

We now briefly present two real life applications and motivate the use of vague regions, vague topological predicates, and vague spatial operations. Vague concepts offer a greater flexibility for modeling properties of spatial phenomena in the real world than determinate concepts do. Still, vague concepts comprise the modeling power of determinate concepts as a special case.

The first example is taken from the animal kingdom and demonstrates the need of different *vague intersects* predicates and the use of a *vague intersection* operation. We view the living spaces of different animal species and distinguish kernel areas where they mainly live and boundary areas like peripheral areas or corridors where they in particular hunt for food or which they cross in order to migrate from one kernel area to another one. We now consider some relationships of their living spaces and ask:

- Which animals (partially) share their living spaces?
- Which hunters penetrate into the living space of other animals?
- What are the areas where two species can only meet by accident?

For two animal species u and v, the interesting situations for the queries are shown below. They all relate to different kinds of intersection which amount to three different kinds of topological predicates (introduced in the next section). The first query asks for kernel/kernel intersections, the second query for kernel/boundary intersections but not kernel/kernel intersections, and the third query exclusively for boundary/boundary intersections. The situation on the left is definitely an intersection. In contrast, the situation in the middle is a vague intersection which, however, is a stronger case than the situation on the right. Other examples of topological relationships and their use will be presented in the next section.

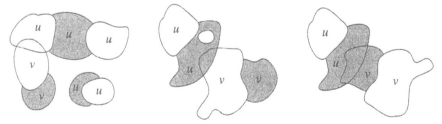

The task to compute the common living spaces of two animal species asks for the intersection of two vague regions. The intersection of two kernels is certainly a kernel, and the intersection of an exterior part with anything else is an exterior part. The open question is now the intersection of a kernel with a boundary and the intersection of two boundaries. Since boundaries are vague, we cannot make a unique statement whether these intersections belong to the kernel parts or to the boundary parts. It only remains to regard these intersections as boundary parts.

The second example demonstrates that concentric models like those presented in [CF96, CG96, Sc96] are captured by our concept. Consider a lake l which has a minimal water level in dry periods (kernels) and a maximal water level in rainy periods. Dry periods can entail puddles. Small islands in the lake which are less flooded by water in dry and more (but never completely) flooded in rainy periods can be modeled through holes surrounded by a boundary. If an island like a sandbank can be flooded completely, it belongs to the boundary part.

4   An Exact Model of Vague Regions

In this section, we give a formal account of vague regions. We first define vague spatial operations in Section 4.1. After that we define predicates in Section 4.2. There we will see that a concept, such as *inside*, is not anymore simply a question of *true* and *false*, but rather needs a vague kind of booleans containing a value like *maybe*. That is, we actually employ a three-valued logic as the range of (standard) predicates. Similarly, numeric operations given in Section 4.3 seem to require a concept of vague numbers (given, for example, by intervals). Since this entails rather extensive changes to the type of real numbers and on its operations, we instead define different exact versions of numeric operations capturing various aspects of vagueness. In general, the problem is how to integrate vague regions with other types and operations of a data model. We will pick up this issue again in Section 6.

For the definition of vague regions we make use of a suitable model for determinate regions as sketched in the previous section. One possible candidate is the point set model the relevant parts of which are given in the Appendix. We can choose any other model as long as it offers the following operations (let R denote the type of regions and $\mathbb{R}$ the set of real numbers):

$$\oplus : R \times R \to R \qquad \qquad \textit{(union)}$$
$$\otimes : R \times R \to R \qquad \qquad \textit{(intersection)}$$
$$\ominus : R \times R \to R \qquad \qquad \textit{(difference)}$$
$$\ominus : R \to R \qquad \qquad \textit{(complement)}$$
$$dist : R \times R \to \mathbb{R} \qquad \qquad \textit{(minimum distance)}$$
$$area : R \to \mathbb{R} \qquad \qquad \textit{(area)}$$

Moreover, R together with the operations $\oplus$ and $\otimes$ must form a boolean algebra. The order predicate of the corresponding boolean lattice is then given by $r \subseteq s \Leftrightarrow r \cup s = s$ ($\Leftrightarrow r \cap s = r$).

We define a *vague region* v as a pair of disjoint regions (k, b) where k gives the *kernel* of v and b denotes the *boundary* of v. We employ the following notation: $v^\kappa = k$ and $v^\beta = b$. Finally, the *exterior*, or *outside*, of v is defined as $v^\varepsilon = \ominus(k \oplus b)$.

4.1 Vague Spatial Operations

In order to define operations, such as **union**, **intersection**, and **difference** of two vague regions u and v, it is helpful to consider the possible relationships between the kernel, boundary, and outside parts of u and v. We do this by giving a table for each operation where a column/row labeled by ●, ◉, or ○ denotes the kernel, boundary, or outside part of u/v. Each field of the table denotes a possible combination (that is, intersection) of kernel, boundary, and outside parts of both objects, and the label in each field specifies whether the corresponding intersection belongs to the kernel, boundary, or outside part of the operation's result.

For example, the *union* of a kernel part with any other part is a kernel part since the union of two regions asks for membership in either region and since membership is certain for each kernel part. Likewise, the union of two boundaries or the union of a boundary with the outside should be a boundary, and only the parts of the space which belong to the outside of both regions contribute to the outside of the union.

On the other hand, the outside of the *intersection* is given by either region's outside because intersection requires membership in both regions. The kernel of the intersection only contains parts which definitely belong to the kernel of both arguments, and intersections of boundary parts with each other or with kernel parts make up the boundary of the intersection.

The definition of difference is motivated by the definition of complement. Clearly, the complement of the kernel should be the outside, and the complement of the outside should be the kernel, but what about the boundary part? Anything inside the vague part of an object might or might not belong to the object, so we cannot definitely say that the complement of the vague part is the outside. Neither can we say that the complement belongs to the kernel. So the only reasonable definition is to define the complement of the boundary to be the boundary itself:

Now the result of removing a vague region v from another vague region u can be defined as the intersection of u with the complement of v. That is, removing a kernel part means intersection with the outside which always yields outside, and removing anything from the outside leaves the outside part unaffected. Similarly, removing a boundary means intersection with the boundary and thus results in a boundary for kernel and boundary parts, and removing the outside of v (that is, nothing) does not affect any part of u.

difference	●	◉	○
●	○	◉	●
◉	○	◉	◉
○	○	○	○

Next we formally define these operations simply by using regions operations, that is, we express the notion of vague regions using well-understood exact regions. Let u and v be two vague regions. Then we define:

u **union** v $\quad := (u^{\kappa} \oplus v^{\kappa}, (u^{\beta} \oplus v^{\beta}) \ominus (u^{\kappa} \oplus v^{\kappa}))$
u **intersection** $v := (u^{\kappa} \otimes v^{\kappa}, (u^{\beta} \otimes v^{\beta}) \oplus (u^{\kappa} \otimes v^{\beta}) \oplus (u^{\beta} \otimes v^{\kappa}))$
u **difference** $v \quad := (u^{\kappa} \otimes (\ominus v^{\kappa}), (u^{\beta} \otimes v^{\beta}) \oplus (u^{\kappa} \otimes v^{\beta}) \oplus (u^{\beta} \otimes (\ominus v^{\kappa})))$
complement $v \quad := (\ominus v^{\kappa}, v^{\beta})$

In the following we use as an abbreviating notation for the intersection of two (determinate) regions simple juxtaposition, and we assign intersection higher associativity than union and difference. That is, the above definition for u **difference** v could also be written more concisely as $(u^{\kappa}(\ominus v^{\kappa}), u^{\beta}v^{\beta} \oplus u^{\kappa}v^{\beta} \oplus u^{\beta}(\ominus v^{\kappa}))$.

It is not difficult to check that the definitions realize the behavior specified by the tables given above. Consider, for example, the **union**-operation. For $w = u$ **union** v we have to show the following three identities:

(1) $w^{\kappa} = u^{\kappa}v^{\kappa} \oplus u^{\kappa}v^{\beta} \oplus u^{\kappa}v^{\varepsilon} \oplus u^{\beta}v^{\kappa} \oplus u^{\varepsilon}v^{\kappa}$
(2) $w^{\beta} = u^{\beta}v^{\beta} \oplus u^{\beta}v^{\varepsilon} \oplus u^{\varepsilon}v^{\beta}$
(3) $w^{\varepsilon} = u^{\varepsilon}v^{\varepsilon}$

For proving (1) we first observe that $\oplus$ is idempotent. We can therefore duplicate the first term $u^{\kappa}v^{\kappa}$. Then using the fact that $\otimes$ distributes over $\oplus$ we can factorize both u^{κ} and v^{κ} and obtain:

$$w^{\kappa} = (u^{\kappa}(v^{\kappa} \oplus v^{\beta} \oplus v^{\varepsilon})) \oplus (v^{\kappa}(u^{\kappa} \oplus u^{\beta} \oplus u^{\varepsilon}))$$

Since $v^{\kappa} \oplus v^{\beta} \oplus v^{\varepsilon} = 1_R$ and $u^{\kappa} \oplus u^{\beta} \oplus u^{\varepsilon} = 1_R$ and since 1_R is the identity of $\otimes$ we get

$$w^{\kappa} = (u^{\kappa} \otimes 1_R) \oplus (v^{\kappa} \otimes 1_R) = u^{\kappa} \oplus v^{\kappa},$$

which is the definition of the kernel part of **union**. Equation (2) can be shown as follows. For arbitrary regions r and s we know:

$$r \oplus s = rs \oplus r(\ominus s) \oplus (\ominus r)s$$

We can use this identity to rewrite the boundary definition as:

$$u^{\beta}v^{\beta} \oplus u^{\beta}(\ominus v^{\beta}) \oplus (\ominus u^{\beta})v^{\beta} \ominus (u^{\kappa}v^{\kappa} \oplus u^{\kappa}(\ominus v^{\kappa}) \oplus (\ominus u^{\kappa})v^{\kappa})$$

Next we evaluate all complements (note that $\ominus v^\beta = v^\kappa \oplus v^\varepsilon$ or $\ominus v^\kappa = v^\beta \oplus v^\varepsilon$):

$$u^\beta v^\beta \oplus u^\beta(v^\kappa \oplus v^\varepsilon) \oplus (u^\kappa \oplus u^\varepsilon)v^\beta \ominus (u^\kappa v^\kappa \oplus u^\kappa(v^\beta \oplus v^\varepsilon) \oplus (u^\beta \oplus u^\varepsilon)v^\kappa)),$$

and apply distributivity of $\otimes$:

$$u^\beta v^\beta \oplus u^\beta v^\kappa \oplus u^\beta v^\varepsilon \oplus u^\kappa v^\beta \oplus u^\varepsilon v^\beta \ominus (u^\kappa v^\kappa \oplus u^\kappa v^\beta \oplus u^\kappa v^\varepsilon \oplus u^\beta v^\kappa \oplus u^\varepsilon v^\kappa)$$

In the resulting term, only $u^\beta v^\kappa$ and $u^\kappa v^\beta$ appear in both parts of the difference; all other intersections to be subtracted have no effect at all since all intersections are pairwise disjoint. Therefore the result is:

$$u^\beta v^\beta \oplus u^\beta v^\varepsilon \oplus u^\varepsilon v^\beta$$

which is exactly the condition required for w^β. For the proof of relationship (3), first note that in a boolean lattice we have for any two regions r and s: $1_R \otimes s = s$, $1_R \oplus s = 1_R$, and $1_R = r \oplus (\ominus r)$. Therefore, we know that $s = (r \oplus (\ominus r))s = rs \oplus (\ominus r)s$, and it follows that $r \oplus s = r \oplus rs \oplus (\ominus r)s$. We also know that $r \oplus rs = r(1_R \oplus s) = r$, so that $r \oplus s = r \oplus (\ominus r)s$. Since $(\ominus r)s$ is another way of denoting the difference $s \ominus r$, we get: $r \oplus (s \ominus r) = r \oplus s$. Now we have by definition that

$$w^\varepsilon = \ominus(w^\kappa \oplus w^\beta) = \ominus(u^\kappa \oplus v^\kappa \oplus ((u^\beta \oplus v^\beta) \ominus (u^\kappa \oplus v^\kappa))) = \ominus(u^\kappa \oplus v^\kappa \oplus u^\beta \oplus v^\beta)$$

By commutativity and de Morgan's law this reduces to:

$$\ominus(u^\kappa \oplus u^\beta) \otimes (\ominus(v^\kappa \oplus v^\beta))$$

which is by the definition of complement equal to $u^\varepsilon \otimes v^\varepsilon$, the condition required for w^ε. The correctness of the other operations is shown in a similar way.

In addition to having the four basic spatial operations on vague regions, it is also sometimes helpful to be able to explicitly deal with their boundary and kernel parts. Thus, we define the following operations:

boundary$(v) := (\emptyset, v^\beta)$
kernel$(v) \quad := (v^\kappa, \emptyset)$
invert$(v) \quad := (v^\beta, v^\kappa)$

In particular, these operations facilitate the computation with parts of vague regions in a purely exact way since the vague spatial operations, applied to vague regions with an empty boundary, behave exactly like the corresponding exact spatial operations. (This can be easily seen from the definitions.)

4.2 Vague Predicates

One of the most basic relationships that can be observed for two regions is whether they intersect or not. Many different cases of intersection can be identified leading to specialized predicates, like *covers* or *meets*, that describe more specific relationships. To define an intersection predicate for two vague regions u and v it is instructive to look at the possible results for the kernel and boundary of $w = u$ **intersection** v. Surely, we want to say that u and v intersect if $w^\kappa = u^\kappa v^\kappa$ is not empty, that is, if the kernel regions of u and v overlap. This is true independent from the value of w^β. Likewise, if the regions of $u^\kappa \oplus u^\beta$ and $v^\kappa \oplus v^\beta$ are disjoint, we can safely say that u and v do not intersect at all. However, if $w^\kappa = 0_R$ and $w^\beta \ne 0_R$, we cannot be sure about the intersection of u and v. This means, we can neither return *true* nor *false*, but we rather have to define the predicate to yield something like *maybe* or *unknown* (comparable to NULL-values known from relational databases).

Therefore, we use a three-valued logic as the range of boolean predicates. The definition of the logical operators parallels the definition of the operations for vague regions (1, 0, and ? are used as abbreviations for *true*, *false*, and *maybe*):

and	1	?	0
1	1	?	0
?	?	?	0
0	0	0	0

or	1	?	0
1	1	1	1
?	1	?	?
0	1	?	0

not	1	?	0
	0	?	1

Now we return to the definition of vague predicates. For example, the definition of intersection is:

$$u \text{ \textbf{intersects} } v = \begin{cases} true & \text{if } u^\kappa v^\kappa \ne 0_R \\ false & \text{if } u^\kappa v^\kappa \oplus u^\beta v^\beta \oplus u^\kappa v^\beta \oplus u^\beta v^\kappa = 0_R \\ maybe & \text{otherwise} \end{cases}$$

The *maybe*-case of intersection can be distinguished further according to whether a kernel/boundary or only a boundary/boundary intersection exists. An example for both situations is shown below:

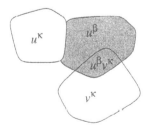

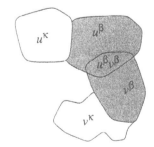

vague intersection

weak vague intersection

We consider the situation depicted on the left to be a stronger indication of intersection than the situation on the right. Accordingly, we define two predicates, **v-intersects** (*vague intersection*) and **w-intersects** (*weak vague intersection*) as follows:

$$u \text{ v-intersects } v = \begin{cases} true & \text{if } u^\beta v^\kappa \oplus u^\kappa v^\beta \neq \mathbf{0}_R \text{ and } u^\kappa v^\kappa = \mathbf{0}_R \\ false & \text{otherwise} \end{cases}$$

$$u \text{ w-intersects } v = \begin{cases} true & \text{if } u^\beta v^\beta \neq \mathbf{0}_R \text{ and } u^\beta v^\kappa \oplus u^\kappa v^\beta \oplus u^\kappa v^\kappa = \mathbf{0}_R \\ false & \text{otherwise} \end{cases}$$

A special case of intersection is also given when u lies inside v. We can safely say that u **inside** v holds if everything of u (that is, kernel and boundary) is inside the kernel of v. If this is not the case, we cannot simply conclude that u **inside** v is *false* since this requires definite knowledge about a part of u being outside any part of v. In other words, whenever $u^\kappa \subseteq v^\kappa \oplus v^\beta$ we are not sure about insideness, and we should define u **inside** v as *maybe*:

$$u \text{ inside } v = \begin{cases} true & \text{if } u^\kappa \oplus u^\beta \subseteq v^\kappa \\ false & \text{if } u^\kappa \not\subseteq v^\kappa \oplus v^\beta \\ maybe & \text{otherwise} \end{cases}$$

As we have done for intersection we can discriminate the *maybe*-case further. If the kernel part of u is completely inside the kernel part of v, then only the boundary of u makes the decision of insideness vague. This is a stronger indication for the inside relationship than in the case that also a part of u's kernel lies in the boundary of v. The possible relationships are shown in the following picture:

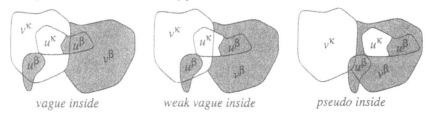

vague inside   *weak vague inside*   *pseudo inside*

We do not consider u to be inside v in the situation on the right because inside implies intersects; in other words, insideness can only hold for intersecting regions. Since the right side can also hold for disjoint regions, speaking of insideness seems to be too optimistic an interpretation. Thus we can define the predicates as follows:

$$u \text{ v-inside } v = \begin{cases} true & \text{if } u^\kappa \subseteq v^\kappa \text{ and } u^\beta \subseteq v^\kappa \oplus v^\beta \text{ and } u^\beta \not\subseteq v^\kappa \\ false & \text{otherwise} \end{cases}$$

$$u \text{ w-inside } v = \begin{cases} true & \text{if } u^\kappa \oplus u^\beta \subseteq v^\kappa \oplus v^\beta \text{ and } u^\kappa \not\subseteq v^\kappa \\ false & \text{otherwise} \end{cases}$$

The complementary predicate for **intersects** is **disjoint**, and its definition is obtained by simply exchanging *true* and *false* in the definition of **intersects**.

Note that we cannot directly express relationships, such as *meets* or *adjacent*, since we currently have no concept of lines and points in our model. However, we can regard weak vague intersections as a kind of adjacency as done by [CF96, CG96].

4.3 Vague Numeric Operations

The definition of numeric operations on vague regions must be based on the corresponding functions for regions. Let us consider, for example, the *area* of a vague region. The area is at least the area of the kernel part and at most the area of kernel plus boundary. So the result of the vague area operation could be an interval given by the minimum and maximum area values. We then, however, have to work with intervals in any further calculations using such an area value. This requires a whole new set of vague arithmetic operations working with intervals. (The situation is similar to the extension to three-valued logic used for predicates.) So in order to keep things simple we instead define *two* operations, **min-area** and **max-area**, and can thus keep ordinary numeric operations.

$$\textbf{min-area}(v) := area\ (v^\kappa)$$
$$\textbf{max-area}(v) := area\ (v^\kappa \oplus v^\beta)$$

The definition of the *distance* between two vague regions u and v is very similar. Again the distance is a vague value: An upper bound is obtained by the distance between the kernel parts of u and v, that is, we are sure that the distance is at most the distance between the kernels. The distance might be smaller, but it is at least as large as the distance between the maximal extensions of u and v, in other words, the minimum distance is given by the distance taking kernel and boundary into account.

$$\textbf{min-dist}\ (u, v) := dist\ (u^\kappa \oplus u^\beta, v^\kappa \oplus v^\beta)$$
$$\textbf{max-dist}\ (u, v) := dist\ (u^\kappa, v^\kappa)$$

The generalization of *area* and *dist* to vague regions is rather straightforward. There are other useful operations on regions, however, for which a generalization to the vague case is not quite so simple or even impossible. Consider, for example, the definition of perimeter. The definition for the exact case is well-known, but what could be the perimeter of a vague region? In a first approach one could be tempted to define minimum and maximum versions similar to the definition of area. This, however, might lead to wrong results. We have indicated that the boundary region can be thought of (at least in some applications) as describing possible locations of the region's contour. But then we cannot give any upper bound on the length of such a curve. In particular, the contour might be much longer than the perimeter of the boundary region, for instance:

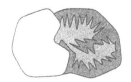

Moreover, we cannot simply take the perimeter of the kernel part as the minimal perimeter. This can be seen as follows. Usually, holes contribute to the perimeter of a region. If now, for example, a kernel part of a vague region v contains a hole which is equal to a boundary part of v, the perimeter of the hole is not counted in the perimeter of the maximal possible region.

In this example, the minimal possible region has a perimeter of $length(C) + length(c)$ whereas the maximal possible region has the minimal possible perimeter of $length(C)$.

Another example is an operation giving the number of connected components for which the generalization to vague regions heavily depends on the semantics of the boundary parts. Since regions need not be connected, a "possible regions"-semantics of boundary parts might well allow several unconnected parts, that is,

 could be a possible region of

Hence, we cannot give an upper bound on the number of components. But, in general, we cannot even give a non-trivial lower bound either (for example, the number of kernel components) since kernel components might be connected by boundary regions. Thus in the example below, the minimal number of components is 1 although there are three kernel regions.

(If we required, however, that a possible region extending into an adjacent boundary is connected, we could give a meaningful definition.)

We deliberately have avoided definitions of operations, such as perimeter and number of components, since we are then not forced to fix a semantics for vague regions. This means, the semantics can be assumed by each application as required and thus makes our model more general.

5 Embedding Vague Regions into Query Languages

In the previous section we have defined operations on vague regions. Next we indicate how these operations can play a part in a spatial query language. We do not give a full description of a specific language. We rather assume a relational data model where tables may contain vague region objects together with a SQL-like query language. For example, if we want to find out all regions where lack of water is a problem for cultivation, we can pose the following query:

select region **from** weather **where** climate = dry

Here we assume a table *weather* having a column named *region* containing vague region values for various climatic conditions given by the column *climate*. A similar query could ask for bad soil regions as a hindrance for cultivation.

Note that the result of both queries is a *set* of vague regions. If we now want to find out about regions where cultivation is impossible due to either reason, we ask for the union of the two region sets. Thus, we first have to cast the sets into single region objects. We therefore use the built-in aggregation function **sum** which, when applied to a set of regions, aggregates this set by repeated application of **union** (in the sense of *fold/reduce* of functional languages). So we can determine regions where cultivation is impossible by:

(**select sum**(region) **from** weather **where** climate = dry)
union
(**select sum**(region) **from** soil **where** quality = bad)

Pollutions are nowadays a central ecological problem and cause an increasing number of environmental damages. Important examples are air pollution and oil soiling. Pollution control institutions, ecological researchers, and geographers, usually use maps for visualizing the expansion of pollution. We can ask, for example, for inhabitable areas which are air polluted (where the kernel part of air pollution denotes heavily polluted areas and the boundary part gives only slightly polluted regions).

select **sum**(pollution.region) **intersection sum**(areas.region)
from pollution, areas
where area.use = inhabited **and** pollution.type = air

Then the kernel part of the result consists of inhabited regions which are heavily polluted, and the boundary consists (a) of slightly polluted inhabited regions, (b) of heavily polluted regions which are only partially inhabited, and (c) of slightly polluted and partially inhabited regions. If we want to reach all people who live in heavily polluted areas, we need the kernel of the intersection together with part (b) of the intersection boundary. How can we get this from the above query? The trick is to force boundary parts (a) and (c) to be empty by restricting pollution areas to their kernel region:

select kernel(sum(pollution.region)) **intersection sum**(areas.region)
from ...

A slightly different query is to find out all areas where people are definitely or possibly endangered by pollution. Of course, we have to use an intersection predicate. More precisely, we want to find those areas for which **intersects** either yields *true* or *maybe*. For this purpose we can prefix any predicate with **maybe** which causes the predicate to fail only if it returns *false*. (Technically **maybe** turns a *maybe* value into *true*.) So the query is:

select   areas.name
from   pollution, areas
where   area.use = inhabited **and**
            pollution.region **maybe intersects** areas.region

We could also express the query by using simply **intersects** and explicitly adding the two cases for **v-intersects** and **w-intersects**. This would be, of course, much longer and less clear.

The following example describes a situation which stresses the conflicting interests of economy and ecology. Assume on the one hand areas of animal species and plants that are worth being protected (nature reserves and national parks are the kernel regions) and on the other hand mineral resources the mining of which prospects high profits. An example for forming a difference of vague regions is a query which asks for mining areas that do not affect the living space of endangered species.

(**select sum**(region) **from** resources **where** kind = mineral)
difference
(**select sum**(region) **from** nature **where** type = endangered)

The kernel of the result describes regions where mining should be allowed. The boundary consists (a) of regions where mineral resources are uncertain and (b) of resource kernels that lie in (non-kernel) regions hosting endangered species. Since national parks are generally protected by the government, it is especially regions (b) conservationists should carefully observe. We can determine these regions by:

(**select kernel(sum**(region)) **from** resources **where** kind = mineral)
intersection
(**select boundary(sum**(region)) **from** nature **where** type = endangered)

The result is a vague region with an empty kernel and a boundary that just consists of the intersection of the mineral resource kernel and the endangered nature boundary.

Next we consider an example from biology already mentioned in Section 3. Assume we are given living spaces of different animal species. The kernel describes places where they normally live, and the boundary describes regions where they can be found occasionally (for example, to hunt for food or to migrate from one kernel area to another through a corridor). First, we can search for pairs of species which share a common living space. This asks for regions which have a non-empty intersection kernel:

select A.name, B.name
from animals (A), animals (B)
where A.region **intersects** B.region

A quite different question, also based on intersection, is whether there are animals that only sometimes enter the kernel region of other animals, for example, to attack them (but usually live in different areas). Here, we ask for an empty intersection kernel and a non-empty boundary/kernel intersection which is exactly the concept of vague intersection. The above query changes to:

select …
where A.region **v-intersects** B.region

Finally, we can also ask for animals that only encounter each other in their boundaries. This would be, for instance, the case for two animal species that are both hunters and usually avoid contact. The corresponding query is obtained by simply using **w-intersects** instead of **v-intersects**.

Assume that we are given a map of land areas (kernels) and mixed areas like shores and banks (boundaries) where the living spaces of animals with their kernel and vague regions are depicted. We can ask for animals that usually live on land and sometimes enter the water or for species that never leave their land area. This can be expressed using the inside predicate. The first example is characterized by an animal's living space being **v-inside** land:

select name
from animals
where region **v-inside** (**select sum**(region) **from** land)

The second example demands plain (that is, strong) **inside**.

A quite different example using insideness relates to the historical development of the Roman Empire, in particular, its expansion. At any moment during this development there were kernel regions representing the areas currently occupied by the Roman conquerors and spheres of influence (vague parts) that were under the control of the Roman Empire but not annexed. If we consider two points in time ($t_1 < t_2$) and hence two vague regions u and v, we could ask whether any of the occupied areas at t_1 have been preserved to t_2. This is the case if the kernel of u is completely inside v, in other words, when u **inside** v **or** u **v-inside** v. If they had to give up kernel regions, this is an example of u **w-inside** v.

Let us finally provide some examples for numeric operations. Oil companies are often interested to determine whether it is worth exploiting a recently discovered oilfield. Hence, they classify oilfields in areas where the existence of oil was proved by soil samples and in areas where the incidence of oil is only assumed. The decision of exploiting then depends on the guaranteed minimal extensions of the oilfields:

select min-area(region) **from** oilfields

An example for applying **max-area** is again pollution where we should be pessimistic and consider the worst case of all possible polluted regions.

The minimum distance between a forest fire and a region of endangered species indicates where protective measures should be performed first. For another example consider an attacked country that might have secure parts (kernel), battle regions (boundary), and even lost parts (outside). To move from one safe area to another one might consider the risk of such a trip be given by the maximum distance between different regions, that is, the difference between secure parts.

6 Conclusions and Future Work

We have defined a data model of regions that is capable of describing many different aspects of vague spatial objects. It is a canonical extension of a determinate region model which facilitates the treatment of vague and exact regions in one model. In particular, this allows a smooth migration from already existing models to vague concepts (at least as far as regions are concerned). Our approach is based on exact spatial modeling concepts which allows to build upon existing work and simplifies many definitions. In particular, we can (re-)use already existing regions implementations to realize vague regions with only minimal effort.

Of course, the current model is limited in some ways, and we are currently investigating extensions along several different lines. First, the presented concept of vagueness can be extended to *other spatial objects*, such as points and lines. For example, a vague line could be thought of as consisting of a kernel part given by a set of (unconnected) curves and a vague part described by a boundary region. An example is a river which may contain fixed segments (determined, for instance, by levees) and a boundary which describes possible flows that depend on water level or season.

A vague point can be simply given by a vague region (with empty kernel) describing possible positions of the point. To define such extensions we first have to extend the basic model of exact regions by lines and points together with operations defined for them. These can then be used to define vague lines and points. We also should consider operations concerning objects of different vague types, for example, the intersection of a vague line with a vague region.

Another direction of extension is the notion of vagueness itself. As yet, there is only one kind of vagueness, but there are many applications which can be best described by having *different degrees of vagueness*. For example, zones of decreasing pollution or regions of different possibilities for certain virus infections. Our model can be easily extended to deal with this kind of applications by having a set of regions labeled with different values of a suitable domain *D* which is subject to certain restrictions. For example, we need operations *max* and *min* to give meaningful extensions of operations like **union** and **intersection**.

Finally, we consider the *integration* of vague regions (and their possible extensions) into other data models. We have already seen how for predicates and numeric operations the vagueness of regions affects the corresponding domains of booleans and real numbers. It is likely that the situation is similar for other domains as well. So the integration of vague regions into any existing data model and query language might cause some trouble since it either requires a redefinition of the data types or a redefinition (and duplication) of operations. That this can be tedious and error-prone has been demonstrated in the description of numeric operations. Note that the problem of "vague infection" is not restricted to standard data types. For example, in [EG94, Er94] graphs have been integrated into a spatial data model. With respect to vague spatial objects, an operation like **subgraph** that computes part of a graph according to a possibly spatial predicate should return a *vague graph*. Now, what are vague graphs, and how can all the graph operations adapted to the vague case? We currently consider the integration an open problem.

References

[Al61]   P. Alexandroff. *Elementary Concepts of Topology.* Dover Publications, 1961.

[Al94]   D. Altman. Fuzzy Set Theoretic Approaches for Handling Imprecision in Spatial Analysis. *Int. Journal of Geographical Information Systems*, vol. 8, no. 3, pp. 271-289, 1994.

[Ar83]   M.A. Armstrong. *Basic Topology.* Springer Verlag, 1983.

[Ba93]   R. Banai. Fuzziness in Geographical Information Systems: Contributions from the Analytic Hierarchy Process. *Int. Journal of Geographical Information Systems*, vol. 7, no. 4, pp. 315-329, 1993.

[Bl84]   M. Blakemore. Generalization and Error in Spatial Databases. *Cartographica*, vol. 21, 1984.

[Bu96]   P.A. Burrough. Natural Objects with Indeterminate Boundaries. *Geographic Objects with Indeterminate Boundaries*, GISDATA Series, vol. 3, Taylor & Francis, pp. 3-28, 1996.

[CF96]   E. Clementini & P. di Felice. An Algebraic Model for Spatial Objects with Indeterminate Boundaries. *Geographic Objects with Indeterminate Boundaries*, GISDATA Series, vol. 3, Taylor & Francis, pp. 153-169, 1996.

[CG96]   A.G. Cohn & N.M. Gotts. The 'Egg-Yolk' Representation of Regions with Indeterminate Boundaries. *Geographic Objects with Indeterminate Boundaries*, GISDATA Series, vol. 3, Taylor & Francis, pp. 171-187, 1996.

[Co96]   H. Couclelis. Towards an Operational Typology of Geographic Entities with Ill-defined Boundaries. *Geographic Objects with Indeterminate Boundaries*, GISDATA Series, vol. 3, Taylor & Francis, pp. 45-55, 1996.

[ECF94] M.J. Egenhofer, E. Clementini & P. di Felice. Topological Relations between Regions with Holes. *Int. Journal of Geographical Information Systems*, vol. 8, no. 2, pp. 129-142, 1994.

[Ed94]   G. Edwards. Characterizing and Maintaining Polygons with Fuzzy Boundaries in GIS. *6th Int. Symp. on Spatial Data Handling*, pp. 223-239, 1994.

[Eg89]   M.J. Egenhofer. Spatial SQL: A Spatial Query Language. Report 103, Dept. of Surveying Engineering, University of Maine, 1989.

[EG94]   M. Erwig & R.H. Güting. Explicit Graphs in a Functional Model for Spatial Databases. *IEEE Transactions on Knowledge and Data Engineering*, vol. 6, no. 5, pp. 787-804, 1994.

[Er94]   M. Erwig. *Graphs in Spatial Databases*. Doctoral Thesis, FernUniversität Hagen, 1994.

[Fi93]   J.T. Finn. Use of the Average Mutual Information Index in Evaluating Classification Error and Consistency. *Int. Journal of Geographical Information Systems*, vol. 7, no. 4, pp. 349-366, 1993.

[Ga64]   S. Gaal. *Point Set Topology*. Academic Press, 1964.

[GS93]   R.H. Güting & M. Schneider. Realms: A Foundation for Spatial Data Types in Database Systems. *3rd Int. Symp. on Large Spatial Databases*, pp. 14-35, 1993.

[GS95]   R.H. Güting & M. Schneider. Realm-Based Spatial Data Types: The ROSE Algebra. *VLDB Journal*, vol.4, pp. 100-143, 1995.

[Gü88]   R.H. Güting. Geo-Relational Algebra: A Model and Query Language for Geometric Database Systems. *Int. Conf. on Extending Database Technology*, pp. 506-527, 1988.

[KV91]   V.J. Kollias & A. Voliotis. Fuzzy Reasoning in the Development of Geographical Information Systems. *Int. Journal of Geographical Information Systems*, vol. 5, no. 2, pp. 209-223, 1991.

[LAB96]   P. Lagacherie, P. Andrieux & R. Bouzigues. Fuzziness and Uncertainty of Soil Boundaries: From Reality to Coding in GIS. *Geographic Objects with Indeterminate Boundaries*, GIS-DATA Series, vol. 3, Taylor & Francis, pp. 275-286, 1996.

[LN87]   U. Lipeck & K. Neumann. Modelling and Manipulating Objects in Geoscientific Databases. *5th Int. Conf. on the Entity-Relationship Approach*, pp. 67-86, 1987.

[Sc95]   M. Schneider. *Spatial Data Types for Database Systems*. Doctoral Thesis, FernUniversität Hagen, 1995.

[Sc96]   M. Schneider. Modelling Spatial Objects with Undetermined Boundaries Using the Realm/ROSE Approach. *Geographic Objects with Indeterminate Boundaries*, GISDATA Series, vol. 3, Taylor & Francis, pp. 141-152, 1996.

[Sh93]   R. Shibasaki. A Framework for Handling Geometric Data with Positional Uncertainty in a GIS Environment. *GIS: Technology and Applications*, World Scientific, pp. 21-35, 1993.

[SV89]   M. Scholl & A. Voisard. Thematic Map Modeling. *1st Int. Symp. on Large Spatial Databases*, pp. 167-190, 1989.

[Ti80]   R.B. Tilove. Set Membership Classification: A Unified Approach to Geometric Intersection Problems. *IEEE Transactions on Computers*, vol. C-29, pp. 874-883, 1980.

[Us96]   E. L. Usery. A Conceptual Framework and Fuzzy Set Implementation for Geographic Features. *Geographic Objects with Indeterminate Boundaries*, GISDATA Series, vol. 3, Taylor & Francis, pp. 71-85, 1996.

[Wa94]   F. Wang. Towards a Natural Language User Interface: An Approach of Fuzzy Query. *Int. Journal of Geographical Information Systems*, vol. 8, no. 2, pp. 143-162, 1994.

[WB93]   M.F. Worboys & P. Bofakos. A Canonical Model for a Class of Areal Spatial Objects. *3rd Int. Symp. on Advances in Spatial Databases*, Springer-Verlag, LNCS 692, pp. 36-52, 1993.

[WH96]   F. Wang & G.B. Hall. Fuzzy Representation of Geographical Boundaries in GIS. *Int. Journal of Geographical Information Systems*, vol. 10, no. 5, pp. 573-590, 1996.

[WHS90]   F. Wang, G.B. Hall & Subaryono. Fuzzy Information Representation and Processing in Conventional GIS Software: Database Design and Application. *Int. Journal of Geographical Information Systems*, vol. 8, no. 2, pp. 143-162, 1994.

[Za65]   L.A. Zadeh. Fuzzy Sets. *Information and Control*, vol. 8, pp. 338-353, 1965.

Appendix

We give a formal definition of the general model for determinate regions that has been informally described in Section 3. An adequate and general method to formally define this model is to use the point set paradigm and point set topology. The point set paradigm expresses that space is composed of infinitely many points and that spatial objects like areal objects are distinguished subsets of space which are viewed as entities. Point set topology [Al61, Ar83, Ga64] allows one to distinguish special topological structures of a point set like its boundary or interior. We start with some basic concepts of point set topology.

Definition. Let X be a set and $T \subseteq 2^X$ be a subset of the power set of X. The pair (X, T) is called a *topological space*, if the following three axioms are satisfied:

(T1) $X \in T, \varnothing \in T$

(T2) $U \in T, V \in T \ \Rightarrow \ U \cap V \in T$

(T3) $S \subseteq T \ \Rightarrow \ \bigcup_{U \in S} U \in T$

T is called a *topology* for X. The elements of T are called *open sets*, their complements in X *closed sets*. The elements of X are called *points*.

When no confusion can arise, T is not mentioned, and X denotes a topological space. In the sequel, let X be a topological space and $Y \subseteq X$.

Definition. The *interior* of Y, denoted by Y°, is the union of all open sets that are contained in Y. The *closure* of Y, denoted by $\bar{Y}$, is the intersection of all closed sets that contain Y. The *exterior* of Y, denoted by Y^-, is the union of all open sets that are not contained in Y. The *boundary* of Y, denoted by ∂Y, is the intersection of the closure of Y and the closure of the complement of Y, that is, $\partial Y = \bar{Y} \cap \overline{X - Y}$.

The relationships between these four topological structures are given by the provable statements (1) $Y^\circ \cap \partial Y = \varnothing$, (2) $Y^\circ \cup \partial Y = \bar{Y}$, (3) $Y^- \cap \partial Y = \varnothing$, and (4) $Y^\circ \cap Y^- = \varnothing$. Obviously we can conclude $X = \partial Y \cup Y^\circ \cup Y^-$.

Since our objective is to model two-dimensional areal objects for spatial applications, we embed them in the Euclidean space (plane) $\mathbb{R}^2$ as an instance of a topological space[4] with metric properties. A problem of applying pure set-theoretic operations to point sets is that undesired geometric anomalies can arise. These anomalies are avoided by the concept of *regularity* [Ti80].

Definition. Y is called *regular closed* if $Y = \overline{Y^\circ}$.

Intuitively, regular closed sets model areal objects containing their boundaries and avoid both isolated or dangling line or point features and missing lines and points in the form of cuts and punctures. Hence, it makes sense to define a *regularization* function *reg* which associates a set Y with a regular closed set, as follows:

---

4. Note that most of the definitions and results in the sequel also hold for general topological spaces.

$$reg(Y) := \overline{Y^\circ}$$

An example of regularization is shown below where the set Y consists of areal, point, and line objects. Some areal objects contain only parts of their boundaries (drawn with broken lines) and have cuts (drawn with broken lines) and punctures. The regularization process eliminates point and line features, cuts and punctures, and includes the missing boundary parts of the areal objects.

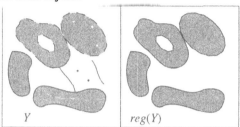

The union of a finite number of regular closed sets is regular closed. The intersection and difference of regular closed sets are not necessarily regular closed. Hence, we introduce *regular set operations* that preserve regularity.

Definition. Let A, B be regular closed sets, and let $\neg A$ denote the (set-theoretic) complement $\mathbb{R}^2 - A$ of A. Then

(i) $A \cup_r B := reg(A \cup B) = A \cup B$
(ii) $A \cap_r B := reg(A \cap B)$
(iii) $A -_r B := reg(A - B)$
(iv) $\neg_r A := reg(\neg A)$

It is obvious that the subspace RCS of regular closed sets together with the regular set operations is a topological space. Regular closed sets and regular set operations express a natural formalization of the dimension-preserving property taken for granted by many spatial type systems and geometric algorithms. The following important theorem holds:

Theorem. RCS with the set-theoretic order relation $\subseteq$ is a Boolean lattice.

This implies that (i) $(RCS, \subseteq)$ is a partially ordered set, (ii) every pair A, B of elements of RCS has a least upper bound $A \cup_r B$ and a greatest lower bound $A \cap_r B$, (iii) $(RCS, \subseteq)$ has a maximal element $\mathbf{1}_r := \mathbb{R}^2$ (identity of $\cap_r$) and a minimal element $\mathbf{0}_r := \varnothing$ (identity of $\cup_r$), (iv) algebraic laws like idempotence, commutativity, associativity, and distributivity hold for $\cup_r$ and $\cap_r$, (v) $(RCS, \subseteq)$ is a complementary lattice, that is, $\forall A \in RCS : A \cap_r \neg_r A = \mathbf{0}_r$ and $A \cup_r \neg_r A = \mathbf{1}_r$.

Definition. A *region* is a regular closed set.

Definition. The type R consists of all regions and has the operations $\oplus$, $\otimes$, $\ominus$, and Θ that are equated with the regular set operations $\cup_r$, $\cap_r$, $-_r$, and $\neg_r$, respectively, and the elements $\mathbf{1}_R = \mathbf{1}_r$ and $\mathbf{0}_R = \mathbf{0}_r$.

On the Assessment of Generalisation Consistency

Vasilis Delis

Computer Technology Institute
Kolokotroni 3, 26221, Patras, Greece
Computer Engineering and Informatics
Department, University of Patras
delis@cti.gr

Thanasis Hadzilacos

Computer Technology Institute
Kolokotroni 3, 26221, Patras, Greece
thh@cti.gr

Abstract

Theory, algorithms, techniques and tools for producing a generalisation of a map have long been available. In this paper we study the inverse problem, namely, given two maps L and M, whether there exists a generalisation G, such that $L=G(M)$. Answering this problem can help with fundamental issues of consistency in multiresolution databases. We view such a database as a collection of map layers depicting the same geographic area at different levels of detail, related through a generalisation hierarchy. From an engineering perspective, multiple representations, of which multiresolution maps is a special case, imply redundancy, which is a threat to the integrity of a database. For integrity control we need a set of tools that ensure that the metric and topological properties of a map layer are retained or monotonically decreased along the generalisation hierarchy. In this paper we study the former, i.e. we propose a framework for the assessment of metric consistency between two map layers.

1 Introduction

In a way, cartography is the art of generalisation (when abstracting the real world to create a map or deriving a small scale map from a large scale one) which has long been a difficult and highly subjective problem. "It is the process of reducing the amount of detail in a map so that the character or essence of the original features is retained at successively smaller scales. In the context of topographic map generalisation this usually involves omission, aggregation, simplification, displacement, exaggeration and symbolization of either individual features or groups of features" [19]. Intuitively it involves a transition between two different perception levels. In undertaking this process, a cartographer makes use of his knowledge on the source representation, the intended purpose of the target representation as well as his knowledge of cartographic conventions and the real world [16]. The rules for such transformations are extremely context dependent: relative size, relative scarcity of the entity, "importance" (for the purpose of the specific generalisation).

A considerable body of literature exists on map generalisation (see [3, 15] for general treatments). Modern tools for digital generalisation, designed and implemented to exploit the latest in computer science, like object orientation [13], expert systems [19], neural networks [23], or case-based reasoning [12], clearly show the need for intricate, context sensitive algorithms which take particular notice of spatial and other relationships among objects and features.

There is a growing demand for current GIS to utilise multiple representations of geographic data at different levels of detail in order to deal successfully with several demanding applications, like efficient browsing over large spatial databases [2], abstract querying of spatial and statistical databases [18] or structured solutions in way-finding [21], but also in order to support map generalisation and conventional cartographic map series production at different scales [10]. What underlies the need for better, closer to humans systems is the concept of abstraction [11]. It is used pervasively in common sense reasoning which often encompasses multiple representation (e.g. hierarchical mental maps).

From an engineering point of view, the concept of multiple representations is a matter of redundancy, so it naturally calls for methods to check and preserve the integrity of the database. In multiresolution spatial (or geographic) databases the various layers that refer to the same geographic area at different scales, are related through generalisation relationships. Since such databases encompass changes in the metric and topological structure of a map layer with the changing resolution at which the map's objects are depicted, it is important to develop methods to assess consistency among the different representations.

The issue of consistency in multiresolution spatial databases is relatively new and most of the work recorded in the literature deals almost exclusively with topological consistency [6, 7, 8, 9]. this work we focus on the geometry of individual objects, ignoring the effects of generalisation on the spatial relationships among objects. Thus, in the sequel of the paper any use of terms like "geometric consistency", "geometric simplification", etc. will refer to metric aspects, excluding any considerations on topological or other spatial relationships. Assessing consistency between two layers assumed to be related through a generalisation transformation means checking whether one of them can be correctly considered to be a generalisation of the other. Geometric consistency therefore refers to whether two representations at different scales are similar with respect to geometric objects' positions (location, size, orientation).

Geometric consistency can be assessed using the *monotonicity assumption* [7]: the geometry of a map through consecutive representation layers must be monotonic, i.e. continuously decrease in detail with the decrease of scale. Like generalisation itself, the assessment of geometric consistency is a subjective and context sensitive problem. Therefore, there is no universal algorithm for this purpose and likewise, assessing whether a map layer M is "geometrically simpler" than another layer L, does not always have a strictly binary (yes, no) answer. Adopting this view, in this work we suggest tools/operations to aid the expert user in building such geometric simplification checking algorithms. We first develop a general framework to define generalisation and a model of spatial entities and then focus on a mathematical definition of "geometrically simpler". This is done in Section 2. Section 3 deals with the problem under a computational perspective. An example is presented on how the tools proposed in Section 2 can be used for the geometric comparison of two layers. Adding semantics and tuning the algorithm is also discussed. In our conclusions in Section 4 possible continuation of this work is outlined.

2 A Map Generalisation Framework

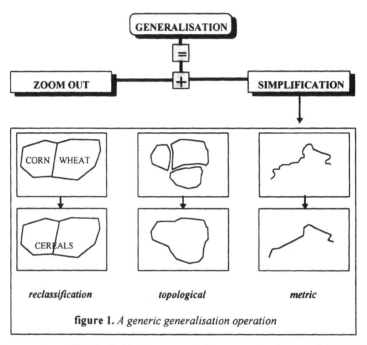

figure 1. *A generic generalisation operation*

Among the most qualified individuals to speak for generalisation would definitely be astronauts and space-travellers for they have a great amount of experience on how the Earth looks like from above, from distances of a few hundreds of meters where everything is visible but smaller, through stages where geographic objects become simpler in detail while some become indistinguishable, right to the other extreme where the only visible remains are simple geometric shapes that correspond to huge geographic areas, drawn on the surface of a pocket-size sphere.

Although childish, the space-trip example indicates two fundamental concepts/operations associated with generalisation, namely continuous zooming and simplification (see fig. 1). This definition of generalisation is neat, with nice properties, but unfortunately not always useful in practice. To continue the space-trip example, a situation where Ireland can not be distinguished from Great Britain but Sicily remains visibly dettached from the remainder of the Italian mainland, is not possible when gazing from a distance and this is theoretically a positive property. Nevertheless, it is often desirable in practice to violate uniformity (under the limitations imposed by a map's scale) when generalising either by exaggerating salient features to be portrayed on the new map and/or simplifying insignificant entities.

Map simplification encompasses three different suboperations pertaining to three usually separated concepts, topology, metric and attribution. A generalised map has to be simpler than its detailed counterpart with respect to all of the above elements (see fig. 1). A common interpretation which we adopt is that a representation A is simpler than a representation B if A captures less detail than B. However, "geometrically simpler" is a fuzzy characterisation. We may accept by convention that a pie (region without holes) is simpler than a donut (region with a hole) but what is the answer for the comparison between a circle and a square? A vast literature exists on the geometric aspects of simplification [3, 15], the majority of which views it in the context of line thinning [5, 14, 1]. Simplification on spatial attributes seems to be relatively neglected in the literature. It falls under a broader category, the generic reclassification operation, treated in [4].

The inverse problem is given two representations at different detail levels, to assess whether a generalisation relationship exists between them, in other words to assess generalisation consistency. In this work we deal with geometric consistency, i.e. we develop a framework to compare two representations at different scales and decide i) whether corresponding geometric objects are similar enough in location and shape and ii) whether the small scale representation contains less detailed objects with respect to the large scale one, in order to be considered as a correct detailed-generalised pair of map layers.

2.1 A Conceptual Model of Spatial Data

Spatial entities are geometric representations of real world entities along with associated attributes. Geometric entities are grouped into three major categories: points (0-dimensional), lines (1-dimensional) and regions (2-dimensional).

A *point* is defined uniquely by its coordinates in a coordinate system on $\mathbf{R}^2$.

An *arc* is a straight line segment with two distinct points, a *start point* and an *end point*. A *line* is a finite sequence of simple arcs $a_1 a_2 ... a_n$ such that the end point of the arc a_i coincides with the start point of the arc a_{i+1}, for every $1 \leq i \leq n\text{-}1$. A line's endpoints are called *nodes*. If a line's nodes coincide, it is called *closed*, otherwise it is called *open*. A *simple line* has a non-crossing interior.

A *region* is a connected subset of 2D space which is characterised by a simple closed line called the *outer boundary*, denoted l_0 and potentially a set of simple closed lines called the *inner boundaries*, denoted l_i for $i>0$, such that i) no two such lines intersect, ii) neither of them is contained in the internal set of the other and iii) all of them are contained in the internal set of the outer boundary. The region then can be expressed by the formula

$$\bigcap_{i>0} (ex(li) \cap in(l_0)) \ ^1 \ (1)$$

---

[1] By Jordan theorem, a simple closed line *l* separates $\mathbf{R}^2$ into two open sets: one is bounded and called the internal set, denoted *in(l)* and the other is unbounded and called the external set, denoted *ex(l)*.

A *map layer* (or simply layer) is a function $L : G \rightarrow D_1 \times D_2 \times ... \times D_n$ where G is a set of geometric objects defined as above and $D_1 \times D_2 \times ... \times D_n$ the Cartesian product of a set of domains. The domain G may in general contain non-overlapping geometric objects (points, lines and regions). This means that two distinct regions can only share a portion of their boundaries or be disjoint and two distinct lines can only intersect at their nodes. Consequently, whenever a line crosses another line or a region's boundary, the former line is split and a node is inserted at the intersection. This definition is a direct interpretation of our intuitive perception of a map as a set of geometric shapes with associated attributes.

2.2 Generalisation and Simplification: Concepts and Definitions

The potential transformations a geometric object may undergo generally in a particular generalisation scenario fall in the following broad categories [17]:

- *preservation*: the object represented in the detailed map is retained in the generalised map with the same dimension and possibly a simplified structure, definitely without introducing more detail in the generalised representation.

- *reduction*: an object may be represented in the generalised map with a lower dimension (a trivial valid case is when a small object in the source map disappears after the generalisation). Thus, a region may become a line or point and a line may also become a point.

- *immersion*: several objects in the detailed map (e.g. a group of small islands close enough to each other) may be aggregated to form one object in the generalised map.

The relation "simpler" between two geometric objects is not a straight-forward assessment, since a generic object could be arbitrarily complex, consisting of areal and lineal parts, etc. However, in practice the entities participating in geographic maps usually have a "bounded" complexity, which allows us to formulate a definition for simplification with a high degree of confidence that it would be practically sufficient. Our definition is based on a simplified version of the mathematical concept of ε-homotopy [17].

Definition 1. A d-buffer of a point p, denoted $d\text{-}buff(p)$, is the disk of diameter d around the centre p.

Definition 2. A d-buffer of a line l, denoted $d\text{-}buff(l)$, is the area formed from the convolution of the line with a disk of diameter d, i.e. the area covered by the disk while sweeping its centre along l.

Definition 3. A d-buffer of a region r, denoted $d\text{-}buff(r)$, is the area given by the following formula:

$$r \cup \bigcup_i d\text{-}buff(l_i) \quad (2)$$

where r is given by formula (1), l_0 is the outer boundary of r and l_i, $i \neq 0$, are the inner boundaries. Intuitively, the d-buffer of a region, is the region itself "fattened" by $d/2$

around the outer side of the outer boundary as well as around the inner side of the inner boundaries.

In the same manner we can define the d-buffer of a set of objects S to be the union of the d-buffers of its elements, i.e.

$$d\text{-}buff(S) = \bigcup_i d\text{-}buff(r_i), \quad r_i \in S \quad (3)$$

Definition 4. All geometric objects x which are topologically contained in the d-buffer of another object y, form a class denoted as the d-class of y. If x belongs to the d-class of y and y belongs to the d-class of x, then x and y are said to be d-equivalent. Analogously, a set of geometric objects S (e.g. a map) is d-equivalent to a set S' if S belongs to the d-class of S' and vice versa (one of the sets can be a singleton). The concept of d-equivalence is a formal interpretation of our intuition on the geometric similarity of objects, and d is the given tolerance.

Now we are ready to give a definition of geometric simplification.

Definition 5. Below we list the definitions of the relation simpler between two geometric objects or between a geometric object and a set of objects. Our approach is bottom-up, that is, we start by the simple cases and inductively define more complex ones.

1. in a trivial case a point p is simpler than any geometric object(set of objects) if the point and the object(set of objects) are d-equivalent (this definition refers to the cases where a group of objects is small enough with respect to d so that it can be represented by a single point in a generalised map).

2. an open(closed) line c is simpler than an open(closed) line c' if c and c' are d-equivalent and the number of arcs c consists of is less or equal than the number of arcs c' consists of (a closed and an open line are not comparable).

3. a line is simpler than a set (possibly a singleton) of regions if the line and the set are d-equivalent and the number of arcs the line consists of is less or equal than the sum of the numbers of arcs the regions' in the set outer boundaries consist of.

4. a simple region r is simpler than a set of simple regions R if r is d-equivalent to R and the number of arcs of r's boundary is less or equal than the sum of the number of arcs of the regions in R boundaries (this definition covers the case where R is a singleton).

5. a set of simple regions R is simpler than another set of simple regions R' if R has a smaller or equal cardinality than R' and there is a surjective function $f: R \rightarrow R'$ such that for every region r in R ($f(r)$ might well be a set of regions in R'), r is simpler than $f(r)$ and $f(R)$ is a partitioning of R'. In the trivial case where R is the empty set, we can accept that R is simpler than R' if R' can be contained in a disk of diameter d (the meaning here is that the group of regions in R' is small enough with respect to d so that it can be eliminated, thus represented by the empty set). Also, R is simpler than $R' \cup R''$, for any set R'' consisting of regions so small to be contained in a disk of diameter d.

figure 2. *a region and its equivalent tree representation*

6. a nice conceptual representation for an arbitrary region is that of a tree whose nodes are simple regions (see figure 2). Any two nodes at the same level are non-overlapping and topologically contained in their father(s). An arbitrary region r is simpler than another region r' if r's corresponding tree has a depth less or equal than the depth of the tree of r' and all the nodes of r at a particular depth are simpler (according to 5.5) than the nodes of r' at the same depth. Notice that if r's depth is less than the depth of r' then the set of nodes at any level deeper than the leaves of r is considered to be the empty set. Similarly, a region r is simpler than a set of regions R if at any depth the set of r's nodes is simpler than the set of nodes of all regions in R at the same depth.

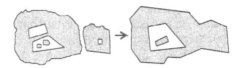

figure 3. *a region generalisation example*

Although the above definition is reasonable for a wide range of cases in practice, still some possible instances are questionable. For example, according to the above, a simple line is more complex than an eyelet if it has more arcs, which may not be always agreed upon by cartographers. Such vague cases should not undermine the validity of our approach. Fuzziness and subjectiveness are inherent in the generalisation problem. We believe that we still make a contribution if we axiomatise a set of cases that the majority of cartographers would agree upon and use them in our study.

3 The Algorithm

So far we have proposed a model of spatial entities and map layers and a set of formal specifications for geometric simplification. The ultimate goal is the definition of an algorithmic component which assesses two layers L and M and decides whether M (the target layer) can potentially be geometrically a simplification of L (the source layer). The definitions of the previous section are our conceptual toolbox. At the implementation level all we need is a buffering tool, an arc counter and a routine that checks for topological containment, all very common to GIS practitioners.

Due to the inherent subjectiveness of generalisation it would be certainly ineffective, if not impossible, to propose a unique algorithm, regardless of the particular context. For example, as already mentioned in Section 2 uniformity in the generalisation of individual features may be purposefully violated in a particular scenario. Therefore, depending on the application, the consistency among particular features (e.g. roads) could be strictly checked (i.e. using a small d value), while greater tolerance (i.e. a larger d) might be used for other features. The extensive research undertaken on generalisation has shown that the synergy between an expert user and a toolbox consisting of generalisation operators is so far the best approach towards the automation of map generalisation. We adopt this approach in this inverse problem (checking instead of deriving a generalisation) and suggest a generic framework and a set of primitive operations to perform this task. These can only provide the basis which the expert user should enrich with case dependent generalisation constraints. Below we present an example of how our toolbox can be used to build a simple algorithm and in following paragraphs we discuss how more meaning can be added.

3.1 Example of a geometric simplification check

In general, a map layer M is simpler than (thus a potential generalisation of) a layer L if every object in L can be either eliminated or represented by a simpler counterpart in M or aggregated with other objects to form a simpler object in M. In addition, every object in M should be associated with an object or an aggregate of objects in L. We assume that the parameter d (the buffer width) is one of the algorithm's input parameters. Its impact on the algorithm will be discussed in a following section. An important point which has to be clarified is that throughout the algorithm none of the manipulations takes into account the zooming factor. This can be interpreted in two equivalent ways: i) when we create buffers and check for containment, we use real coordinates and ignore any legibility constraints (such as pixel granularity and human sight limitations), or ii) we want to compare the two maps in the same scale, so we zoom in the target map in order to bring it to the scale of the source map and then check for simplification.

Our purpose is to create a mapping from objects in the source map to valid simpler counterparts in the target map. The meaning of the mapping is the following: i) if a single source object a is mapped to a target object b then b is a simpler representation of a, and ii) if more than one objects $a_1, a_2, ..., a_n, n>1$, are mapped to the same object b then b is a simpler representation of the spatial aggregation of $a_1, a_2, ..., a_n$. A correct mapping has to be a surjective function. In other words, the target map is a correct geometric simplification of the source map if every object in the latter is mapped to one and only one simpler object in the former, while every object in the target map is the mapped image of at least one object in the source map (see fig. 4).

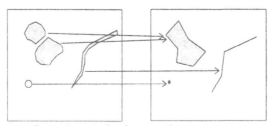

figure 4. *example of a generalisation mapping*

An example

Consider a source large scale map layer consisting of regions and a target small scale layer consisting of regions and points. Assume a mapping f that associates a region x in the source map to an entity y in the target map if x belongs to the d-class of y. We will denote as R_f the relation with all such pairs (x,y).

Step 1.

For every point p in the target map we calculate $f^{-1}(p)$, that is, we compute R^p which is the restriction of R_f for all the points in its range. Eventually a valid generalisation (see figure 5) should be represented by a function f' (a mapping where every region would be associated with exactly one point). It is possible that (depending on the magnitude of d) R^p is not a function, i.e., some regions are mapped to more than one points. In this case we have to restrict R^p to a function f' by keeping for every region only one mapping edge to a point. A straight-forward consequence is that the cardinality of R^p's range should be less or equal than the cardinality of its domain (otherwise some points would be unmapped). Points have a useful property: if a geometric entity x belongs to the d-class of a point p then p is contained in the d-class of x. This means that as soon as R^p is restricted to a surjective function g, then g corresponds to a correct geometric simplification, i.e. for every point p in the target layer, p belongs to the d-class of $g^{-1}(p)$, thus d-equivalence is satisfied (assuming that $g^{-1}(p)$ can possibly consist of more than one entities from the source layer, in an extreme case), that is, p is simpler than $g^{-1}(p)$ according to definition 5.1.

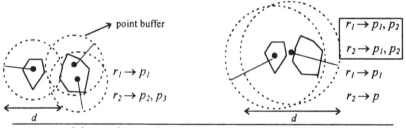

a. *an invalid generalisation, the mapping can not be resolved (p_2 or p_3 remains unmapped)*   **b.** *a valid generalisation*

figure 5. *Resolving R^p mappings to surjective functions*

Step 2.

For every region r in the target map we calculate $f^{-1}(r)$, that is, we compute R^r which is the restriction of R_f for all the regions in its range. As in the point case in step 1, a valid generalisation should be represented by a function f' (a mapping where every region in the source map would be associated with exactly one region in the target map). In a completely analogous to the previous step manner, we try to restrict R^r to a function that represents a correct simplification. Unlike the point case, in the region case restricting R^r to a surjective function g, does not guarantee that g corresponds to a correct simplification, since 1) for every region r in the target map we have to check that r belongs to the d-class of $f^{-1}(r)$ (notice that when considering regions, a belongs to the d-class of b does not imply that b belongs to the d-class of a so this check is necessary to prove d-equivalence between r and $f^{-1}(r)$) and 2) if d-equivalence is verified, boundaries simplification has to be verified according the condition regarding the number of arcs of the regions (see definition 5).

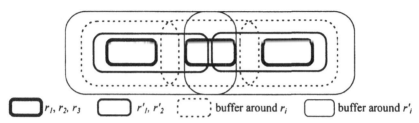

figure 6. *a region-to-region generalisation example*

For the above example $R' = \{(r_1, r'_1), (r_2, r'_1), (r_2, r'_2), (r_3, r'_2)\}$. One can observe that in both possible restrictions of R' to a function (either keeping (r_2, r'_1) or (r_2, r'_2)), the resulting functions are not valid generalisations because d-equivalence is not satisfied.

Therefore, all the different combinations for restricting R' to a function have to be examined. Assume that R''s domain consists of regions r_1, r_2, ..., r_n and each region r_i is mapped under f to $n_i > 0$ target regions. We can restrict R' to a function in $n_1 \times n_2 \times ... \times n_n$ ways, by keeping for every source region only one link to a target region. This is the worst case however, because a speed-up can be gained if we examine only those combinations that do not leave any target region unmapped.

Step 3.

After a successful completion of both steps 1, 2, the algorithm must check whether any unmapped regions in the source layer exist (that is, whether there are regions in the source layer not contained in the d-buffer of any point or region in the target layer). If these unmapped regions are significant (e.g. large enough or important entities not to be eliminated in the generalised layer) then the checking has failed in the sense that there are no generalised representations for these regions in the target layer. Otherwise (e.g. if the unmapped regions are unimportant or small enough, i.e. each of them contained in a disk of diameter d), these regions can reasonably be eliminated in a generalised representation, so the simplification checking was successful.

3.2. The parameter d

The buffer diameter d is the allowed tolerance for the differences in location and shape that two objects representing the same real-world entity at different scales can exhibit. There is a strong locality factor inherent in the generalisation problem, i.e. a generalised representation is required to have the same location and shape as the corresponding detailed one. However in practice there is always a divergence from this rule that the reduction in scale accounts for, due to the inevitable loss of detail.

A reasonable question that arises is, what is the appropriate d value when checking for a generalisation relationship between two maps? Again there is no single answer to this question. Several factors like the sources of the maps (e.g. a map layer originally derived by a satellite image is likely to differ significantly from a layer derived from manual digitization, depicting the same area), the application context (e.g. land ownership applications require more accuracy than environmental applications), etc., may affect the choice. The two extremes are a very large d value (in which case a whole map could be represented by a point, if all map objects fell in the d-buffer of the point), or an infinitesimally small value (in which case only two identical maps would pass the simplification check).

However, a "good" d value is one that guarantees that the results of the simplification checks are not unreasonable. Any value less or equal than the minimum representable distance at the particular scale is a safe value. This is clarified in the example in figure 7, where exceeding the minimum distance m leads to a "bad" simplified representation of a line successfully passing the simplification check.

The magnitude of d also affects quantitative aspects of a simplification checking algorithm, that is, its computational complexity. Considering again the example algorithm presented in the previous section, the larger the value of d, the more objects that are likely to belong to the d-class of a particular object in the target layer, the more ways to restrict R_f to a surjective function. In practice however, testing many detailed candidates for a generalised representation would be very unreasonable and suspicious. So normally, the average number of the mapped generalised images for the objects in the detailed layer should be very close to 1, for any realistic, cartographically proper generalisation.

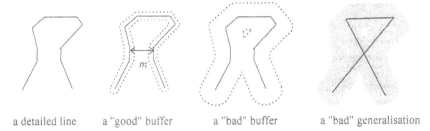

a detailed line   a "good" buffer   a "bad" buffer   a "bad" generalisation

figure 7. *the impact of d's magnitude*

3.3. Improving the semantics of simplification checking algorithms

Generalisation is frequently divided in two distinct processes in the literature [22, 20]: *Geographic information abstraction* mainly concerns managing geographical meaning in spatial databases while *simplification* concerns structuring map presentations. So far we have dealt with geometric simplification using purely geometric techniques, thus completely ignoring the semantics of the information. This however could lead to undesirable situations. Consider the example depicted on figure 8.

figure 8. *an example of an undesirable generalisation*

The source map consists of five regions, representing a park and a lake that crosses it. It may well be possible that a simplification algorithm would accept the generalised representation, even if the lake is now split into heterogeneous, not even connected components. This happens because overlapping real-world entities (in this case the park and the lake) are split into separate components in order to achieve a planar representation in a GIS layer. Thus, each geometric object may be generalised independently, possibly resulting in a non-uniform representation for a single real world entity.

Likewise, a road may be found to be a geometrically perfect simplification of a underground water pipe that follows exactly the road's route, because their lineal representations happened to be very close in shape and location.

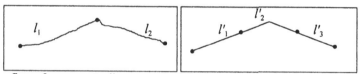

figure 9. *a semantically correct simplification missed by the algorithm*

As a third example, consider the simple case where a lineal feature (e.g. a river) is depicted on a detailed and a generalised layer, but at the implementation level divided into a different number of line objects respectively (see fig. 9).

According to the algorithm proposed in Section 3.1, none of the l_i falls in the d-class of any of the l'_i, so the algorithm would reject this case.

All the above examples indicate that a purely geometric simplification checking that ignores semantic information that a map layer's geometric objects convey, is bound to yield meaningless or undesirable results. We suggest two strategies based on two orthogonal concepts, objects and layers (in the sense that spatial objects potentially include attributes from several layers while layers may include several types of objects), to alleviate these effects.

A general strategy is to decompose both the source and target maps in several layers of thematic information, check the corresponding thematic layers in pairs and then combine the results. Our purpose is to be able to compare representations of a geographic entity and not of a part of it, in other words to create layers in which real world entities are atomically portrayed. Recalling the park-lake example, we could split both the detailed and generalised layer in two other layers, depicting parks and lakes respectively, and then perform the simplification checking. By applying this rule we ensure that the simpfication algorithm would manipulate the lake or park regions atomically, thus improving the quality of its results.

Alternatively, manipulating objects instead of pure geometric figures (that is, taking descriptive attributes into account) allows for meaningful checks. For example, roads should not be compared with water pipes, even if their geometric representations were very similar. Also, the use of object ids can ensure atomic treatment of real world entities. For example, considering the third case described above, by associating with each l_i and l'_i the id of the object that these lines are part of, we can force the simplification algorithm to compare not just simple lines from the source and target layer, but these aggregates of lines that have a common object id, which in our case represent real world entities.

4 Conclusion

In this work we made the first step on a long open road. The incentive came from the recent research need to support multiple representations of geographic information in order to deal effectively with several demanding applications. In a GIS context, such representations are map layers related through map generalisation relationships. Since multiple representations is a form of redundancy, consistency (in this case generalisation consistency) must be assessed and maintained.

In this work we develop a framework for the assessment of geometric simplification (a component of the general map generalisation problem). We do so by giving formal but practical definitions for the "geometrically simpler" relationship, which we subsequently use as building blocks for algorithms that check the geometric similarity of map layers. Our framework can be combined with recent results on the topological similarity of map layers, thus contributing to the development of a generic strategy for the assessment of generalisation relationships.

We are currently considering the implementation of the techniques presented in this paper in the ARC/INFO GIS and we believe that continuation of the research on how semantic information can be exploited and incorporated in the generalisation problem, which was only alluded to here by means of examples, would be a very fruitful research direction.

Acknowledgements

The first author gratefully acknowledges a PhD studentship from Bodosakis Foundation.

References

[1] Beard, K., (1991), Theory of the Cartographic Line Revisited: Implications for Automated Generalisation, *Cartographica* 18(4), pp. 32-58.

[2] Bertolotto, M., De Floriani, L., Puppo, E., (1993), Multiresolution Topological Maps, (eds.) Molenaar, M., De Hoop, S., *Advanced Geographic Data Modelling - Spatial Data Modelling and Query Languages for 2D and 3D Applications,* Netherland Geodetic Commission, Publications on Geodesy - New Series, 40, pp. 179-190.

[3] Buttenfield, B., McMaster, R., (1991), *Map Generalisation: Making Rules for Knowledge Representation,* Longman Scientific and Technical, London.

[4] Delis, V., Hadzilacos, Th., (1996), On Map (Re)classification, *Seventh International Symposim on Spatial Data Handling,* Delft, Netherlands.

[5] Douglas, D., Peucker, T., (1973), Algorithms for the Reduction of the Number of Points Required to Represent a Digitized Line or its Caricature, *Canadian Cartographer* 10(2), pp. 112-122.

[6] Egenhofer, M. and Sharma, J. (1992) Topological Consistency, *Fifth International Symposium on Spatial Data Handling,* Charleston, S.C., pp. 335-343.

[7] Egenhofer, M., Clementini, E., Di Felice, P., (1994), Evaluating inconsistencies among multiple representations, *6th International Symposium on Spatial Data Handling,* Edinburgh, Scotland, pp. 901-920.

[8] Egenhofer, M. and Tryfona N. (1996a) A Computational Model to Determine Consistency Among Aggregates and Their Parts, Technical Report, *National Center for Geographic Information and Analysis,* Maine, Orono, U.S.A.

[9] Egenhofer, M. and Tryfona N. (1996b) Multiresolution Spatial Databases - Consistency Among Networks, Technical Report, *National Center for Geographic Information and Analysis,* Maine, Orono,U.S.A.

[10] Frank, A. and Timpf, S. (1994) Multiple Representations for Cartographic Objects in a Multiscale Tree - An Intelligent Graphical Zoom, *Computers and Graphics,* 18(6), pp. 348-376, Springer Verlag, New York, NY.

[11] Giunchiglia, F., Walsh, T., (1992), A Theory of Abstraction, *Artificial Intelligence,* vol. 56, no 2-3, pp. 323-390.

[12] Keller, S., (1994), On the Use of Case-based Reasoning in Generalisation, *6th International Symposium on Spatial Data Handling,* Edinburgh, Scotland.

[13] Lagrange, J., Ruas, A., (1994), Geographic Information Modelling: GIS and Generalisation, *6th International Symposium on Spatial Data Handling,* Edinburgh, Scotland, pp. 1099-1117.

[14] Muller, J. C., (1990), The Removal of Spatial Conflicts in Line Generalisation, *Cartography and Geographic Information Systems,* 17(2), pp. 141-149.

[15] Muller, J., Lagrange, J., Weibel, R., (1995), *GIS and Generalisation: Methodology and Practice,* Taylor & Francis, London.

[16] Nyerges, T., (1991), Representing Geographical Meaning, in (eds.) (Buttenfield and McMaster, 1991), pp. 59-85.

[17] Puppo, E., Dettori, G., (1995), Towards a Formal Model for Multiresolution Spatial Maps, *Proceedings of SSD'95,* Lecture Notes in CS, 951, pp. 152-169, Springer.

[18] Rigaux, P., and Scholl, M., (1995) Multi-scale Partitions: Application to Spatial and Statistical Databases, *Proceedings of SSD'95,* Lecture Notes in CS, 951, pp. 170-183, Springer.

[19] Robinson, G., Lee, F., (1994), An automatic generalisation system for large-scale topographic maps, in (ed.) Worboys, M. F., *Innovations in GIS - 1,* Taylor and Francis, London.

[20] Ruas, A., Plazanet, C.,   (1996), Strategies for Automated Generalisation, *Seventh International Symposim on Spatial Data Handling*, Delft, Netherlands.

[21] Timpf, S., Volta G., Pollock, D. and Egenhofer, M., (1992) A Conceptual Model of Wayfinding Using Multiple Levels of Abstraction, *Lecture Notes in Computer Science 639*, pp. 348-376, Springer Verlag, New York, NY.

[22] van Smaalen, J., (1996), A Hierarchical Rule Model for Geographic Information Abstraction, *Seventh International Symposim on Spatial Data Handling*, Delft, Netherlands.

[23] Werschlein, T., Weibel, R., (1994), Use of Neural Networks in Line Generalisation, *Proc. EGIS '94*, Paris, pp.76-85.

Spatial Access Methods

New Linear Node Splitting Algorithm for R-trees

C.H. Ang and T.C. Tan

Department of Information Systems & Computer Science
National University of Singapore
Republic of Singapore, 119260
E-mail: angch@iscs.nus.sg, tantc@iscs.nus.sg

Abstract. A new linear-time node splitting algorithm for R-trees is proposed. Compared with the node splitting algorithm that requires quadratic time and is used in most implementations of R-tree, it is more superior in terms of the time required to split a node, the distribution of data after splitting, as well as the area of overlapping. Most important of all, it has a better query performance. The claim is substantiated by an analysis of the algorithm and a set of empirical results.

1 Introduction

The volume of data to be handled by many information systems is ever increasing. In order to process the data efficiently, data collected are organized into huge disk files that can be accessed quickly with the use of appropriate index structures. The B-tree is commonly used when the data to be manipulated are points from one dimensional space while octree [5], multi-priority tree [8] and R-tree [3] can be used for higher dimensional space.

The R-tree, a data structure that is regarded as the multi-dimensional version of the B-tree, is known for its ease of implementation, its space efficiency, and its adaptability to higher dimensionality. It has a structure that is quite similar to that of the B-tree. It usually uses a separate file to organize all the indexes to the objects that are being stored. The objects can be retrieved via their identifiers (ids) in the index file. The indexes are being stored in disk blocks termed *nodes*. Information stored includes the object ids, the minimum bounding rectangles (*MBR*s) of the objects or groups of objects, and the pointers that put all nodes in an R-tree into place. There are many entries in an R-tree node. Each entry of a leaf node is of the form (R, P) where R is the MBR of the object with id P. For a non-leaf node, R is the MBR that encloses all the objects that can be reached by following the node pointer P.

To perform an insertion in an R-tree, a new object is inserted into the subtree whose MBR needs the least amount of enlargement. When a node N is already filled to its capacity M, an inclusion of a new object will cause an overflow and hence the node has to be split. A new node is then allocated and some of the data in the overflowed node are moved over to this new node. The MBR of N

is updated and a new index is inserted into the index node which contains N's index. Different ways of splitting a node may be used to achieve different goals, such as the time efficiency in the R-tree creation, the minimal overlapping of the MBRs, or the minimal total coverage of the MBRs.

Since the MBRs in an R-tree may overlap, a query rectangle that intersects with any of these MBRs requires the traversal of the related subtrees. As such, using R-tree to answer window queries may be slow. Various ways to improve R-tree's performance have been proposed; the R^+-tree [7, 6], the R^*-tree [1], and the Hilbert R-tree [4] are just some of them. These variants of the R-tree either avoid overlapping altogether by splitting the objects into many parts, or pack the objects in different ways. Where node splitting is concerned, the variants of R-tree are more or less using the same algorithm.

In this paper, we propose a new linear time node splitting algorithm that outperforms the quadratic algorithm. The paper is organized as follows. Section 2 reviews the available node splitting algorithms and Section 3 describes the new algorithm. We then apply our algorithm to some examples taken from other papers for a quick comparison. In addition, several experiments are conducted, the performance are assessed based on several criteria, and the results are reported and analyzed in Section 4. Section 5 is the conclusion.

2 Some Existing Algorithms

Each node in an R-tree has a capacity M and, except for the root, has to be filled with no less than m entries. The value of m is usually between $\frac{1}{3}$ to $\frac{1}{2}$ of M. In his seminal paper [3] on R-tree, Guttman proposed several ways to split a node: the exhaustive method, the quadratic algorithm, and the linear algorithm. All methods proposed attempt to minimize the total area covered by the MBRs of the 2 nodes resulted from splitting and they are briefly described below:

1. *Exhaustive method:*
   It tries all possible ways in splitting the node in order to find the optimal solution. Due to the exponential time required, this method is impractical and will be excluded from our discussion henceforth.

2. *Quadratic algorithm:*
   It consists of two phases. In phase 1, a pair of objects is selected to be the seeds with one in each of the two nodes (groups) resulted from the splitting. These are the two objects that would have produced the largest white space if they were to be placed in the same node.

   In phase 2, for each of the remaining objects, calculate the difference, upon the inclusion of that object into each of the two groups, between the increases in the areas of the MBRs of the two groups. The object that causes the largest difference will be chosen and assigned to the group with the smaller MBR area expansion. We say that the object is 'nearer' to this selected group.

Thus in each iteration of phase 2, only one object is being assigned to the group of objects that is 'nearer'. At the end of each iteration, if all the remaining unassigned objects can be included into the smaller node such that the node will then contain exactly m elements, then this is done so. This is to make sure that each node will contain the minimum number of entries. With such bulk assignment of objects, it is obvious that the goal of minimizing the total coverage may be compromised.

3. *Linear algorithm:*

It consists also of two phases. In phase 1, the two objects that are separated farthest apart along the direction of certain dimension are chosen as the seeds. The remaining objects will subsequently be assigned one by one to the group that is nearer, in the same sense as described above. This algorithm also maintains the minimum occupancy of each node.

According to the empirical results obtained by Guttman, it seems that the performance of the linear algorithm is not too far off from that of the quadratic algorithm in terms of the space requirement of the resulting R-trees as well as the query performance using them, whereas in [1] the quadratic algorithm is shown to perform much better than the linear one.

Let us look at some examples to see how the above algorithms work. In Figure 1, the overflowed node containing rectangles R1 to R4 is to be split into 2 nodes. Using the linear algorithm, R1 and R2 are chosen as seeds since their separation in the x direction is the largest. Since R3 is nearer to R2 and R4 is nearer to R1, the resulting two nodes contain {R1, R4} and {R2, R3}. The MBRs of these two nodes overlap. If the overflowed node is split into {R1, R3} and {R2, R4}, then there will be no overlap.

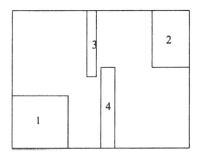

Fig. 1. Overflowed node with 4 rectangles.

In Figure 2, the node containing 5 rectangles is to be split. Using the quadratic algorithm, R1 and R2 are chosen as the seeds since they would have created the largest white space if they were to be placed into the same node. Since R5 is nearer to R1, R3 is nearer to {R1, R5}, and R4 is nearer to R2, the node is split into two nodes {R1, R3, R5} and {R2, R4} which again have their MBRs overlapping each other. If the overflowed node is split into {R1, R5} and {R2, R3, R4}, or {R3, R5} and {R1, R2, R4}, then there will be no overlap.

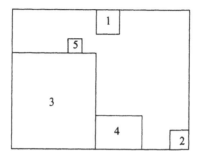

Fig. 2. Overflowed node with 5 rectangles.

It has been pointed out in [1] that inserting a new object into a node which needs the least expansion of its MBR is biased towards the node that contains more rectangles. It is generally true that the more rectangles a node contains, the larger its MBR is, and hence the least enlargement is needed to include a new rectangle. As a result, a large node tends to get larger. Therefore, during node splitting, the first few assignments made in redistributing the rectangles have great influence on the distribution of the remaining rectangles. The uneven distribution of the rectangles resulted from the application of the node splitting algorithm proposed by Guttman is apparent from the empirical results shown in section 4.

3   New Linear Algorithm

In order to optimize the search performance of the R-tree, we should aim to minimize the overlapping areas of the bounding rectangles involved. This will reduce the chance of the query rectangle intersecting with an overlapped area which requires several search paths to be followed.

In order to minimize the overlap between the two new nodes produced when a node splits, all rectangles are pushed as far apart as possible towards the boundary of the MBR of the overflowed node. For ease of discussion, let us first consider only rectangles in two dimensional space. Each rectangle S can be represented by a 4-tuple (xl, yl, xh, yh) where (xl, yl) and (xh, yh) are the coordinates of the bottom-left corner and the top-right corner of S respectively.

Let the MBR of the overflowed node N be $R_N=(L, B, R, T)$. We use four lists $LIST_L$, $LIST_B$, $LIST_R$, and $LIST_T$ to remember those rectangles that are nearer to a border than to its opposite. These lists are the main structures used in the New Linear Algorithm which is described as follows:

Algorithm NewLinear

$LIST_L \leftarrow LIST_R \leftarrow LIST_B \leftarrow LIST_T \leftarrow \emptyset$

For each rectangle $S = (xl, yl, xh, yh)$ in the overflowed node N with $R_N = (L, B, R, T)$,

    If $xl - L < R - xh$

        then $LIST_L \leftarrow LIST_L \cup S$

        else $LIST_R \leftarrow LIST_R \cup S$

    If $yl - B < T - yh$

        then $LIST_B \leftarrow LIST_B \cup S$

        else $LIST_T \leftarrow LIST_T \cup S$

If $\max(|LIST_L|, |LIST_R|) < \max(|LIST_B|, |LIST_T|)$

    then split the node along the x direction

else If $\max(|LIST_L|, |LIST_R|) > \max(|LIST_B|, |LIST_T|)$

    then split the node along the y direction

else /* tie breaker */

    If overlap$(LIST_L, LIST_R) <$ overlap$(LIST_B, LIST_T)$

        then split the node along the x direction

    else If overlap$(LIST_L, LIST_R) >$ overlap$(LIST_B, LIST_T)$

        then split the node along the y direction

    else split the node along the direction with smallest total coverage

Note that each rectangle S appears once in either $LIST_L$ or $LIST_R$ depending on whether it is horizontally closer to the left border or the right border of R_N, and once in either $LIST_B$ or $LIST_T$ depending on whether it is vertically closer to the bottom border or the top border of R_N. The split will be carried out horizontally or vertically depending on the evenness of the distribution. The direction that ensures a more even distribution will be used to split the node. In the event of a tie, the least overlap is favoured, and if again there is a tie, the one with the minimum coverage is chosen.

If there are n rectangles in the overflowed node, then it requires $O(n)$ to scan the list of rectangles once in order to create the four lists. Note that $|LIST_L| + |LIST_R| + |LIST_B| + |LIST_T| = 2n$. Once the direction of splitting the node is determined, the two lists related are the contents of the two nodes resulted from the split.

It is quite clear that this algorithm does not require any seed to grow the two new nodes. Moreover, it can be easily extended to handle the node splitting for data taken from k-dimensional space. What is needed is to have two lists to keep track of those objects that are closer to one of the two borders in each dimension. The given list of rectangles to be redistributed is still scanned once. The direction chosen to split the overflowed node is the one that ensures a more even data distribution. Since there are kn entries in the $2k$ lists, the time complexity of the algorithm is $O(kn)$, or simply $O(n)$ as k is usually a small value.

In the list of rectangles to be redistributed, if there are a few rectangles that are outliers while others form a cluster, then the redistribution will be lopsided. In this case, the outliers can be reinserted so that they may be absorbed by some of the surrounding nodes. If the outliers keep falling back to the same overflowed node, then they are to reside in a separate node which, of course, will have a very low utilization. This is really a rare case and we would like to report that such situation has not been encountered so far in our experiments.

4   Comparisons and Empirical Results

To have a quick comparison between the performances of various node splitting algorithms, let us visit the previous examples again. Referring to Figure 1, we have $LIST_L=\{$R1, R3$\}$ and $LIST_R=\{$R2, R4$\}$ in the x direction, and $LIST_B=\{$R1, R4$\}$ and $LIST_T=\{$R2, R3$\}$ in the y direction. Since both lists contain the same number of rectangles and only the split in the x direction does not produce two nodes that overlap, so $LIST_L$ and $LIST_R$ are used and the overflowed node is split in the x direction.

As for Figure 2, we have $\{$R1, R5$\}$ and $\{$R2, R3, R4$\}$ in the y direction, and $\{$R3, R5$\}$ and $\{$R1, R2, R4$\}$ in the x direction. There is no overlap in either way of splitting, so either one can be used. Since the split in the y direction will produce a smaller total coverage, it is preferred and chosen.

In addition, two examples have been used in [1] to show the deficiency of the quadratic algorithm and the variant proposed by Greene in [2]. They are reproduced in Figures 3 and 4 for comparison. In both examples, the new linear algorithm produces better or at least equally good results.

In Figure 3b, the quadratic node splitting algorithm is used with the restriction that at least 30% of the space in a node must be utilized, ie, $m = 0.3M$. Although there is no overlapping of the resulting nodes, the distribution of rectangles is not even. The split as carried out by R^*-tree produced the same result as shown in Figure 3e. When m is increased to $0.4M$, although the distribution is now more even as shown in Figure 3c, the nodes overlap significantly. On the other hand, Greene's method is able to split the node more evenly with smaller overlap. Yet none of these methods can produce a better result than that of the new linear algorithm which produces two nodes containing the same number of rectangles with no overlapping.

In Figure 4b, Greene's split produces two nodes which overlap and are unbalanced. The new linear algorithm is able to produce the same result as that of R^*-tree which uses a more sophisticated and time consuming node splitting method.

Besides having a lower time complexity, the new algorithm produces a more even distribution of data when a node is split and the overlap between the MBRs bounding the two nodes is also smaller. To demonstrate its strength, a series of experiments has been conducted to compare the performance of the

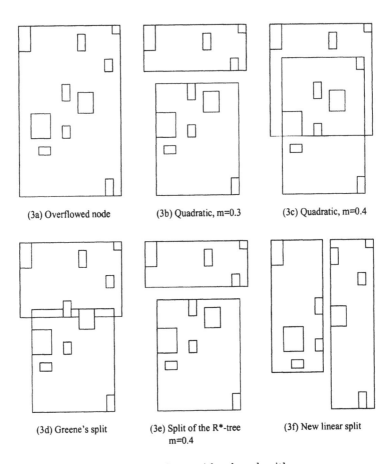

(3a) Overflowed node   (3b) Quadratic, m=0.3   (3c) Quadratic, m=0.4

(3d) Greene's split   (3e) Split of the R*-tree   (3f) New linear split
m=0.4

Fig. 3. Comparisons with other algorithms.

linear, the new linear, and the quadratic algorithms. We do not include those splitting algorithms used in the variants of R-tree in our comparisons because they are basically the quadratic algorithm employing different criteria in phase 2 to distribute the rectangles.

We note that the time needed to create an R-tree is mainly determined by the I/O activity involved and, to a certain extent, by the efficiency of the node splitting algorithm used to grow the tree. A node splitting algorithm that fails to ensure an even data distribution will usually produce a tree with more nodes that are either quite full or less than half-filled. This may cause more node splittings subsequently and thus more I/O's and time will be incurred.

The programs used in the experiments are all written in C and are run on Digital AlphaServer 8400 5/350 with 4 processors. The first experiment is to measure the time taken in creating R-trees when different splitting algorithms are used. Five R-trees are built by inserting n rectangles that are randomly

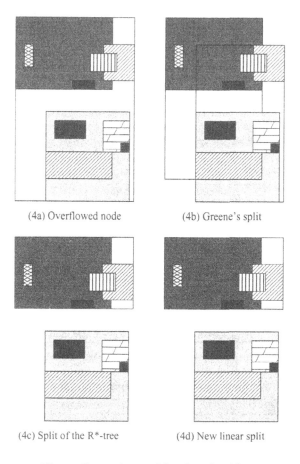

(4a) Overflowed node

(4b) Greene's split

(4c) Split of the R*-tree

(4d) New linear split

Fig. 4. Comparisons with other algorithms.

generated, with n ranging from 10,000 to 100,000 in steps of 10,000, and the average time is recorded. Table 1 shows the results of this experiment.

Note that the splitting algorithm is called only when a node overflows, an ocassion that arises only once after many insertions. Yet as can be seen from Table 1, the new linear algorithm manages to reduce the R-tree creation time by more than 40%. This is really a very significant improvement. For small R-trees, the creation times are reduced by more than 50%. As the size of the R-tree increases, its depth also increases and more I/O's are required to insert a rectangle. As a result, the saving in the time used in splitting the overflowed nodes becomes less dominant.

Since node splitting will invariably produce two nodes of different sizes most of the time, we would like to define some metrics to measure the quality of the split. Let us call the node with fewer rectangles $L1$ and the one with more rectangles $L2$. We are interested in the percentage of the space that is covered by $L1$ and $L2$ within the MBR of the overflowed node, the percentage of the overlap between them, and the number of rectangles in each of them. The statistics are shown in Table 2.

Number of Rectangles	Quadratic	New Linear
10,000	8.88	4.98
20,000	16.25	7.60
30,000	23.78	11.38
40,000	31.82	16.17
50,000	42.20	22.02
60,000	51.45	31.13
70,000	61.95	33.97
80,000	70.16	39.53
90,000	77.61	40.18
100,000	90.11	49.43

Table 1. R-tree creation time (sec.)

	Quadratic,m=0.33M	Quadratic,m=0.5M	New Linear	Linear
L1's Coverage (%)	49.20	50.05	48.07	61.53
L2's Coverage (%)	55.74	58.84	49.76	64.03
Overlap (%)	7.37	10.91	0.36	27.73
Number of rectangles in L1	47.18	50.00	47.18	43.56
Number of rectangles in L2	52.82	50.00	52.82	56.44

Table 2. Rectangle distribution and overlap

From Table 2, it is noted that the original linear algorithm, even though it is simple to implement and fast to execute, does not produce satisfactory results. The nodes cover too big an area, overlap too much, and the rectangles are not evenly distributed.

Compared with quadratic algorithm with $m = 0.33M$ and $m = 0.5M$, the new linear algorithm always produce nodes that cover area of almost equal size, with minimum total coverage, minimum overlap and the rectangles are evenly distributed. Note that even though the quadratic algorithm is able to produce nodes of equal size when $m = 0.5M$, this is achieved at the expense of the extent of overlap and the total coverage.

The last experiment is designed to assess the quality of the resulting R-trees in terms of query performance. An R-tree could be created in shorter time with the use of the new linear node splitting algorithm, but what is more important is whether the R-tree created is efficient for query processing.

There are point query, intersection query, containment query, etc. A typical query may ask for all rectangles that intersect with a given query rectangle to be retrieved. Since a point query is easy to satisfy and containment query is not used frequently, we therefore focus on the performance of intersection query.

In the last experiment, 100 rectangles are generated randomly and used to perform intersection queries on the two R-trees which contain 100,000 rectangles and are created using the quadratic algorithm and the new linear algorithm respectively. The number of node accesses is then recorded. The same experiment is repeated 10 times and the data collected can be found in Table 3. It is noted that on the average, the number of node accesses has been reduced by 16%. This is again a significant improvement.

Test	Quadratic	New Linear
1	833	743
2	949	731
3	960	768
4	1028	820
5	922	868
6	912	752
7	996	821
8	892	834
9	964	796
10	965	804
11	951	804

Table 3. Number of node accesses for 100 intersection queries

5 Conclusion

In this paper, we have described a simple linear time algorithm to split an overflowed node in an R-tree. It is shown in many aspects to be more efficient than the quadratic algorithm that is currently in use. With the use of the new node splitting algorithm, the overlap between the nodes resulted from the split is reduced significantly with no compromise in the evenness of data distribution.

Two experiments are carried out to find the running time required to create an R-tree using various node splitting algorithms as well as to gauge the quality of the splits. Since variants of R-tree are using basically the same quadratic node splitting algorithm proposed by Guttman with some modifications, the metric that is used to measure the quality of the splitting will not be much different from that of the R-tree, and their time complexity will not change asymptotically. As such, it is not necessary to involve them in the comparisons as far as the efficiency of the node splitting algorithm and the quality of the split are concerned. The third experiment shows that the R-tree created using the new linear node splitting algorithm is more superior in query processing as well.

Although the algorithm proposed is optimal, there are rooms for some minor improvements which are to be worked on in future. Here are some examples:

1. To break the tie between two lists of equal number of rectangles, in addition to opting for a node with smaller overlap, we may use other criteria such as having smaller total coverage or smaller extent in certain dimension.

2. We may choose the direction that minimizes the overlap as the splitting direction instead of one that minimizes the difference in the list sizes. We have tried this out and observed a further reduction in the overlap with no noticeable effect on the evenness of data distribution.

3. We may tinker with the assignments of the rectangles that cause the overlap so as to see whether the overlap can be reduced. These rectangles, instead of assigning them to the nearest border, may be assigned to the opposite if the overlap can be reduced.

Acknowledgement

We would like to thank Lee Wee Fong, Lee Sue Chin, and Lim Joo Pheng for their help in the implementation and Ooi Beng Chin for his valuable comments.

References

1. Norbert Beckmann, Hans-Peter Kriegel, Ralf Schneider, and Bernhard Seeger: The R^*-tree: An efficient and robust access method for points and rectangles. Proc. of the SIGMOD Conference, Atlantic City, NJ. (May 1990) 322-331
2. D. Greene: An implementation and performance analysis of spatial data access methods. Proc. of the 5th International Conference on Data Engineering. (1989) 606-615
3. Antonin Guttman: R-tree: A Dynamic Index Structure for Spatial Searching. Proc. of the SIGMOD Conference, Boston. (June 1984) 47-57
4. Ibrahim Kamel and Christos Faloutsos: Hilbert R-tree: An improved R-tree Using Fractals. Proc. of the 20th VLDB Conference, Santiago, Chile. (May 1994) 500-508
5. Hanan Samet: The Design and Analysis of Spatial Data Structures. Addison-Wesley (1989)
6. T. Sellis, N. Roussopoulos, and C. Faloutsos: The R^+-tree: A dynamic index for multi-dimensional objects. Proc. of 13th VLDB Conference, England, (1987) 507-518. Also available as Computer Science TR 1795, University of Maryland, College Park, MD (1987)
7. M. Stonebraker, T. Sellis, and E. Hanson: An Analysis of rule indexing implementations in data base systems. Proc. of the First International Conference on Expert Database Systems, Charleston, SC. (April 1986) 353-364
8. Ching-Der Tung, Wen-Chi Hou, and Jiang-Hsing Chu: Multi-Priority Tree: An Index Structure for Spatial Data. Proc. of 1994 International Computer Symposium. (1994) 1285-1290

The S-Tree: An Efficient Index for Multidimensional Objects

Charu Aggarwal[1], Joel Wolf[1], Philip Yu[1] and Marina Epelman[2]

[1] IBM T.J. Watson Research Center, Yorktown Heights, New York
[2] Massachusetts Institute of Technology, Cambridge, Massachusetts

Abstract. In this paper we introduce a new multidimensional index called the S-tree. Such indexes are appropriate for a large variety of pictorial databases such as cartography, satellite and medical images. The S-tree discussed in this paper is similar in flavor to the standard R-tree, but accepts mild imbalance in the resulting tree in return for significantly reduced area, overlap and perimeter in the resulting minimum bounding rectangles. In fact, the S-tree is defined in terms of a parameter which governs the degree to which this trade-off is allowed. We develop an efficient packing algorithm based on this parameter. We then analyze the S-tree analytically, giving theoretical bounds on the degree of imbalance of the tree. We also analyze the S-tree experimentally. While the S-tree is extremely effective for static databases, we outline the extension to dynamic databases as well.

1 Introduction

In this paper we consider spatial structures which are useful for representing multidimensional objects. Typical 2-dimensional objects include cartographic maps of countries. Typical 3-dimensional objects include medical images, such as MRI brain scans. These applications require the storage of data objects which may either be point sets or have non-zero measure. Such multidimensional or spatial data objects are not generally well represented by traditional structures which use a 1-dimensional ordering of key values, such as B-trees or ISAM indexes.

A number of interesting methods have been proposed for the efficient representation of spatial objects. A good survey of these data structures may be found in [18]. For handling point set data, k-d-B-trees [16] may be used, but these are not useful for handling objects with non-zero measure. Other commonly used methods include k-d-trees [3], hB-trees [14] and $cell$-trees [7]. Another interesting class of methods involves the transformation of rectangles into points of higher dimensionality [9]. Other methods include $linear$ $quad$-$trees$ [1, 6], z-$ordering$ [15] and $space$ $filling$ $curves$ [5, 10].

R-$trees$ are an excellent method for dealing with data containing both point sets and objects of non-zero measure. R-trees were first proposed by Guttman [8] as a multidimensional extension of B-trees. The resulting data structure has the form of a height balanced tree in which each leaf node contains between m and M index records, and each internal node except for the root contains between m

and M children. Each leaf node of this R-tree contains record entries which are of the form $(I, object\text{-}identifier)$. Here I is the *minimum bounding rectangle* of the data object (that is, the smallest box parallel to the axes in K-dimensional space which contains the object). The structure of an internal node entry is slightly different in that the entries are of the form $(I, pointer)$. In this case *pointer* contains the address of a lower node in the R-tree, and I is the minimum bounding rectangle of the set of data objects in the leaf node entries which are descendants of that node. This data structure tends to provide excellent space utilization and good computational performance. (In the above we make the modest assumption that the size in bytes of an object-identifier is close to that of a pointer. Otherwise, the leaf and internal nodes might conceivably have different fanouts. We will continue to make this assumption throughout the paper. The value of M is a design choice, and may not correspond to precisely 100% page utilization in order to accommodate future insertions.)

A number of variants of R-trees have been proposed which can provide better performance depending upon the nature of the data. Some of these variants are R^*-trees [2], R^+-trees [19] and *Hilbert-packed R-trees* [12]. The last of these methods concentrates on static representations for which the initial packing is the crucial factor affecting the performance of the data structure. A space-filling Hilbert curve is employed to provide a linear order on K-dimensional space while maintaining reasonable locality. Because of the way in which they are packed, Hilbert-packed R-trees are initially, in fact, fully packed with branch factors M, except for possibly one node at each level.

One important feature of real data is that it is often skewed. Consequently, there are typically regions of very high density as well as large regions of very low density. Many of the above methods fail to take advantage of this essential feature. Some interesting work in this direction has been done in [4, 11].

In this paper we present a new kind of data structure in order to represent spatial data. These trees are not necessarily height balanced, but as a result they are better able to deal with data skew. Because of this we shall call them S-trees (for *Skew tolerant trees*). These trees are not direct extensions of B-trees, although we shall see that they do share some similarity with them. The key property of S-trees is that they trade off mild imbalance in the resulting tree in return for significantly reduced area, overlap and perimeter in the resulting minimum bounding rectangles. In fact, the S-tree is defined in terms of a parameter which governs the degree to which this trade-off is allowed.

This paper is organized as follows. Section 2 gives an overview of the basic S-tree and describes its key properties. Section 3 gives basic algorithms for the construction of S-trees. A theoretical S-tree analysis is presented in Section 4. In Section 5 we outline several important S-tree algorithmic variants. While S-trees were initially geared towards static databases, we outline some operations for accommodating dynamic databases in Section 6. Section 7 gives results of our experiments evaluating the performance of S-trees. A conclusion is given in Section 8.

2 Overview

In order to motivate the discussion of S-trees, we first discuss two types of queries which will typically be required in spatial database applications.

(1) **Region Query:** Given an *aligned rectangle* (that is, a box in K-dimensional space parallel to the axes), find all data objects which intersect it.

(2) **Point Query:** This is a special case of the region query in which the aligned rectangle is a single point.

The goal is to minimize the number of index pages that a given query must access. Thus, to achieve the best possible performance in an index structure for multidimensional objects the following three design objectives must be considered.

(1) **Area:** The areas of the minimum bounding rectangles corresponding to the different nodes must be kept small, because the probability that a particular node may be accessed increases with increasing area.

(2) **Overlap:** An equally critical factor may be the overlap between different minimum bounding rectangles. This is because the overlap usually occurs in densely populated regions of the database, which also happen to be frequently accessed. So this overlap must be kept small as well.

(3) **Perimeter:** For region queries, it is desirable that the minimum bounding rectangles be as square as possible. Equivalently, it is important that the perimeter of the minimum bounding rectangles also be as small as possible.

Each of these design objectives have been discussed before, in the context of R-trees and R^*-trees [2, 12]. They play important roles in the design of S-trees as well.

The S-tree we construct will not necessarily be height balanced. In fact, associated with each S-tree is a parameter p which we refer to as the *skew factor* of the tree. This factor p lies in the range $(0, 1/2]$, and will be a good indicator of how well balanced the tree is required to be. Sacrificing the balanced property means that on average the path from the root to a leaf node will increase. However, as we shall see, the advantages of being able to pack the tree properly may outweigh the disadvantages: There is a trade-off between the degree of balance in the tree and the flexibility we have in meeting the three design objectives above. A low value of p results in potentially greater imbalance but greater design flexibility, and hence lower areas, overlaps and perimeters. A skew factor near $1/2$ implies that the tree will be quite well balanced, but may not do as well with respect to some or all of these criteria.

We need a little additional notation. A node is said to be *penultimate* if all of its children are leaf nodes. The *branch factor* of a node is defined as the number of children of that node. In other words, this is the fanout. The *leaf number* N_A of a node A is defined as the total number of data objects in the leaf descendents of that node. Denote the area of an aligned rectangle I by $\mathcal{A}(I)$.

The properties of an S-tree with a skew factor of p can now be listed as follows:

(1) The structure of the leaf and internal nodes in S-trees are identical to the internal record structure of R-trees. In other words, each record in a leaf node contains entries of the form $(I, object\text{-}identifier)$, where I is the minimum bounding rectangle of the K-dimensional data object. Similarly, each internal node contains entries of the form $(I, pointer)$, where $pointer$ contains the address of a lower node in the S-tree, and I is the minimum bounding rectangle of the set of data objects in the leaf descendants of that node.

(2) Each node has a branch factor of at most M. As with R-trees, the value of M is a design choice, and may not correspond to precisely 100% page utilization. (Choosing a lower utilization affords flexibility in handling future insertions, for example.) For the basic S-tree algorithm, each node which is not a leaf node or a penultimate node has a branch factor of exactly M. And variants of the basic algorithm do better on these nodes as well. Thus an S-tree differs significantly from a standard R-tree. (We point out again, however, that a Hilbert-packed R-tree is initially nearly fully packed.)

(3) Any pair of nodes having the same parent are referred to as sibling nodes. For any pair of sibling nodes B_1 and B_2, we must have:

$$p \leq \frac{N_{B_1}}{N_{B_2}} \leq \frac{1}{p} \tag{1}$$

We refer to this condition as the *skewness guarantee*, since it provides a worst-case bound on the skewness of descendents of that node. If p is close to $1/2$, this guarantee is fairly tight. On the other hand, if p is small this bound is loose. Note that it is not possible to give such a guarantee in the case of an R-tree even if the tree is perfectly balanced. This is because each node may have a variable number of children. (The Hilbert-packed R-tree, on the other hand, is initially nearly perfectly packed: Except for at most one node in each level, the leaf numbers of any two siblings are identical.)

The above properties give good insight into the performance of an S-tree. The fact that all nodes except for the leaf nodes and the penultimate nodes have the highest possible branch factor helps significantly in reducing the length of the path from the root to the leaf node, and consequently also the number of accesses which are required in order to reach a leaf node.

The S-tree is designed to meet our design objectives as closely as possible within the framework of trees satisfying the above conditions. Of course, these objectives may conflict with each other, and we need a method for determining their relative priorities. Specifically, we will try to minimize the total area of the minimum bounding rectangles while constraining the allowable overlap by a parameter o. Formally, the *overlap factor* is defined to be the area covered by the intersection of the bounding rectangles divided by the sum of the corresponding areas. This overlap factor o lies in the range $[0, 1]$. The algorithm will naturally tend to create relatively square, low perimeter minimum bounding rectangles. We also encourage this by adjudicating ties appropriately. In the next section we will discuss the initial construction of a basic S-tree with a skew factor p and an overlap factor o.

3   Construction of the Basic S-tree

Fix a skew factor p in $(0, 1/2]$ and an overlap factor o in $[0, 1]$. (We will discuss appropriate choices of robust and effective factors in a later section.) In this section we discuss how the basic S-tree is constructed for an initial set of multidimensional data objects. The process consists of two stages:

(1) **Binarization:** In this stage we construct a binary tree such that the entries in the leaf nodes correspond to the spatial data objects. Thus all non-leaf nodes have branch factors of 2. Let A be an internal node of the binary tree with children B_1 and B_2. We ensure that the leaf numbers N_{B_1} and N_{B_2} of each of these children are each at least $p \cdot N_A$. Among all partitions examined which satisfy this condition we will choose the one which minimizes the sum $A(I_{B_1}) + A(I_{B_2})$ of the areas of the minimum bounding rectangles subject to the constraint that $A(I_{B_1} \cap I_{B_2}) \leq o \cdot (A(I_{B_1}) + A(I_{B_2}))$. (If no partition examined satisfies this constraint we will choose the one which minimizes the ratio $A(I_{B_1} \cap I_{B_2})/(A(I_{B_1}) + A(I_{B_2}))$ instead.)

(2) **Compression:** In this stage we convert the binary tree into a tree for which all but the leaf and penultimate nodes have branch factors of M. We achieve this by iteratively compressing pairs of non-leaf parent and child nodes in the tree into single nodes. The leaf nodes are not affected by this operation.

We will show that the skewness guarantee holds for the tree constructed during the binarization stage, and that the guarantee is maintained during each iteration within the compression stage. We now describe each of these stages in detail.

3.1   Binarization

The binary tree is constructed in a top-down fashion. (By comparison, the packing method for Hilbert-packed R-trees can be classified as bottom-up instead.) Thus, we start with the full set of data objects and partition them into two sets which satisfy the properties described above. Then we partition each of these two sets of data objects in turn using the same method, and so on recursively until we arrive at the individual leaf nodes.

It is therefore sufficient to describe the binarization process for an arbitrary node A. If $N_A \leq M$ we make A a leaf node and stop. Otherwise, we consider the minimum bounding rectangle I_A, and choose a dimension k for which this rectangle is longest. (Choosing this dimension will tend to cause relatively square minimum bounding rectangles.) We then represent each of the N_A data objects by their geometrical center, and order them by increasing values of their kth coordinate. Candidate partitions (B_1, B_2) considered will have the property that B_1 contains the first q data objects according to this order, B_2 contains the remaining $N_A - q$ data objects, and $p \cdot N_A \leq q \leq (1 - p) \cdot N_A$. *Sweeping* over these $(1 - 2p) \cdot N_A$ choices, we choose the one for which

$$A(I_{B_1}) + A(I_{B_2}) \tag{2}$$

is minimized subject to

$$\mathcal{A}(I_{B_1} \cap I_{B_2}) \leq o \cdot (\mathcal{A}(I_{B_1}) + \mathcal{A}(I_{B_2})) \tag{3}$$

if such a partition exists, and the one which minimizes

$$\frac{\mathcal{A}(I_{B_1} \cap I_{B_2})}{\mathcal{A}(I_{B_1}) + \mathcal{A}(I_{B_2})} \tag{4}$$

otherwise. Ties can be adjudicated in favor of rectangle pairs whose total perimeter is minimized. Clearly the rectangles I_{B_1} and I_{B_2} can be computed incrementally as the sweep progresses. Also, we certainly have $N_{B_1} \geq p \cdot N_A$ and $N_{B_2} \geq p \cdot N_A$, as desired. Figure 1 shows a sample sweep operation. (The bounding rectangle is longest in the x-dimension, and we are assuming that $p = 1/4$. We have $N_A = 20$. The bounds then imply that $N_{B_1} \geq 5$ and $N_{B_2} \geq 5$, or equivalently $N_{B_1} \leq 15$, resulting in the potential cuts shown.)

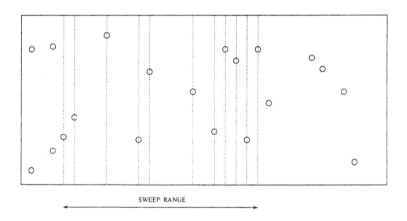

SWEEP RANGE

Fig. 1. Sweep Operation for Creation of Binary Tree

Notice that if $o = 1$ the constraint vanishes, and we are picking the partition with minimal total area. On the other hand, if $o = 0$ the constraint is only satisfied by partitions with no overlap. So we are picking either the partition having the smallest total area without overlap, or the one with the smallest relative overlap. Pseudocode for the binarization algorithm is given below.

Algorithm Binarization (Input: Parameters p and o, data objects)
Create an ordered list L consisting of a single node
        corresponding to all data objects

while L is not empty **do**

    Pick the first node A on list L and delete A from L

    If $N_A > M$ **then do**

        Find longest dimension k of I_A

        Solve sweep algorithm optimization using p and o to determine

                child nodes B_1 and B_2

        Add nodes B_1 and B_2 to back of list L

   end

end

3.2 Compression

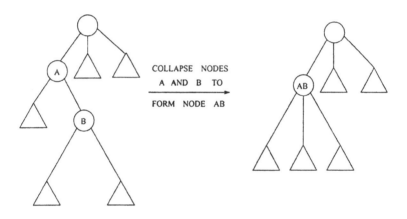

Fig. 2. Collapsing Two Nodes into One

Before we describe the algorithm itself, we discuss what it means to collapse two nodes into one. First of all, two nodes A and B can be collapsed into one node when they satisfy a parent-child relationship. Suppose, for example, that node A is the parent of node B. The new node AB has the same ancestor as node A. The children of node AB are the union of the children of node A (excluding B) and the children of node B. This situation is illustrated in Figure 2. Note that the child node B has a branch factor of 2, and therefore the branch factor of the new node AB is 1 more than the branch factor of the original node A.

In fact we shall always collapse a parent node with a child node having a branch factor of 2, so that the branch factors of the resulting new node will always increase by 1 at a time. In this way we can be certain that collapsing will never result in a node with a branch factor exceeding M. The idea is to pick a node, collapse it with one of its children to form a new node, and keep iterating the process until either the branch factor of that node is exactly M, or all children of that node are leaf nodes. Of course, given the parent node we need to choose the child to collapse it with. The rule is simple: We always choose the child with the highest leaf number. The motivation behind this rule is that it restricts the amount of skewness in the tree, and helps us in establishing the *skewness guarantee* in the resulting S-tree. Notice also that this rule automatically guarantees that the chosen child will *not* be a leaf node, because all children which are leaf nodes will have leaf numbers less than or equal to M, while all children which are not will have leaf numbers greater than M. It remains to choose the order in which we choose the parent nodes, and this is top-down. To accomplish this, we start with an ordered list of the non-leaf nodes obtained via breadth-first search starting at the root, removing collapsed parent nodes but replacing them with the newer nodes unless their branch factors are full. We continue the process until this list of potential parents is empty. At its conclusion, only the penultimate and leaf nodes can have branch factors of less than M. Pseudocode for the compression algorithm is given below.

Algorithm Compression (Input: Binary tree T)
Create an ordered list L of the non-leaf nodes of T using breadth-first search
while L is not empty **do**
    Pick the first node A on list L and delete A from L
    If A has non-leaf children **then do**
        Let B be a non-leaf child of A with the highest leaf number
        Collapse A and B to form node AB
        If branch factor of $AB < M$ **then** add node AB to front of list L
    end
end

We note in passing that a reasonable implementation of the binarization algorithm would place the non-leaf nodes in breadth-first order to begin with, so that no reordering of the nodes is necessary.

The compression procedure actually helps rebalance the tree created in the binarization stage. The steps of the compression algorithm shown in Figure 3 illustrate this point. Assume $M = 4$. The first tree in the figure is the result of the binarization stage. Note that nodes A through F are non-leaf nodes, ordered by breadth-first search, while the remaining 7 nodes are leaf nodes. In the next two steps we collapse node A into node AB, and then into node ABC, which has a full branch factor of 4. In the final step we collapse node D into node DF. Although this penultimate node has a branch factor of 3, there are no non-leaf children. Similarly node E has a branch factor of 2, but has no non-leaf children. So the list becomes empty and the compression process ends. We see from the figure that although the initial binary tree is skewed, the final compressed S-tree is relatively well balanced.

Figures 4 and 5 illustrate the difference between R-trees and S-trees. We have actually adapted both the example and the packing shown in Figure 4 from

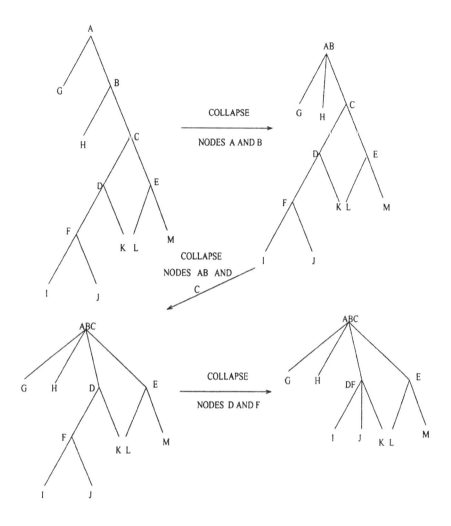

Fig. 3. Compression of the Binary Tree

Guttman's original paper [8]. The R-tree packing shown is obtained if $M = 3$. On the other hand, the S-tree packing shown in Figure 5 results from a skew factor of $p = 1/4$, and an overlap factor of $o = 1$, with $M = 3$ as before. Notice that the S-tree algorithm takes advantage of the skewness of the underlying data objects, and actually results in a packing with less area, perimeter *and* overlap. Of course, given the skewness factor chosen, the improvement in overlap is somewhat (but not entirely) serendipitous.

It is of some interest to analyze the time complexity of the algorithm for

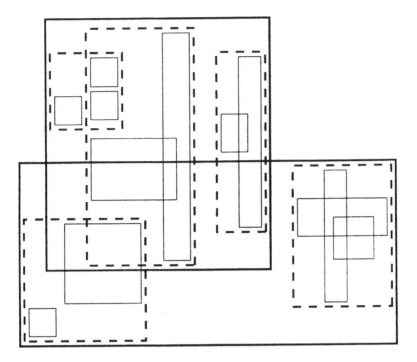

Fig. 4. An R-tree Packing

constructing the S-tree. It turns out that the bookkeeping associated with the construction of the minimum bounding rectangles predominates. In other words, ordering the lower-left corner for N data objects in each of K dimensions takes $O(K \cdot N \cdot \log N)$ steps, as does the ordering of the upper-right corner and the ordering of the geometrical centers.

Theorem 1. *The S-tree construction algorithm requires $O(K \cdot N \cdot \log N)$ steps, where N is the number of K-dimensional spatial data objects to be packed.*

Proof. We analyze the computational complexity of the binarization and compression stages.

As a worst case assumption for the binarization stage, assume that whenever a node is partitioned it is split in the most skewed way possible by the sweep algorithm, giving the left child with a fraction p of the data objects, and the right child a fraction $1 - p$ of the data objects. (It is easy to see that no binarization stage can take longer.) Note that the process of splitting a node requires us to first choose the dimension along which to split and then use the linear sweep algorithm. Consequently, the total time required for this is $O(K + N)$. Thus if

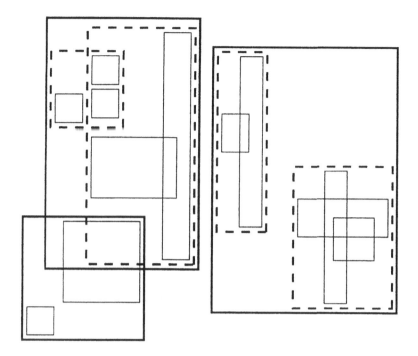

Fig. 5. An S-Tree Packing

$B(N)$ denotes the time required to perform binarization on N objects, we get

$$B(N) = p \cdot B(p \cdot N) + (1 - p) \cdot B((1 - p) \cdot N) + O(K + N) \qquad (5)$$

Solving this recurrence, we get

$$B(N) = (K + N) \cdot \log_{1/p} N \qquad (6)$$

which is dominated by our initial sorting complexity. The same is true for the compression step, which requires at most N compressions each of which takes at most $O(\log N)$ time. This completes the proof.

4 Theoretical S-Tree Analysis

In this section, we shall analyze the S-tree representation in terms of its skewness and page utilization. We shall first proceed to show that the skewness guarantee is indeed maintained in the S-tree construction. Note that it is not possible to give such a skewness guarantee in the case of an R-tree, because each child may have a variable number of children. In the case of S-trees, all internal nodes except for penultimate nodes have the desired page utilization. Thus it may turn out that even though the S-tree is not perfectly height balanced, the average length of the path from the root node to the leaf node may be less.

Theorem 2. (Skewness Guarantee:) *An S-tree constructed with skew factor p must satisfy the skewness guarantee*

$$p \leq \frac{N_{B_1}}{N_{B_2}} \leq \frac{1}{p} \tag{7}$$

for any pair of siblings B_1 and B_2.

Proof. First we show that the tree constructed in the binarization stage satisfies the skewness guarantee. Then we shall show that a collapse operation in the compression stage does not disturb the skewness guarantee. Since the S-tree is constructed by using a series of collapse operation steps on the initial binary tree, an inductive argument will establish that the final S-tree must also satisfy the skewness guarantee. The key in this proof is that when we collapse a node with one of its children we choose the child with the highest leaf number.

Initialization Step: Let A be a node in the binary tree with children B_1 and B_2. We have

$$p \leq \frac{N_{B_2}}{N_A} \tag{8}$$

We also have

$$p \leq \frac{N_{B_1}}{N_A} = \frac{N_A - N_{B_2}}{N_A} = 1 - \frac{N_{B_2}}{N_A} \tag{9}$$

These equations may be summarized as

$$p \leq \frac{N_{B_2}}{N_A} \leq 1 - p \tag{10}$$

Taking reciprocals and simplifying using the fact that $N_A = N_{B_1} + N_{B_2}$, we get

$$\frac{p}{1-p} \leq \frac{N_{B_1}}{N_{B_2}} \leq \frac{1-p}{p} \tag{11}$$

Setting $r = p/(1 - p)$, we note that $p < r$. Thus

$$p < r \leq \frac{N_{B_1}}{N_{B_2}} \leq \frac{1}{r} < \frac{1}{p} \tag{12}$$

as desired.

Inductive Step: Let A be a node which is collapsed with one of its children B. Denote the other children of A by $C_1, ..., C_k$. Since the collapse operation occurs in a top-down fashion, the node B has never been collapsed, and must itself have two children. Call these children D_1 and D_2. The children of the collapsed node AB are $C_1, ..., C_k$ plus D_1 and D_2. Now it follows from the induction hypothesis that

$$p \leq \frac{N_{C_i}}{N_{C_j}} \leq \frac{1}{p} \tag{13}$$

for any i and j between 1 and k. It also follows from the inductive hypothesis that

$$p \leq \frac{N_{D_1}}{N_{D_2}} \leq \frac{1}{p} \tag{14}$$

Finally, it follows from the inductive hypothesis that

$$p \leq \frac{N_B}{N_{C_i}} \leq \frac{1}{p} \tag{15}$$

for any i between 1 and k. It remains to show that the skewness guarantee holds for an arbitrary node C_i and an arbitrary node D_j, where i is between 1 and k, and j is either 1 or 2. But we have

$$p \leq \frac{N_B}{N_{C_i}} \cdot p \leq \frac{N_{D_j}}{N_{C_i}} \leq \frac{N_B}{N_{C_i}} \leq \frac{1}{p} \tag{16}$$

The first inequality holds because node B was chosen for collapsing over alternative node C_i. The second inequality is holds because of the binarization stage condition on p. The third inequality is trivial. The fourth inequality is by induction, as noted above. This completes the proof.

In the binarization stage of building an S-tree, we achieve a better packing by allowing some amount of imbalance. It is therefore instructive to analyze the average increase in path length that this imbalance might cause. (We have noted that the compression stage rebalances the tree to some extent, but it does not do so completely.) We shall analyze the average path length increase in both a worst case and average case binary tree.

First we deal with the worst case binary tree scenario. A worst case tree would be created via binarization if at every decision point the algorithm allocates a fraction p of the leaves to one of the children and a fraction $(1-p)$ of the leaves to the other child. We can assume, without loss of generality, that the child getting the smaller share of leaves is the left child, and the child getting the larger share is the right child. (For a path length analysis we need not specifically concern ourselves with the distribution of data objects. So we focus on leaves instead.)

Theorem 3. *Let T be a binary tree with the property that for each node A with left child B_1 and right child B_2, the ratio of the number of leaves beneath B_1 to the number of leaves beneath A is p. Then the ratio of the average path length to a leaf in T to that of a perfectly balanced binary tree with the same number of leaves is given by $\log_{\frac{1}{p^p \cdot (1-p)^{(1-p)}}} 2$.*

Proof. Let $T(N)$ be the average path length to the root in a worst case binary tree containing N leaves. To compute $T(N)$, consider a randomly chosen path from the root to a leaf. With probability p the path will pass through the left child of the root, and with probability $1-p$ the path will pass through the right child of the root. But the binary tree corresponding to the left child contains $p \cdot N$ leaves, and the binary tree corresponding to the right child contains $(1-p) \cdot N$ leaves.

Thus, adding 1 for the edge from the root, we obtain the following recurrence relation:

$$T(N) = p \cdot T(p \cdot N) + (1 - p) \cdot T((1 - p) \cdot N) + 1 \qquad (17)$$

It is now straightforward to prove by induction on N that the solution to this recurrence relation is given by the value $T(N) = \log_{\frac{1}{p^p \cdot (1-p)^{(1-p)}}} N$. Since the average path length in a perfectly balanced tree with N leaves is given by the value $\log_2 N$, the theorem follows.

Note that when $p = 1/2$, the ratio becomes 1, as it should. We can use this theorem to plot the percentage increase in path length to a random leaf node in a worst case binary tree as a function of the skew factor p. Instead we simply note that this percentage would be unacceptable for small values of p, though one must remember that this result is for the binarization stage only. The compression stage will help rebalance the tree.

In any case it is more realistic to analyze the average path length increase in a probabilistically average case binary tree. Such a tree would be created via binarization if at every decision point the algorithm allocates a fraction q of the leaves to one of the children and a fraction $(1-q)$ of the leaves to the other child, where q is uniformly distributed between p and $(1 - p)$. We have the following analogue of Theorem 3.

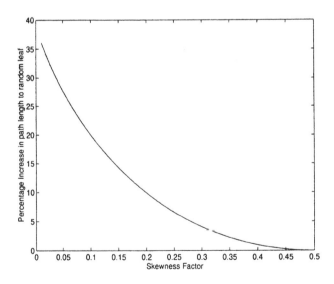

Fig. 6. Effect of Skewness in Average-Case Tree after Binarization Stage

Theorem 4. *The ratio of the average path length to a leaf in an average case binary tree T constructed with skew factor p to that of a perfectly balanced binary tree with the same number of leaves is given by $\frac{-(1-2p)\cdot ln(2)}{(1-2p)\cdot ln(1-p)+p^2\cdot ln((1-p)/p)-(1-2p)/2}$.*

Proof. Let $T(N)$ be the average path length to the root in an average case binary tree containing N leaves. Let $f_p(q)$ denote the probability density function for the case of a uniform distribution between p and $(1-p)$. In other words, $f_p(q) = 1/(1-2p)$. By an argument analogous to that given in the proof of the previous theorem we have the following recurrence relation:

$$T(N) = \int_{q=p}^{(1-p)} (q \cdot T(q \cdot N) + (1-q) \cdot T((1-q) \cdot N)) f_p(q) dq + 1 \qquad (18)$$

We "guess" that $T(N)$ has the form $K \cdot \ln N$, and find the value of K at the same time by substituting the above expression into the recurrence relation. This gives

$$K \cdot \int_{q=p}^{(1-p)} (q \cdot \ln q + (1-q) \cdot \ln(1-q)) dq + (1-2p) = 0 \qquad (19)$$

Solving the above equation we get:

$$K = \frac{-(1-2p)}{(1-2p) \cdot \ln(1-p) + p^2 \cdot \ln((1-p)/p) - (1-2p)/2} \qquad (20)$$

Since we know that the average path length in a perfectly balanced binary tree is given by $\log_2 N$, the result follows.

Figure 6 plots the percentage increase in path length to a random leaf node in an average case binary tree as a function of the skew factor p. Even when p is very small, the average length of a path from the root to a randomly chosen leaf degrades by at most 40%. The compression stage will rebalance the tree further. In any case, as we shall see, this degradation is more than compensated by the better flexibility in packing.

5 Variants of the Basic S-Tree Algorithm

In this section we briefly describe some alternatives to the basic S-tree algorithm. Some of these pertain to the binarization scheme and some to the compression scheme. In each case we will describe the motivation, the trade-offs, and the effects on the theoretical properties of the algorithm. As will be seen in the next section, some seem to be significant from a performance perspective, while others do not.

5.1 Binarization

We describe a few alternative approaches to the sweep algorithm in the binarization scheme.

The most important problem with the basic S-tree algorithm is that the leaf nodes are not guaranteed to be fully packed. Since M is typically large, the number of leaf nodes dominates the total size of the tree. So this utilization problem is potentially serious, both in terms of performance and tree size. Fortunately, an easy solution is available: In the sweep loop we can change the increment from 1 to M, ensuring that the first rectangle, say, always has a multiple of M data objects underneath it. So we sweep over $(1 - 2p) \cdot N_A/M$ choices instead of $(1 - 2p) \cdot N_A$ choices when minimizing the objective function (2). Notice that nearly all nodes will now have a multiple of M data objects, and, in particular virtually all of the leaf nodes will be fully packed. We also reduce the computation time of the binarization algorithm by a factor of approximately M. On the negative side, we sacrifice the quality of the optimization problem slightly.

There is a natural generalization which allows one to compromise between choosing increment 1 and increment M. Let Δ be an integer between 0 and $M - 1$. In the sweep we examine all choices in which the first rectangle has $i \cdot M - j$ data objects underneath it, where i is a positive integer and $0 \leq j \leq \Delta$. When $\Delta = 0$ we obtain the special case of increment M just described. When $\Delta = M - 1$ we get the original increment 1.

Along somewhat different lines, we can perform the sweep in all dimensions, not just the longest. This may improve the solution in terms of the area and overlap, at the expense of perimeter (and computational time, if not in computational complexity). Another approach might be to use Guttman's linear or quadratic split algorithm [8], suitably modified to obey the rules on partition size. (These two algorithms were originally proposed to handle insertions in R-trees. Because it is quadratic in the number of data objects, the latter split algorithm may be prohibitively expensive, at least for use at high levels in our binarization process. But some combination of the linear split algorithm at high levels of the tree and the quadratic split algorithm at low levels may constitute a reasonable strategy.)

Note that none of the above alternatives have any effect on the skewness guarantee.

5.2 Compression

We have seen that choosing increment M causes the leaf nodes to be nearly fully utilized. The original compression algorithm forces all but the penultimate non-leaf nodes to be fully utilized as well. But the penultimate nodes still present a modest problem. In fact, the top-down nature of the compression scheme actually serves to disconnect small portions of the S-tree at the lowest levels. So one is left with occasional penultimate nodes with very modest branch factors. In order to alleviate this we consider the notion of applying a precompression stage: We compress bottom up one level, as follows: Identify each node A arising from the binarization stage with the property that the number of leaf descendents of A is less than or equal to M, but the number of leaf descendents of the parent of A is greater than M. These will become the penultimate nodes of the final S-tree, so collapse all non-leaf nodes underneath each such A. Then apply the compression stage to the remaining tree, exactly as before. A little analysis will show that the skewness guarantee can be maintained *provided* all leaf nodes are equally packed. This will be nearly the case if we choose our increment to be M. But if we use the original increment of 1 we must sacrifice this theoretical property.

6 Other S-tree Operations

The S-Tree is essentially a different data structure from the R-Tree but it is similar enough so that operations such as insertions, deletions, search and nearest neighbor queries remain exactly the same. The efficiency of the packing algorithm under S-stree also makes periodic reorganizations more practical. Although the skewness guarantee cannot be maintained and the 100% page utilization is also lost at the top level nodes, a good initial packing is a substantial advantage in achieving an efficient representation even in the face of dynamic insertions and deletions. Reorganizations are efficient because, as noted, the complexity of the algorithm is no worse than that of sorting the points along each dimension.

For insertion and deletion we use an approach similar to the Hilbert-packed R-Tree method described in [11]. Insertions and deletions can be done by using the notion of *cooperating siblings*. The idea of cooperating siblings was first used in the context of Hilbert-packed R-Trees in [11]. In this paper we show how to use this idea in the context of S-Trees. The essential idea of cooperating siblings is that whenever a node overflows or underflows, some of its sibling nodes are chosen and the children of all of these siblings are regrouped so that there is no resulting overflow or underflow. In the case of the Hilbert-packed R-Tree, the choice of which nodes to choose as the cooperating siblings is really easy because of the fact that we can simply order all the spatial objects by Hilbert number. Consequently, this defines a natural ordering among the sibling nodes at various levels of the tree so that adjacent nodes can be chosen as cooperating siblings. In the case of the S-Tree we choose the cooperating siblings of a node which overflows/underflows by using a simple greedy algorithm which tries to choose the most closely clustered nodes. The number of cooperating siblings (including the overflowing node itself) to be considered is a parameter, denoted by s, as in [11]. (In the case of an underflow, it is $s + 1$.) Thus the algorithm for choosing the cooperating siblings is as follows:

(1) Let A be the overflowing (underflowing) node.
(2) Greedily choose at most $s - 1$ (s in case of an underflow) other siblings of the node A such that at each step the minimum bounding rectangle of the cooperating siblings has as small an area as possible.

Next we need to show how to redistribute the children of a set of cooperating siblings. Note that the initial tree does not have the strong guarantee that the space utilization of a node should be at least $s/(s+1)$. Consequently, an overflow operation is not necessarily an s to $(s+1)$ split, as it is in the case of the Hilbert-packed R-Tree. Here the children of the cooperating siblings are redistributed into a new set of siblings which may even be smaller than s in cardinality. However the new set of siblings will have the guarantee that each node has a space utilization of at least $s/(s+1)$. In other words, we redistribute the children of these cooperating siblings into the minimum number of groups which avoids any overflows and also results in a space utilization of $s/(s+1)$ for each of the new siblings formed. This can again be done by using a modified version of the sweep algorithm, which regroups the children of the cooperating siblings so that the area/overlap characteristics are as good as possible. We shall give more detailed descriptions and empirical results on the insertion/deletion method in subsequent work.

7 Experimental Results

In this section we describe the results of simulation experiments designed to test the performance of S-trees relative to Hilbert-packed R-trees. (Based on the analysis given in [12], it would appear that Hilbert-packed R-trees perform better than the other R-tree variants.) All experiments performed for this paper involved synthetically generated 2-dimensional data. Much of our general methodology follows that described in [12], though we also introduce skew into the data.

We assume without loss of generality that all data and all queries are contained within the unit square. In other words, both the x and y values of each corner of every bounding and query rectangle lie between 0 and 1. In our experiments we typically packed trees containing $N = 50,000$ data objects. These objects consisted of both points and rectangles, based on a parameter F in the range $[0,1]$. In other words, $F \cdot N$ data objects were points, and the remaining $(1 - F) \cdot N$ data objects were legitimate rectangles. We considered various values of the *density* D of the underlying data objects: The density is defined as $D = N \cdot a$, where a is the average area of the data objects, and thus corresponds to the average number of data objects covering a random point in the unit square. Following [12], we chose values of D between 0.25 and 2.00.

Now we describe the method used to define the underlying data objects with varying degrees of skew. We partitioned the unit square into 100 subsquares by subdividing each side into 10 equal sections. (This choice was somewhat arbitrary, and could be easily modified.) We ordered these 100 subsquares randomly. We then assigned centers for the bounding rectangles of the underlying data objects to the randomly ordered subsquares according to a Zipf-like distribution with $\nu = 100$ distinct values and *data skew parameter* θ. (Briefly, a Zipf-like distribution is given by $p_i = c/i^{(1-\theta)}$, where $c = 1/(\sum_{i=1}^{\nu} 1/i^{(1-\theta)})$ is a normalization constant. Setting $\theta = 0.00$ corresponds to a pure Zipf distribution, which is highly skewed: The most popular subsquare contains 19% of the data objects. Setting $\theta = 1.00$ corresponds to a uniform distribution. Setting $\theta = 0.50$

corresponds to moderate skew, the most popular subsquare containing 5% of the data objects. For further details on Zipf and Zipf-like distributions, see [13].) Within each subsquare the centers were chosen via uniform distributions along both axes.

For the point data objects, the center also corresponds, of course, to the corners of the degenerate bounding rectangle. For the legitimate rectangles, we chose sides in a manner which conformed to the underlying data object density. We did this by picking the side in each of the two axes independently as $2\sqrt{D/((1-F)\cdot N)}$ times a randomly chosen number from a uniform distribution on $[0,1]$. (However, we cropped the side if it would otherwise extend beyond the unit interval in either direction.)

We used a similar technique to vary the average size of a side in a query. However, we chose the centers from a uniform rather than a skewed distribution. We would not anticipate that forcing the centers to have some skewed distribution will have a negative impact on the S-tree performance, particularly if there is correlation between the query and data object centers. But we have not done this study yet.

For our experiments we generally chose a branch factor of $M = 40$. This number appears to be reasonable, based on typical page sizes and the size of the various object-identifiers and pointers in 2 dimensions. We also chose a skew factor of $p = 0.3$ and an overlap factor of $o = 1$. Recall that the latter choice eliminates the overlap constraint. This choice was driven by empirical results, for example those shown in Figure 7. (Here, and in all other graphs discussed in this section, the metric used is the average ratio of the number of S-tree index pages touched by a query to the number of Hilbert R-tree index pages touched, expressed as a percent. So the horizontal line at $y = 100$ corresponds to a tie.) The figure shows the effect of various values of p, where $o = 1$. One sees that choosing a skew parameter somewhere in the range 0.2 to 0.3 appears close to optimal: There is enough flexibility to build a good tree, but the imbalance is not too great. It turns out that the performance of the S-tree is relatively insensitive to the overlap parameter, unless o is very small. In this case the performance seems to degrade somewhat. In Figure 7 we have used $M = 40$, $\theta = 0.00$ and $D = 1.5$. But other experiments yielded similar results. We show point queries, queries with average sides equal to 5% of the unit interval, and queries with average sides equal to 10% of the unit interval. These queries correspond to 0.00%, 0.25% and 1.00% of the areas of the unit square, respectively.

We briefly mention the S-tree variants which had the best performance: Not surprisingly, we found that a sweep increment of M worked better than an increment of 1. We also chose $\Delta = 0$, since results did not improve for other values. These choices guarantee nearly full utilization of the leaf nodes. We looked at the longest dimension only during each sweep. Here the modest improvement did not justify the extra cost. In the compression stage we did employ the lowest level bottom up compression. Given the sweep increment choice this meant that we essentially retained our skewness guarantee.

What is striking about our results is the robustness of the S-tree algorithm. It *never* performed worse than the Hilbert *R*-tree schemei for reasonable parametric choices, though for queries with large sides the performance of the two was quite close. On the other hand, on queries with small sides the *S*-tree did significantly better, and did best of all on point queries. See, for example, Figures 8 and 9. The former is for $M = 40$ and $\theta = 0.50$ representing the case with medium data skew, while the latter is for $M = 30$ and $\theta = 0.00$, representing the case with heavy data skew. Both figures are nearly convex, asymptoting at approximately 100%. It is natural to expect that most queries will have sides towards the left of these figures, since queries with large sides return massive amounts of data, and may be best done without an index in any case.

Figures 10, 11 and 12 show the sensitivities to varying one of the major parameters. In the first we varied the data skew parameter (θ), keeping $M = 40$ and $D = 1.5$. In the second we varied the data density, keeping $M = 40$ and $\theta = 0.00$. In the third we varied the branch factor, keeping $D = 1.5$ and $\theta = 0.00$. We show three different average sides in each case, and their relative performance is as would be expected. There are no particularly obvious effects to report, except in Figure 11. Here one sees that performance of the *S*-tree consistently improves with smaller data object densities.

8 Conclusions

In this paper we have introduced a new multidimensional index structure, known as the *S*-tree. While similar in spirit to the *R*-tree, the new structure allows for mild imbalance in the tree in return for significantly reduced area, overlap and perimeter in the resulting minimum bounding rectangles, and thus improved performance.

In this paper we have concentrated on the initial *S*-tree design strategy, which is appropriate for static databases, of course. As noted, however, the *S*-tree can handle insertions and deletions as well. In a future paper we plan to devote more space to the dynamic algorithms, and experimentally study *S*-tree performance under such conditions.

We have described only experiments for 2 dimensions in the current paper. However, experiments with higher dimensions have yielded similar results. We will discuss these in greater detail in the future. We shall also examine the performance of the *S*-tree on real databases.

References

1. Aref, W., Samet, H.: Optimization Strategies for Spatial Query Processing. Proceedings of the VLDB Conference. (1991) 81–90.
2. Beckman, N., Kriegel, H., Schneider, R., Seeger, B.: The R*-Tree: An Efficient and Robust Method for Points and Rectangles. Proceedings of the ACM SIGMOD Conference. (1990) 322–331.
3. Bentley, J.: Multidimensional Binary Search Trees Used for Associative Searching. Communications of the ACM. **18(9)** (1975) 509–517.

4. Faloutsos, C., Kamel, I.: Beyond Uniformity and Independence: Analysis of R-Trees using the Concept of Fractal Dimension. Proceedings of the ACM PODS Conference. (1994) 4–19.

5. Faloutsos, C., Roseman, S.: Fractals for Secondary Key Retrieval. Eighth ACM SIGACT-SIGMOD-SIGART Symposium on Principles of Database Systems (PODS). (1989) 247–252.

6. Gargantini, I.: An Effective Way to Represent Quad Trees. Communications of the ACM. 25(12) (1982) 905–910.

7. Gunther, O., Bilmes, J.: Tree Based Access Methods for Spatial Databases: Implementation and Performance Evaluation. IEEE Transactions on Knowledge and Data Engineering. 3(3) (1991) 342–356.

8. Guttman, A.: R-Trees: A Dynamic Index Structure for Spatial Searching. Proceedings of the ACM SIGMOD Conference. (1984) 47–57.

9. Hinrichs, K., Nievergelt, J.: The Grid File: A Data Structure to Support Proximity Queries on Spatial Objects. Proceedings of the WG'83. (1983) 100–113.

10. Jagadish, H.: Linear Clustering of Objects with Multiple Attributes. Proceedings of the ACM SIGMOD Conference. (1990) 332–342.

11. Kamel, I., Faloutsos, C.: Hilbert R-Tree: An Improved R-tree using fractals. Proceedings of the 20th VLDB conference. (1994)

12. Kamel, I., Faloutsos, C.: On Packing R-Trees. Proceedings of the 2nd International Conference on Information and Knowledge Management. 490–499.

13. Knuth, D.: The Art of Computer Programming, Vol. 3: Sorting and Searching. (1973).

14. Lomet, D., Salzberg, B.: The hB-Tree: A Multiattribute Indexing Method with Good Guaranteed Performance. ACM TODS. (1990) 15(4) 625–658.

15. Orenstein, J.: Spatial Query Processing in an Object-oriented Database System. Proceedings of the ACM SIGMOD Conference. (1986) 326–336.

16. Robinson, J.: The K-D-B Tree: A Search Structure for Large Multidimensional Dynamic Indexes. Proceedings of the ACM SIGMOD Conference. (1981) 10–18.

17. Roussopoulos, N., Leifker, D.: Direct Spatial Search on Pictorial Databases using Packed R-Trees. Proceedings of the ACM SIGMOD Conference. (1985)

18. Samet, H.: The Design and Analysis of Spatial Data Structures. Addison Wesley. (1989)

19. Sellis, T., Roussopoulos, N., Faloutsos, C.: The R+ tree: A Dynamic Index Structure for Multi-Dimensional Objects. Proceedings of the 13th International Conference on VLDB. (1987) 507–518.

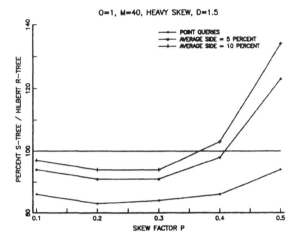

Fig. 7. Varying Skew Factors

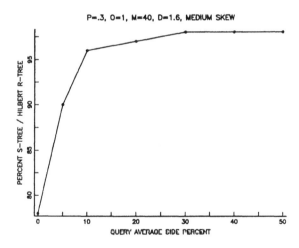

Fig. 8. Varying Query Sizes

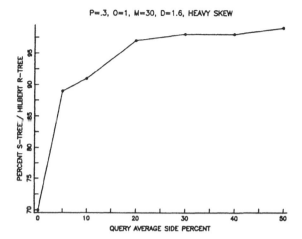

Fig. 9. Varying Query Sizes

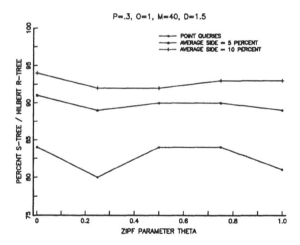

Fig. 10. Varying Zipf Parameters

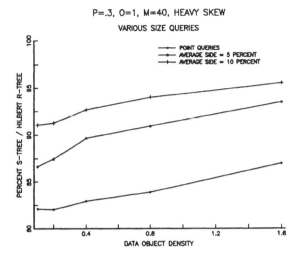

Fig. 11. Varying Data Densities

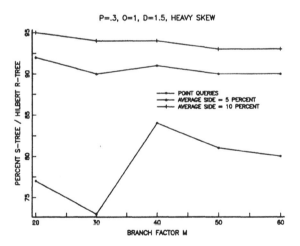

Fig. 12. Varying Branch Factors

Keynote Paper

Perspectives in GIS Database Architecture

Dick Newell

Smallworldwide plc
1 High Street, Chesterton
Cambridge
CB4 1WR
United Kingdom
dick.newell@smallworld.co.uk

So far, the efforts of the mainstream database vendors have done little to impact the progress of GIS implementation. Traditionally, spatial data has been considered different in some respect to other corporate data. Although there have been many attempts to accommodate all data, spatial and aspatial in a commercial database, few if any meet the scalability and performance requirements of a large corporate GIS. All of these systems contain inelegant compromises to overcome the deficiencies of the basic platform that is used. Indeed a major part of the R&D budgets of GIS vendors is spent in database management development, either developing proprietary architectures, or providing proprietary work arounds to industry standard engines. There are now a number of trends emerging, including the DBMS vendors themselves providing a spatial capability, the enhancement of relational databases to provide object-oriented interfaces, the emergence of three-tier architectures as an extension to the two-tier client-server architecture of traditional systems, replication technology as one solution to the distributed database problem, data warehouses and many others. The presentation attempts a broad overview of the field, including a description of R&D developments in the author's own company, including successful architectures that use an industry standard relational DBMS.

Author Index

Springer
and the
environment

At Springer we firmly believe that an
international science publisher has a
special obligation to the environment,
and our corporate policies consistently
reflect this conviction.
We also expect our business partners –
paper mills, printers, packaging
manufacturers, etc. – to commit
themselves to using materials and
production processes that do not harm
the environment. The paper in this
book is made from low- or no-chlorine
pulp and is acid free, in conformance
with international standards for paper
permanency.

Lecture Notes in Computer Science

For information about Vols. 1–1192

please contact your bookseller or Springer-Verlag